ADVANCES IN
Applied Microbiology

VOLUME 14

CONTRIBUTORS TO THIS VOLUME

Bernard J. Abbott
Nancy N. Gerber
William E. Gledhill
John J. H. Hastings
V. W. Jamison
H. A. Lechevalier
Mary P. Lechevalier
Barney J. Magerlein
A. Kathrine Miller
D. Perlman
John N. Porter
R. L. Raymond
Oldrich K. Sebek
G. L. Solomons

ADVANCES IN

Applied Microbiology

Edited by D. PERLMAN

School of Pharmacy
The University of Wisconsin
Madison, Wisconsin

VOLUME 14

 1971

ACADEMIC PRESS, New York and London

COPYRIGHT © 1971, BY ACADEMIC PRESS, INC.
ALL RIGHTS RESERVED
NO PART OF THIS BOOK MAY BE REPRODUCED IN ANY FORM,
BY PHOTOSTAT, MICROFILM, RETRIEVAL SYSTEM, OR ANY
OTHER MEANS, WITHOUT WRITTEN PERMISSION FROM
THE PUBLISHERS.

ACADEMIC PRESS, INC.
111 Fifth Avenue, New York, New York 10003

United Kingdom Edition published by
ACADEMIC PRESS, INC. (LONDON) LTD.
24/28 Oval Road, London NW1 7DD

LIBRARY OF CONGRESS CATALOG CARD NUMBER: 59-13823

PRINTED IN THE UNITED STATES OF AMERICA

CONTENTS

LIST OF CONTRIBUTORS ix
PREFACE xi

Development of the Fermentation Industries in Great Britain

JOHN J. H. HASTINGS

I.	Prelude to Industrialization	1
II.	The Vintage Years of Industrial Alcohol	8
III.	The First Closed Deep Fermentation	12
IV.	Production of Vitamins	19
V.	Citric Acid and Surface Culture	23
VI.	The Entry of Antibiotics	25
VII.	Enzymes	38
VIII.	Gaps in the Program	41
IX.	The Influence of Continuous Fermentation	42
	References	45

Chemical Composition as a Criterion in the Classification of Actinomycetes

H. A. LECHEVALIER, MARY P. LECHEVALIER, AND NANCY N. GERBER

I.	Introduction	47
II.	Pigments in the Classification of Actinomycetes	50
III.	Antibiotics in the Classification of Actinomycetes	54
IV.	Deoxyribonucleic Acid Base Composition in the Classification of Actinomycetes	55
V.	Cell Wall and Whole-Cell Sugar Composition in the Classification of Actinomycetes	57
VI.	Lipid Composition in the Classification of Actinomycetes	60
VII.	Conclusion: A Tentative System of Classification of Actinomycetes	62
	References	69

Prevalence and Distribution of Antibiotic-Producing Actinomycetes

JOHN N. PORTER

I.	Introduction	73
II.	Isolation and Enumeration of Actinomycetes	74
III.	Classification of Actinomycetes Isolated from Soils	81
IV.	Antibiotic-Producing Actinomycetes	83
V.	Identification and Classification of Antibiotics Produced by Actinomycetes	86
VI.	In Retrospect	89
	References	90

Biochemical Activities of *Nocardia*

R. L. Raymond and V. W. Jamison

I.	Introduction	93
II.	Hydrocarbon Oxidation	94
III.	Nonhydrocarbon Oxidation	105
IV.	Transformation of Steroids and Sterols	111
V.	Summary	119
	References	120

Microbial Transformations of Antibiotics

Oldrich K. Sebek and D. Perlman

I.	Introduction	123
II.	Types of Microbial Transformations	126
III.	Techniques Useful in Studying Microbial Transformations of Antibiotics	143
IV.	Summary	145
	References	146

In Vivo Evaluation of Antibacterial Chemotherapeutic Substances

A. Kathrine Miller

I.	Historical Background	151
II.	*In Vivo* Test Procedures	152
III.	Special Tests	171
IV.	Epilog	179
	References	179

Modification of Lincomycin

Barney J. Magerlein

I.	Introduction	185
II.	Lincomycin-Related Antibiotics	186
III.	Chemical Modification of Lincomycin	188
IV.	Microbial Modification of Lincomycin and Clindamycin	224
V.	Summary	227
	References	227

Fermentation Equipment

G. L. Solomons

I.	Introduction	231
II.	Auxiliary Equipment	233
III.	Process Control Instruments	235
IV.	Sterilization of Culture Medium	238
V.	Analytical Instrumentation	240
VI.	Biochemical Analysis of Fermentation Cultures	244
	References	246

The Extracellular Accumulation of Metabolic Products by Hydrocarbon-Degrading Microorganisms

BERNARD J. ABBOTT, AND WILLIAM E. GLEDHILL

I.	Introduction	249
II.	Products from Aliphatic Hydrocarbons	252
III.	Products from Cyclic Hydrocarbons	318
IV.	Concluding Remarks	376
	References	378

AUTHOR INDEX ... 389
SUBJECT INDEX ... 406
CONTENTS OF PREVIOUS VOLUMES ... 409

LIST OF CONTRIBUTORS

Numbers in parentheses indicate the pages on which the authors' contributions begin.

BERNARD J. ABBOTT, *Esso Research and Engineering Company, Linden, New Jersey* (249)

NANCY N. GERBER, *Institute of Microbiology, Rutgers University of New Jersey, New Brunswick, New Jersey* (47)

WILLIAM E. GLEDHILL, *Monsanto Company, St. Louis, Missouri* (249)

JOHN J. H. HASTINGS, *St. Catherine's Lodge, Merriott, Somerset, England* (1)

V. W. JAMISON, *Sun Oil Company, Marcus Hook, Pennsylvania* (93)

H. A. LECHEVALIER, *Institute of Microbiology, Rutgers University, The State University of New Jersey, New Brunswick, New Jersey* (47)

MARY P. LECHEVALIER, *Institute of Microbiology, Rutgers University, The State University of New Jersey, New Brunswick, New Jersey* (47)

BARNEY J. MAGERLEIN, *Research Laboratories, The Upjohn Company, Kalamazoo, Michigan* (185)

A. KATHRINE MILLER, *Merck Institute for Therapeutic Research, Rahway, New Jersey* (151)

D. PERLMAN, *The School of Pharmacy, University of Wisconsin, Madison, Wisconsin* (123)

JOHN N. PORTER, *Lederle Laboratories Division, American Cyanamid Company, Pearl River, New York* (73)

R. L. RAYMOND, *Sun Oil Company, Marcus Hook, Pennsylvania* (93)

OLDRICH K. SEBEK, *Upjohn Research Laboratories, Kalamazoo, Michigan* (123)

G. L. SOLOMONS, *The Lord Rank Research Centre, High Wycombe, Buckinghamshire, England* (231)

PREFACE

In this volume we have attempted to focus attention on those facets of applied microbiology which may have some immediate as well as long-range economic potential. In these days of economic recession it is desirable to bring forth some optimistic proposals for the future, and a number of the contributions to this volume meet this challenge.

Hastings in his own inimitable fashion summarizes centuries of fermentation activity in Great Britain and conveys the attitude that there will always be a "silver lining in most of the dark clouds" that beset the fermentation industry in that country. Solomons evaluates new developments in design of fermentation equipment which will result in economic advantage and make possible the continuation of the fermentation industries in the face of the economic competition mentioned by Hastings. New sources of antibiotics are discussed by Porter, by Lechevalier et al., and by Miller, whose efforts, although centered on chemotaxonomy and antibiotic evaluation, still suggest that it is worthwhile to examine microbial metabolites for potential chemotherapeutic activities. Magerlein's summary of the successful extension of the chemotherapeutic potential of lincomycin is a confirmation of the value of interdisciplinary research teams in promoting new projects.

The amazing potential of microbial systems in transforming hydrocarbons, sterols, and antibiotics is surveyed in contributions by Raymond and Jamison and by Sebek and Perlman. Abbott and Gledhill's complete survey of the activities of hydrocarbon-degrading microorganisms complements Raymond and Jamison's intensive examination of the potential of genus *Nocardia*.

The sophistication and depth of these individual contributions indicates strength in certain aspects of applied microbiology. We hope that these essays will encourage many to join the groups examining the potential of microorganisms. We have confidence that there are rewards enough for all.

D. PERLMAN

ADVANCES IN
Applied Microbiology

VOLUME 14

Development of the Fermentation Industries in Great Britain

JOHN J. H. HASTINGS[1]

Formerly Director of Patents and Registration
Lilly Industries, Limited, Great Britain

I.	Prelude to Industrialization	1
II.	The Vintage Years of Industrial Alcohol	8
III.	The First Closed Deep Fermentation	12
IV.	Production of Vitamins	19
V.	Citric Acid and Surface Culture	23
VI.	The Entry of Antibiotics	25
VII.	Enzymes	38
VIII.	Gaps in the Program	41
IX.	The Influence of Continuous Fermentation	42
	References	45

I. Prelude to Industrialization

It is easy to say that industrial fermentation began in Great Britain with the twentieth century, and to start the story from there. But to understand the birth pangs of our modern practice one must go back into the history and culture and climate of our country, and examine the raw materials, the arts and crafts of construction, and the men that started us on our way.

We were invaded many times before William the Conqueror came in 1066: first by the Romans, who tried to establish vineyards round their country villas, and failed, except maybe in the fair fields of Kent, because there were few years when our climate would produce anything but small sour grapes unfit for wine. Then came the Danes and the Saxons, who did little to add to our fermentation knowledge. They drank the ale and the mead and the metheglyn which we already knew, but they were seamen and fighters first and husbandmen last of all.

Then came William the Norman, tyrannical, systematic, and determined to reduce Britain to an orderly province of his fair Normandy, portioning out the land to his barons and recording every acre in his Domesday Book. And on his heels came the great monastic houses to add to the handful of missionaries and religious orders that had begun to make us a Christian country. They indeed brought knowledge and much experience that helped to develop our agriculture. They tried, as the Romans did, to grow vineyards, and failed as often.

---

[1]*Present address:* St. Katherine's Lodge, Merriott, Somerset, England.

But they also fostered the kinds of grain that would grow on our rich lowlands, barley and wheat, and oats that would flourish even in the hill country. They brought rye, but too often the deadly ergot would occur, and although rye lingered on through the Middle Ages, it has long since disappeared as a major crop in these islands. No matter how loudly patriotic Scots may sing "Comin' through the rye" on Burns night, they have very little chance of doing it today.

So wheat and barley and oats became our staple grains, and the common alcoholic beverage was ale or beer. This was not the end of our resources in raw materials for fermentation. In the western fringes, where the native Celtic population had retreated before one invasion after another, honey from the great areas of heathland and forest was a prime source of liquor, so that the Welshman had his metheglyn and the Cornishman his mead. Along the foothills of the Welsh border and in the valleys of the West country, the small sour apples of the time became the raw material for cider, and they are still cultivated for this purpose today. Across the centuries, our skill in producing fermented liquor grew from a variety of raw materials suited to our climate to products that have been traditional since Norman times.

It was all on the very small scale. Nearly every monastery had its malthouse and its brewhouse, while every small inn brewed its own ale. Every great mansion did the same, and even the farms made their own brews for their men. It was not until the dissolution of the monasteries by Henry VIII that breweries began to grow in the towns and cities to supply ale and beer to several inns in the neighborhood. The master brewers grew in wealth and skill, and became respected citizens in the community, so much so that it became a way of life that even the younger sons of the great landed gentry might take up without loss of face.

There were few occupations that the sons of the aristocracy might follow. The eldest son would of course learn to rule the family estate like a minor princeling, and there are many examples of the wisdom with which these estates were kept up. The younger sons had little choice. They might take a commission in the army or even in the navy. They might enter the political arena and buy a seat in parliament. If they showed intelligence but no zest for the armed forces, they might take holy orders. And finally, if they showed no capabilities in any of these directions, they might enter brewing. For a brewery was a small kingdom of its own, with craftsmen who respected their betters as their forebears had done before them. The brewer himself did little but rule and issue commands, while the beer flowed out and the money flowed in to help him to live the life of his peers. He could hunt, he could wine and dine with his fellows, he could have his season in town when London called, he could marry into any of the great families of the kingdom, and many a fallen estate was restored by the money brought in by ale.

These were the men in control of the fermentation industry when the twentieth century dawned. True, some of them became gifted amateurs in the profession, but it was an art and not a science that they controlled, and they viewed new ideas almost with horror, for brewing had been carried on for hundreds of years with very little change, and why should man want to alter a drink that was his pleasure and necessity. The origin of malted grain was lost in the mists of time, and the right temperature at which to extract it had been found by a hundred thousand trials. The choice of barley from our limited range of grain was almost a necessity, since the spent husks acted as a natural filter that enabled the maximum yield of sweet liquor to be drained from the grain after mashing with hot water. Oats, with their still coarser husks, gave a poorer yield and a harsher flavor, and found their place in the dark-colored stouts and porters that took the fancy of the workers in the towns and seaports.

As for materials of construction, English oak was the choice for fermentors and barrels throughout the centuries, until in the end a veritable shortage made it necessary to turn to Memel oak from the eastern Baltic to supplement our needs. The coopers who made round vessels from such timber were among the most highly skilled craftsmen in the land. The trouble with wood was that when it became fouled with bacterial contamination it was almost impossible to clean, in the absence of modern disinfectants, and the effect on fermentation could be tragic. Slate and stone came into substantial use to build rectangular fermentors, and could be scrubbed to a high degree of cleanliness, thus reducing the risk of spoiled brews. Cast iron had its supporters, but of all the metals copper was the choice. It could easily be worked and riveted, and showed itself extremely suitable for boiling the malt extract, or mash, with hops before cooling to fermentation temperature. The boiling vessel is called the copper to this day.

As for the fermentation itself, the most remarkable feature of British ales and beers is that they are all made with top-fermentation yeasts, as distinct from the bottom-fermentation yeasts long used for European lager-type beers. This no doubt arose in the first place from our casual small-scale methods, where there was no effective temperature control other than the shelter of a building. The new yeast crop would rise to the surface and be skimmed off as a cream which could be stored in a cool place and used to start the next brew. The top-fermenting strains survived by this process of selection. If the yeast deteriorated, the simplest cure was to throw it away and borrow some from the nearest friendly brewery that had a healthy crop, and favors are still exchanged in this way. The necessity for skimming the yeast crop from the surface of the fermentation meant that the fermentor had always to be open to the atmosphere, and it says much for the vigor of our top-fermentation yeasts that often they are maintained healthy and with minimal contamination for many years.

Strange as it may seem, there was little serious attempt in England itself to produce potable distilled liquor from fermented grain. From the beginning of the seventeenth century, brandy was considered the distilled liquor par excellence in this country, and in the absence of home-grown wines to produce it, the source was France, from which it was imported legally and illegally in vast quantities. The imposition of duty by the government made smuggling one of the most widely practiced sports of the times.

In Scotland the position was very different. As a separate kingdom it had had from time immemorial close ties with France. There was a large sea-trade between the two countries, and a community of interests when both were at loggerheads with England. There is no doubt that France introduced the process of distillation to Scotland, but not to make brandy from wine. Scotland had its rich barley growing counties on the eastern seaboard, and its clear cold streams in the hills for cooling water. There were many hidden valleys where such a thing as an excise officer might never venture. What more natural that the canny Scots, short of foreign exchange for brandy, should conduct a do-it-yourself campaign with their own raw material, and with that grim determination that is their character, succeed beyond their wildest dreams, so that today Scotch whiskey is the choice for the connoisseur throughout the world, and the most important export of that craggy kingdom (Ross, 1970).

Just as the English brewer has resisted change, so has the Scotch distiller. He still uses an antiquated pot-still to make the best malt whiskies, though these choice products are usually diluted with grain spirit from continuous stills to make many of the famous brand names that are known on every side. And when a pot-still wears out, it is reproduced as a Chinese copy down to the last rivet, so that no trace of a difference in flavor will appear. Long ago the Scotch distiller learned to produce malts with a high diastatic power so that every trace of carbohydrate in the malt could be converted into fermentable sugar. One might therefore imagine that he would aim for maximum conversion efficiency of sugar to alcohol, with skillfully controlled pure yeast fermentations. Not so. The fermentors that feed the pot-stills are frequently contaminated with bacterial infections that live in some degree of harmony with the yeast. They take their small share of sugar and produce no alcohol, but they convey their own special flavor that enables the expert taster to name the distillery from which the elixir came.

In the early days of the antibiotic industry, when it was found that some of these products could inhibit many strains of bacteria without affecting yeasts, the idea was put to the largest of the distillery combines that the addition of certain antibiotics to the fermentation could improve the yield of alcohol. The proposal was received with great coldness. To risk losing their special flavors for the possibility of 1 or 2% more alcohol was the last thing they were prepared to do.

In the later years of the nineteenth century, distilleries became very closely controlled by law. Excise duty on potable spirits had begun its spiral of ever increasing revenue, and the first place where loss of revenue must be prevented was the point of manufacture. The great Excise Act of 1880 laid down stringent regulations on every stage of the process. It provided permanent excise officers at every distillery, and a license to distill could only be obtained under the most onerous conditions. Excise locks were positioned on every piece of equipment at points where spirit might be withdrawn, strict calculations based on volume and specific gravity were required to be made on every fermentation to determine the alcohol that might and should be produced, and the distiller was required to obtain a minimum specified yield of alcohol as shown by these calculations. The distilled spirit had to be stored in a bonded warehouse until the necessary duty was paid to release it from bond and permit its sale. A still more onerous and costly provision was based on the principle that an excise officer could not be in two places at the same time. He could not maintain a watch on the fermentor room and oversee the operation of the still. The law therefore required the stills to be locked out of use while the fermentors were prepared and the inlets to the fermentors to be locked when the stills were brought into use. This meant that each half of the distillery was only in use for approximately half the available time, thereby doubling the amount of capital required to obtain a given output. Existing distillers were on the whole satisfied with the Act, since it meant that any new competitors would require substantial financial resources to enter the field.

In the case of fermented beverages such as beer, excise tax was based on the original gravity of the liquid before fermentation, on the thesis that a given quantity of dissolved solids would produce a proportional yield of alcohol. Tables were drawn up from which to calculate the original gravity of the brew from the final gravity of the beverage before and after laboratory distillation to remove alcohol. After they were brought into use they were shown to be incorrect, but it was many years before revised tables were legalized and brought into effect.

So the stage was set for the twentieth century to begin, with fermentation practice entirely devoted to production of fermented or distilled beverages with the traditional adjunct of edible vinegar made by the trickle-tower process. The industry used almost entirely indigenous materials and methods inherited by generations of brewers that had gone before. It functioned under men who, to say the least, were conservative and more often than not antipathetic to any scientific approach to their practices. For more than a generation the discoveries of Pasteur had passed over their heads, and in 1900 a microscope in a brewery was a rarity, more to be kept as a showpiece for great occasions than to be used as a working tool. Not until 1916 did Rayleigh's work of the previous century on polarized light result in a first published paper on the use of an interference refractometer in a brewery laboratory (Adler and Leurs, 1916).

Nevertheless, change was on the way, and in the end both the brewing and the distillation industries made great contributions to research in fermentation. On the continent, many scientists of repute had followed the Liebig-Pasteur controversy with interest, and in a number of universities were actively engaged in further studies. The microbiology and culture of yeast received close attention, the enzymes of malted grain and their ability to change the constituents to fermentable products were studied with care. The hunt was on to determine the course of breakdown of sugars to alcohol, and new analytical methods, of use not only in research but also in fermentation practice, were frequently reported in the literature. British brewers gradually became aware that many of their day-to-day problems were susceptible to scientific examination. Few breweries dreamed of establishing their own laboratories, and in any case there were few men capable of staffing them. In the first quarter of the century the great era of brewing consultants was established. Most of them were in the first place highly competent analysts who set up their own laboratories and trained their staff of assistants to a high degree of accuracy. Their knowledge of bacteriology and biochemistry was rather limited, most of it having been gathered on the way. Indeed, there was almost nowhere in the country where a student could obtain any training in industrial microbiology or biochemistry. But men like Arthur Ling, Chaston Chapman, and Lloyd Hind and a number of others made valuable contributions to working practice and analytical control. All capable men, their success was assured in an industry that in places was ignorant almost beyond belief, and their very success as consultants made brewers who could hardly recognize a yeast cell appreciate the value of their methods. The breweries who made the most successful use of their advice were among the first to set up their own laboratories, often manned by technical assistants tempted from the consultants.

The so-called brewery chemist was required to be analyst, biochemist, and microbiologist all in one, but he rated very low in the hierarchy of the brewery, and was all too often regarded as an imposed nuisance by the brewery staff. His only recognized virtue was that he was cheaper than paying large consulting fees. But here and there personalities began to emerge, so skilled in their work, so much in love with their job, so logical and helpful in making deductions from their findings, that they were bound to impress those managements wise enough to listen. Such a man, for example, was the great Scottish brewery chemist, J. S. Ford, who established the outstanding laboratories at Younger's Brewery in Edinburgh. Not only did he serve his company well, but he carried out considerable research of general application and trained many fine workers to follow him the the field of brewing—men like Adam Tait and Louis Fletcher, whose research is widely recorded in the brewing journals.

Quite early in this stage of development, men of goodwill in the industry joined together through the Institute of Brewing to establish a research foundation, and in the end almost every brewery in Britain was persuaded to contribute according to its size, and continues to do so. Initially the Institute had no laboratories of its own, but it accomplished three objectives. It was able to finance research projects in various fields of brewing in those universities most fitted to do so; it encouraged the teaching of subjects relevant to brewing both as full-time and as part-time courses; and by holding regular sectional meetings of its members in all parts of the country, scientists, including those it sponsored, were able to meet practical brewers and discuss the work they were undertaking. Centers such as the University of Birmingham, the College of Technology in Manchester, and the Heriot-Watt College in Edinburgh all began to show an increasing interest in the field, both in teaching and in research. Wye Agricultural College in Kent became the center of botanical studies in the breeding of new varieties of hops, and in parallel, Sir John Lawes

Institute at Rothamsted devoted significant efforts to growing barleys suitable for malting. Eventually the Institute, with the full support of the industry, was able to establish the Brewing Industry Research Foundation in extensive laboratories at Nutfield in Surrey, which now house a highly qualified staff undertaking extensive research in applied and fundamental fields.

The potable spirit industry had no such organization, but when the manufacture of bakers' yeast came substantially under its wing as a large-scale commercial operation, the Distillers Company set up its own research station at Great Burgh near Epsom in the early 1920s, under the direction of Dr. Vargas Eyre. The company had already taken the lead in the production of industrial alcohol and had every incentive to undertake research into yeast and alcohol production. Again, one of the difficulties was to find men with the basic training in microbiology and biochemistry to carry out such research. Like the Institute of Brewing, the Distillers Company, which represented by far the greater part of the Scotch whiskey industry, was wise enough to encourage teaching of such subjects in universities and technical colleges, and gave substantial support where appropriate. It was in this way that the two industries helped to provide the nucleus of technical men who were able when the time came to take part in great developments in the fermentation industry.

II. The Vintage Years of Industrial Alcohol

It is true to say that the first major development in industrial fermentation came right at the beginning of the century, when industrial alcohol became a practical proposition, but this was no great technical problem. The growing organic chemical industry had begun to clamor for alcohol both as a solvent and as an intermediate. The introduction of the Coffey continuous still had made a great impression on the whiskey industry, enabling cheap, pure grain spirit to be used for blending with strongly flavored and more costly pot-still whiskeys which were in short supply in an expanding market. If whiskey distillers could use such a product, why should it not be made available for chemical manufacture? Excise duty was the problem, but eventually a case was made out and accepted that the chemical industry, which the government wished to encourage, should not be made to bear this burden provided that the spirit was denatured before use to make it incapable of illegal diversion to the potable market.

The original form of denaturing consisted in adding a proportion of wood naphtha containing methyl alcohol and a small percentage of pyridine-like substances, the intention being to make the spirit so nauseating to the taste that it could not be swallowed without vomiting. This treatment is still practiced and is used also for that sold to the general public which contains a dye for identification. Such a product is suitable for many industrial purposes, and is widely used, but eventually chemical processes were devised where this type of denaturant could interfere with the reaction or result in a product with undesirable odors. At a later date it was wisely decided to grant indulgences in special cases to permit the spirit to be denatured with other substances, usually one of the reactants in the process for which it was intended. For example, in the manufacture of ethyl acetate, acetic acid is usually accepted as a suitable denaturant.

Now alcoholic fermentation is the easiest fermentation known to man. Any natural-sugar-containing solution, such as fruit juice, when exposed to airborne yeast cells, will start alcoholic fermentation on its own, and this has long been known to ancient peoples throughout the world. Once Pasteur had made his famous observations, it was child's play to bring the process under technical control. Here was an organism able to multiply very rapidly, strongly competitive with other microorganisms in the consumption of sugars, able to synthesize the protein of its cell structure from simple nitrogen compounds such as ammonium sulfate, and welcoming other mineral salts which were not hard to identify. Leave it quiescent in such a sugar solution, and it would convert almost every trace of sugar to alcohol. Agitate and aerate the solution continuously, and it would abandon alcohol production to convert almost all the dissolved fermentable matter to more yeast cells. So two major industrial fermentation processes grew up together. Not only did the chemical industry get what it required in the way of industrial alcohol, but a handful of large yeast plants was able to supply a distribution network that satisfied the need for bakers' yeast throughout the country.

To make the cheapest industrial alcohol by fermentation, the cheapest source of fermentable sugar is required, and very quickly attention was turned from grain to blackstrap molasses from the West Indies, from Cuba, and indeed from any sugar-producing area of the world where transportation in oceangoing tankers was the available route.

As is well known, blackstrap molasses represents the concentrated juice of the sugar cane after all crystallizable sugar has been extracted, and it normally contains 50–55% of a mixture of invert sugar and sucrose, with all the mineral salts originally in the juice. These are adequate for yeast fermentation with the possible exception that a small addition of ammonium sulfate may sometimes be needed.

Industrial distilleries therefore sprang up on the main tidal estuaries of the country for two very good reasons. In the first place they needed seaport facilities for the delivery of molasses at very low cost. In the second place, they had to dispose of millions of gallons of spent wash from the distillation, an effluent very high in biological oxygen demand and a menace if discharged in any inland area. By discharging into a tidal flow they therefore reduced this nuisance to a minimum and satisfied existing legislation with regard to pollution. It was thus no accident that the main industrial distilleries and yeast plants were erected near the mouth of the Thames, the Humber, the Clyde, on Merseyside and on the Bristol Avon near its discharge to the river Severn.

The operational restrictions of the 1880 Excise Act were still imposed, but when the First World War started the government quickly recognized that these restrictions had not been specially designed to enable them to lose the war. Many of the restrictions were swept aside in order to allow all sections of the distillery to operate simultaneously and continuously so that maximum output could be achieved. One of the problems was that slow moving molasses tankers made almost a sitting target for German U-boats, but enough molasses came through to keep up supplies of alcohol to those factories engaged in the production of munitions and other vital necessities of war.

When peace came the restrictions were reimposed, but the authorities were now well aware of the importance of the industry to the economic life of the nation, and every possible interpretation of the law was applied which would ease the problems of the manufacturers without increasing the risk of loss of revenue by illegal diversion of spirits. It should be realized that by this time the standard rate of duty had risen enormously from what it was at the beginning of the century, and evasion of duty could have significant attraction to people prepared to evade the law. It must be said that the authorities cooperated wholeheartedly with manufacturers in finding safe and acceptable solutions to every problem of this nature.

It seemed that the production of industrial alcohol by fermentation was set fair for a long and prosperous future. Blackstrap molasses accumulated with every harvest in the sugar-producing areas. It was stored in such tanks as were available, but bulk storage tanks were an expensive investment for waste products of little worth. The sugar producers had to dispose of the accumulated molasses before each crop of cane came along, or discard it altogether. In consequence it was frequently available at give-away prices, and the main cost to the user was in carrying it away. What was more, at the beginning of the 1930s the Caribbean was faced with a glut of sugar and could not even dispose of the whole of its main product. This meant economic disaster until the blow was softened by producing a new type of molasses. The whole cane juice was treated by acid inversion to degrade the sucrose to hexose and then concentrated to give a golden brown syrup that would not crystallize and contained 70–75% total sugars. This was called "high test" molasses and earned a substantially higher price than blackstrap because of the resulting reduction in freightage costs. Two tankers of "high test" had approximately the same fermentable sugar content as three tankers of blackstrap, and required correspondingly less storage space at the distilleries. Moreover, it was an acceptable method of avoiding duty on what would otherwise have been an importantion of pure sugar. The high test molasses contained only about 2% of inorganic salts, compared to about 12–13% ash in blackstrap, and was a much happier medium to adjust for consistent and high-yielding fermentations.

By this time a number of highly efficient multicolumn continuous stills had been developed by such makers as Lummus and E. B. Badger in America and Usines de Melle in France, and the recovery of highly purified spirit began to approach the theoretical. Demand for absolute alcohol increased, produced mainly by the benzene azeotropic method in a single column fitted with a decanter, and it was possible to produce absolute alcohol at a price which allowed it to be blended with petrol to give a motor fuel having a high antiknock rating.

This was the golden age of fermentation alcohol for industry, but few could have foreseen how short it was to be. There were three factors that brought it to an end. At the beginning of the Second World War, the first factor was only a shadow of the type that threatens every fermentation process. Synthesis of ethyl alcohol from petroleum was already a reality, though still a doubtful economic competitor with the fermentation product. In the second place, the war was hardly under way when the government put an almost total ban on the importation of molasses for alcohol production. It could not afford to risk three tankers loaded with molasses in convoy when one cargo of industrial alcohol from the United States or other available sources would satisfy the same end. British distilleries closed down or were used for other war purposes, and the technical staff and operatives diverted to other duties. When peace came, there was still the prospect of returning to the position as it was in 1939, but now the petrochemical industry began to grow on a scale never contemplated 6 years before, and synthetic alcohol became a real threat, even if fermentation efficiency and cost could be held at the previous level. To make the situation worse, the third blow was struck at the fermentation process. A practice that had only been used on a limited scale in the United States before the war, of mixing molasses with cattle feed, became widespread almost as soon as the war was over. Both administrations supported the farm vote in various ways, and the featherbedding which this produced enabled American farmers to buy molasses at a price with which no distiller could compete. Demand began to match supply, and for the first time in history there was a seller's market for molasses. There was now no question of competing with synthetic alcohol, and the death knell had sounded for the fermentation process. One distillery after another was closed down, and although one or two distilleries were retained to use bargain lots of molasses picked up in odd corners of the world, the golden age was over.

It was another example of the fact that supplies of raw materials at give-away prices seldom last. There is a not entirely apochryphal story of a small island republic that lies not far to the west of the United Kingdom and is famous for growing potatoes. These were classed as grade one for human consumption, grade two for animal feeding, grade three fit for neither man nor beast and a dead loss to the farmers. Unfortunately grade three potatoes formed a high proportion of the crop. The government, seeing this, decided in its wisdom to put up one or more industrial alcohol distilleries — not large plants, not competitive with the scale of production abroad, but sufficient to provide the old country with its own requirements of industrial alcohol and to use up grade three potatoes at dirt-cheap prices as raw material for the fermentation. After the first distillery went into operation with much acclaim, the price of grade three potatoes on the open market climbed steadily until it was higher than the price for grade one, and that was the end of a beautiful dream.

These situations can, however, change almost overnight. What, for example, will be the effect when Britain succeeds in joining the European Common Market? Will it find itself within a wall of beet sugar producers, and cane sugar from the West Indies excluded from its purchasing ability? This could mean another glut of unsaleable sugar in an area that depends on little else for its livelihood, with resulting poverty and destitution. The answer may lie in industrial alcohol once again.

III. The First Closed Deep Fermentation

Shortly before the First World War began, Chaim Weizmann was undertaking research on synthetic rubber in Manchester University, and the process that he visualized required economic supplies of normal butyl alcohol. He knew of the work of Fernbach, who had demonstrated that certain strains of *Clostridium* would ferment starch under anaerobic conditions to give a mixture of acetone and normal butyl alcohol. Weizmann obtained such an organism, though it is possible that he did not isolate it himself, since the isolation of such anaerobes is not a matter for the unskilled, and he was no bacteriologist. But he learned to culture it in suitable media, and found that it would ferment nearly 4% of starch, of which approximately 30% was converted to a mixture of volatile solvents. These consisted of about 60% of normal butyl alcohol, 35% of acetone, and 5% of other compounds having a boiling range intermediate to the two major components. There were small amounts of ethyl alcohol, isopropyl alcohol, and mesityl oxide.

Early in this story the First World War began, and Weizmann was quick to appreciate the new situation. Acetone was an essential solvent in the production of certain types of explosives such as cordite for small arms ammunition. Strange as it may seem, up to 1914 Britain had imported almost all of her supplies of this solvent from Germany, and had no acetone production of her own. Weizmann had no doubts in his own mind that his fermentation would solve the problem, and he pushed all thoughts of normal butyl alcohol into the background.

For months he struggled through the hierarchy of Whitehall along with many other inventors, some of them equally brilliant, some of them insane, but finally he succeeded in getting an interview with Lloyd George himself. The Prime Minister was tremendously impressed. It is said that when he was asked about munitions problems in the House of Commons, he waved a tube of Weizmann's culture in the air, saying "Gentlemen, acetone, this is the answer." At any rate, Weizmann was given carte-blanche to gather a team of active young scientists around him, and it is interesting to note that one of them was a young Ph.D. from Manchester named Thomas Kennedy Walker, who later became Professor T. K. Walker of Manchester University and one of the leading authorities in the world on industrial fermentation processes.

The team was given plant facilities in the south of England with a miscellaneous collection of equipment, including of all things wooden fermentor vessels in which to operate an anaerobic fermentation with an organism terribly sensitive to infection. Very little was known about the behavior of the organism even under laboratory conditions, let alone on the large scale, and it was only the single-minded devotion of the team under the inspiration of Weizmann's leadership that brought the project to fruition. It was this small band of workers and those they trained after them, who produced a generation of craftsmen in industrial bacteriology, with a skill that made most medical bacteriologists look like plumbers. Men like Fred Arzberger in the States, who devised outstanding techniques for the isolation of anaerobes, and Charles Grover of the Distillers Company, whose manipulation and study of single microbial cells was pure artistry. What a pity that today everyone in the life sciences wants to become a molecular biologist, and few graduates can transfer a pure culture without getting it contaminated.

After many struggles the team at last began to produce acetone, but seldom in very good yield, and before long a very serious situation arose. The country was on the verge of starvation due to submarine attack on our food convoys, and we could no longer spare the corn required for the fermentation. There was only one sensible answer, to move the fermentation westward, first to Canada, where the team acquired two brilliant young workers in the shape of D. A. Legg and Clarence Hancock, who were later to become outstanding exponents of industrial fermentation. Then, with the entry of the United States into the war, the whole operation was moved down to the corn belt in Indiana. At the end of the war the plant was sold to commercial enterprise, and Commercial Solvents Corporation undertook the operation of the process.

Weizmann was not forgotten in Britain. He was offered almost every honor and reward in the government's power. His reply made history. All I want is a home for my people in Israel. The answer to this was the Balfour Declaration. The new state of Israel became reality, and Chaim Weizmann was its first President.

Back in Indiana all went well for a time. The organism steadily fermented up to 3.8% of starch and gave a 30% yield of solvents with a ratio of one part of acetone to two parts of butanol (the registered trademark of C.S.C. for normal butyl alcohol). In the last stages of the war the accumulation of butanol had been a headache. There was little outlet for it, and it was a toss-up whether to discard it or to put it into storage for future possible use. But immediately after the war came the development of cellulose lacquers for spraying car bodies and other articles of commerce. Butanol and its ester, normal butyl acetate, were heaven-sent solvents for these lacquers, and the fermentation process rode on a wave of success. Commercial Solvents Corporation were far from satisfied. They were aware of the coming threat of synthetic acetone from petroleum, and synthetic isobutyl alcohol was a growing competitor. The fermentation process yielded little more than 1% of mixed solvents in the fermented medium, and recovery by distillation was expensive and complicated, particularly as batch distillation was employed for final fractionation to give individual solvents of the purity required for the market.

An extensive research department was set up, with Weizmann retained as adviser and consultant. In one of the many projects for improving efficiency, men like Arzberger and Legg persistently isolated new strains of the organism from natural sources, in an attempt to find a strain that would ferment higher concentrations of starch than the 3.8% that had so far proved to be a maximum. Organism after organism was tested, indexed, and put on the shelf as no better than its predecessors.

Suddenly someone saw the light. Why could they not switch to molasses, as the alcohol producers had already done. Surely there was a *Clostridium* that would ferment sugars instead of starch. As it turned out, the sensible thing was done first. Previously isolated cultures were pulled down from the shelf, and each was tested in a medium where starch was replaced by molasses sugars. Thus came to light the famous C.S.C. No. 8 culture, the first of a long and improving series. It would ferment at least 6% of sugar and gave a concentration of nearly 2% of solvents in the fermented medium. Distillation costs were nearly halved at a single stroke, and cheap molasses replaced corn which had fluctuated unpleasantly in price from year to year.

The American situation is described at some length because of its later impact on the British situation. In England, an attempt was made soon after the first war to start up an acetone-butyl alcohol plant at Kings Lynn, using the old starch-fermenting organism. Shortly after the plant reached full-scale operation there was a violent explosion and the whole factory was destroyed. The cause was never determined with certainty. There are two major hazards in the process. A large proportion of the fermentable sugar is converted to a mixture of hydrogen and carbon dioxide. Although 60% of this mixture consists of the latter, when mixed with the right proportion of air it is violently explosive and readily detonated. In addition, acetone vapor itself is a second dangerous hazard, the explosive limits in air lying between 3% and 80% of acetone vapor.

No attempt was made to reinstate the factory. Some years later the Distillers Company made a reassessment of the process on a pilot-plant scale, but concluded that it was uneconomic in the circumstances prevailing at that time. They were no doubt right. But by

1935 the position had changed. Cultures had been discovered that would ferment up to 7.5% sugar, and high-test molasses had arrived. The cost of carriage by sea to an English port was if anything less than the long overland haul to Indiana and Illinois, where the two main U.S. plants were now in operation.

Commercial Solvents Corporation made an intensive study of the situation, and decided to erect a plant at Bromborough, across the river Mersey from Liverpool. Harbor facilities for the largest tankers of the time were available at the adjacent docks at Port Sunlight, where a large molasses-receiving tank farm was installed. The factory was designed to produce both industrial alcohol and the acetone-butyl alcohol complex, with separate fermentor houses and distillation units for the two processes. Commercial Solvents Corporation had sufficient confidence in the standard of scientific teaching in this country to man the technical staff entirely with British-trained personnel, under the leadership of John Hastings, who had been fortunate enough to have studied under T. K. Walker in Manchester.

Apart from minor teething troubles the solvents process moved smoothly into operation at the end of 1935, using immunized strains of *Clostridium saccharo-acetobutylicum* developed in the research laboratories at Terre Haute, Indiana. These at one time had rescued the Indiana plant from what appeared to be total disaster. After several years of almost uneventful operation the Indiana fermentations were struck by a phage infection that caused a total loss of production. With typical American vigor, two decisions were promptly made. First, to put the whole research team onto the problem of a remedy, and second, to build an entirely new plant 200 miles away at Peoria in Illinois where operations might possibly be carried out free from infection. Long before the British plant was built the problem was well under control. The research team had developed superb techniques for the immunization of bacterial cultures from phage, and although they were long kept secret for industrial reasons, it is certain that they were well in advance of the general knowledge at the time. Meanwhile, the Peoria plant had come into highly successful operation.

The acetone organism is satisfied with very simple nutrition. For many years the practice was to add only ammonium sulfate and chalk, together with a little calcium superphosphate, to the solution of molasses, the purpose of the chalk being to neutralize acidity as the nitrogen was used by the organism. These were the only mineral needs, apart possibly from trace elements naturally present in the molasses itself.

In view of the sensitivity of the organism to infection, fermentation is of course carried out in closed deep fermentors that are sterilized under steam pressure, as is the medium itself. Pure cultures are usually grown up through five stages, the last two in the plant itself in successively larger steel vessels before being added to the final fermentor. As the organism begins to multiply the medium rapidly becomes more acid, reaching a trough about pH 5.0, after which the pH normally rises rapidly and the sugar is converted into solvents. If the medium is badly adjusted, the fermentation may stay in a pH trough and utilize the sugars without making any of the desired products. In view of this, Bromborough workers devised a method of adding aqueous ammonia semicontinuously on a logarithmic scale to match the growth of the organism, and in this case, only superphosphate was added to the molasses medium. The method gave such satisfactory control that there was never any further problem of remaining in the pH trough.

When one considers that nearly 70% of the sugar converts to hydrogen and carbon dioxide, and that each large-scale fermentation can contain between 3 and 4 tons of sugar per 10,000 gallons of medium, it will be seen that the volume of gas produced is quite substantial. To circulate it in the closed fermentation system in order to maintain anaerobic conditions is a relatively simple matter, but in the end it must be blown to waste unless a use can be found for it. Fortunately when the carbon dioxide is scrubbed out the remaining hydrogen has a very high degree of purity. In the United States this hydrogen was used in synthetic chemical processes, and in England for the hydrogenation of low-melting edible oils, thus assisting the overall economics of the fermentation process.

When the Second World War came, although the production of fermentation alcohol was brought almost to a standstill, production of acetone was given the highest priority and the country's fermentation capacity was rapidly increased by sequestering suitable alcohol fermentors. Because there was very little auxiliary fermentation equipment to go with them, a successful semicontinuous method of fermentation was devised. This was helped considerably by the aqueous ammonia technique. A large inoculum was prepared equal to one-third of the volume of the final fermentor, and grown until the organism was reaching the peak of logarithmic multiplication. At this point it was possible to introduce the total ammonia requirement of the complete fermentation and to add the remaining two-thirds of the medium. Growth continued without any significant lag phase, and after a very short interval, it was possible to transfer one-third of this volume to a second fermentor, again adjust the ammonia addition, and repeat the process indefinitely as long as sterility was maintained. Fermentation in the remaining two-thirds of the medium was extremely rapid, complete conversion to solvents taking about 12 hours.

It was in fact the speed of this last step that led to a still simpler operation. A complete fermentor grown as a culture and prepared under sterile conditions was used for the simultaneous inoculation of three fermentors that were made up to full volume with unsterilized molasses solution. Molasses is always heavily populated with casual microorganisms, but these are largely dormant in the concentrated material, and the high population of actively multiplying acetone organisms, already reaching a maximum at the time of mixing with unsterile medium, ensured that the fermentation went through to a successful conclusion.

Although the yield efficiency was usually slightly lower than achieved under completely aseptic conditions, there was a most substantial saving in steam utilization at a time when fuel economy was of the utmost importance. Moreover, an enormous improvement in throughput was obtained, the average fermentation cycle being 30–32 hours compared with the 90 hours required under more orthodox batch conditions.

It is interesting to note that the government, taking a long-term view of the war, decided that importation of molasses for acetone production would eventually have to stop, and sought ways of making acetone from entirely indigenous materials. The route decided on was to make calcium carbide from our own coal and limestone, to generate acetylene from the calcium carbide, and to convert acetylene to acetone by the well-known catalytic process. Canada gave tremendous aid in this respect by helping to design the plant, by training a team of our own technicians at Shawinigan Falls to undertake the job, and by sending the best of their own technical men to Britain to supervise and operate the start-up. Among these was Dr. Hugh Sutherland who eventually became President of the Shawinigan Chemical Company.

Unfortunately, while our men were in Canada, Norway was invaded and conquered by Hitler. Norway had provided almost the whole of our calcium carbide for acetylene welding, and the plant intended for acetone had to be hurried into production to keep our engineering industry going. By 1941 it was in operation and saved a very critical situation. No acetone was made by this route, and a great deal had to be imported from America. On one occasion a ship loaded with 50-gallon drums of acetone was torpedoed in our western approaches.

Those drums which did not explode floated and came ashore all round the coasts of the British Isles. They were salvaged and collected at the Bromborough plant, for redistillation where necessary. It says much for the quality of American drum manufacture that nearly all the drums were unbreached and contained acetone in perfect condition, though the drums themselves were battered into inconceivable shapes.

After the war, when the Bromborough plant tried to restore normal commercial conditions, it found itself facing the same serious position with regard to synthetic competition and molasses supplies that had affected industrial alcohol production. A highly efficient fermentation was developed using home-produced beet molasses, giving higher fermentation yields than ever before, but in the end competition became too great and the process was finally closed down in the early 1950s. It had been highly successful in its day, largely due to the technical influence of E. J. Farrar, who had developed his skill as an industrial bacteriologist under the careful eye of T. K. Walker.

Before it closed down, however, the acetone fermentation played an interesting part in the development of fermentation for the production of the vitamin riboflavin.

IV. Production of Vitamins

There have been many attempts to use waste products from fermentation processes for animal feed, some of them highly successful. In Scotland, screened solids and dried solids from grain alcohol fermentations, known as "draff" and "dreg" respectively, have found a useful place for this purpose. Mycelium from antibiotic fermentations has been used, but not so successfully, since it is usually too high in phosphates and unless dried or used when fresh it can autolyze most unpleasantly. In areas where disposal of spent liquors from fermentation processes is a major problem, total evaporation to recover both soluble and insoluble solids is often found necessary.

With increased awakening to pollution of rivers in the United States, the acetone-butyl process found itself seriously up against this problem, the Wabash and Peoria river no longer being available for such an effluent. Total evaporation of the solvents fermentation residue gave a golden powder rich in available protein and eminently acceptable as an animal feed supplement to enrich carbohydrate material. Best of all, analysis showed that the dried "slop" contained a very useful amount of riboflavin, up to 80 $\mu g/gm$, together with a number of other growth factors in lesser amount, such as pantothenic acid and thiamine. This considerably added to the value of the dried material, since it could be used more sparingly in a mixed feed and not merely to provide protein. After very successful trial marketing there was no hesitation in installing large multiple evaporation units to take care of all available liquid residues from the fermentation. The product soon took a useful place in the animal feed spectrum.

In England, the process was on the drawing board almost at the beginning of the Second World War, but although the country was crying out for animal feed and much of it had to be imported by sea, the government regarded the project as of secondary importance and put it on the shelf. In consequence thousands of tons of rich animal feed were discarded into the tidal basin of the river Mersey throughout the war, among the wrecks of ships that had tried to bring animal feed into the port of Liverpool and had failed on the very doorstep when the Luftwaffe found them there.

When the project was pulled down from the shelf it was too late. Pure synthetic riboflavin was already available, and the solvents residue could not compete as a source of the vitamin in spite of its added attractions. Research in this field had not however stood still. More than one microorganism had been discovered that would produce riboflavin in a suitable medium. Bromborough workers made an extensive microbiological study and finally selected a strain of *Eremothecium ashbyii* as their organism of choice. The complete process was worked out, including the method of recovery, and showed so much success in pilot-plant tests that for full-scale working the process was transferred to another fermentation plant in the group before the Bromborough factory closed down.

A comparatively simple technique enabled crystalline riboflavin to be recovered in high yield, but the majority of feedstuffs manufacturers found it preferable to purchase the total evaporate from the fermentation, blended to contain a standard 5% of riboflavin in the product. In this way they obtained free a protein-rich carrier and the first stage of dilution of the riboflavin in the feed was undertaken for them. Blending pure crystalline riboflavin in the coarsely particulate feed at a rate of a few milligrams per pound is not an easy operation and requires at least three successive mixing stages to obtain effective and accurate distribution. The acceptability of the dried concentrate enabled the fermentation process to compete with synthetic material long after the latter was marginally lower in cost, but as feedstuffs manufacturers began to modernize their mixing plants to introduce other trace ingredients, interest in the fermentation product declined until it was no longer really profitable.

Almost unnoticed, another fermentation process for the production of a vitamin crept into being during this period. This was for the production of ascorbic acid. The process is not a complete fermentation in the sense of using a microorganism to obtain an end product from the natural constituents of the medium in which it is cultivated, but it is most important historically as the forerunner of many later processes where microbial enzymes are employed to carry out an intermediate step in a series of chemical reactions to obtain an end product. Such a process is now well recognized as the cheapest and most efficient in the preparation of a number of synthetic hormones, although Britain has not been among the leaders in this field.

The use of fermentation in the preparation of ascorbic acid lies of course in the oxidation of D-sorbitol to L-sorbose. The process has been carried out commercially for a number of years, and has been studied by many workers, particularly in Japan. More than 90% conversion of sorbitol to sorbose is commonly obtained, and manufacturing competition is very strong (Spencer and Gorin, 1968).

No vitamin fermentation has attracted so much attention in recent years as that for the production of vitamin B_{12}. There were three important steps in the development of this process. First came the brilliant work of Lester Smith and his team in the Glaxo laboratories, when they extracted the pure vitamin from liver (Lester Smith, 1948). The second was the almost chance discovery of the vitamin in antibiotic residues by Merck workers in America. They had in fact isolated the crystalline vitamin almost simultaneously with the Glaxo discovery, and it was their knowledge of this field that helped them to ask the right question when a puzzle arose in their other studies.

At the time when the nature of the vitamin was revealed, an interesting but still controversial discovery was made in the field of antibiotics. Penicillin, when added to animal feed in exceedingly small amounts, was found to produce significant increase in growth rate and in feed utilization. Several other major antibiotics were quickly shown to give a similar result. Merck workers had no difficulty in demonstrating this effect with aureomycin, but when they used fermentation residues containing a measurable amount of aureomycin, the effect was again significantly greater than that expected from the equivalent of aureomycin itself. The right men asked the right questions and found the right answer—vitamin B_{12}. The concentration in the fermentation was minute, but on the basis of their previous knowledge they devised the most ingenious multistage recovery process that successfully yielded a crystalline product and ended with a series of worldwide patents on their work.

Quite naturally, the fermented medium from every type of antibiotic fermentation was examined for this product, and in Britain spent streptomycin broth was soon shown to contain about 0.5 µg/ml. This was more than sufficient for effective recovery, and suitable extraction plant was quickly installed.

The remarkable effectiveness of this vitamin in the remission and cure of pernicious anemia immediately created a world demand, and in the early 1950s the pure substance sold at around £200 per gram. Many lesser clinical conditions showed at least subjective improvement, though in some cases it was difficult, and still is, to establish scientific yardsticks to demonstrate effectiveness beyond all doubt. In the case of pernicious anemia, there was no question. The total quantity of vitamin available from spent antibiotic fermentation liquors proved inadequate to meet the demand, and a search was made in other quarters. Effluent from the activated sludge sewage disposal process was shown to contain recoverable quantities, and plants were set up in a number of European countries.

Finally the research followed the pattern of the advance of riboflavin, and microorganisms were isolated that would give fifty times the concentration of vitamin found in spent antibiotic broth. Strains of *Propionibacterium* were introduced in Britain suitable for fermentation for the vitamin alone, and before long media were developed that would yield 25 µg/ml in the fermented broth. This greatly simplified extraction procedure, and brought the cost of manufacture tumbling down. When the international price fell to £3 per gram, the bottom dropped out of the market for those manufacturers using residual processes, and for most of the major manufacturers who could compete at this price the commercial attractiveness had gone. It was an area where volume of sales could not be increased sufficiently to compensate for the fall in price, as it is doubtful that the true world requirement has ever exceeded 300 kg per annum. In consequence there are very few serious manufacturers today. From the point of view of fermentation technology this is a pity, because the process has shown itself to be ideally suited to modern techniques of continuous fermentation, but even with a world monopoly the fermentation unit would be small indeed.

Like every modern wonder drug, there has been considerable misuse of vitamin B_{12}, which fortunately is quite harmless even where it does no good. A great deal has been poured into multivitamin preparations regarded by the medical profession as of very doubtful value, but this has done nothing to obscure the outstanding merit of the product in treatment of the dreadful ailment of pernicious anemia. It is a contribution to the medical armory that is likely always to be needed and in its own way ranks in significance with the major antibiotics.

V. Citric Acid and Surface Culture

If one is to consider the advance of British fermentation industry in terms of groups of products, this is the point at which to review organic acids, because the early technique employed for one of them had a profound influence at the beginning of the antibiotic era which was still to come.

Little needs to be said here about the ancient art of vinegar-making, as traditional to the Englishman as his fish and chips. The process has not in living time provided acetic acid of industrial quality for chemical manufacture and other processes. Synthetic routes starting with the oxidation of alcohol and later of petroleum have long been the source of glacial acetic acid at low cost. In terms of tonnage this is one of the leading organic chemicals in the country, but fermentation has no part to play.

Lactic acid fermentation started on a really substantial scale in the second decade of the century, and the history of lactic acid in Britain is the history of Bowmans of Warrington, a family company that devoted itself to this purpose. Every microbiologist knows that lactic acid-producing organisms are among the most widespread in nature, but many of them are temperamental and not easy to maintain viable in pure culture. It is not too difficult, however, to isolate and select useful commercial strains that will convert sugar to lactic acid in high yield providing most of the acid is removed as it is formed by precipitation as the calcium salt. The main problem in lactic acid production is not the fermentation but the recovery of the free acid. Most of the modern research in this field has been undertaken by Bowmans. Dark colored syrups have been found acceptable for a number of commercial purposes, but for edible lactic acid and the production of high quality organic esters as used in synthetic lacquers, a colorless acid free from impurities is desirable. For this purpose the Bowman company developed an excellent solvent extraction process that has been well described in literature (*Chem. Ind.*, 1959). This gives the highest quality of lactic acid available in commerce. Although lactic acid has no single major use, it has such a variety of applications that the total demand makes its production a worthwhile part of the fermentation industry.

Even before the First World War, attention was being given to the question of producing fumaric, gluconic, and citric acids and the possibility of using molds for their manufacture. While production of the first two has since been developed on quite a useful scale, it was citric acid that attracted the most immediate attention when the first war ended. The soft drinks and confectionery industries were both very substantial potential customers, and up to this time had obtained all their supplies from the citrus fruit industry, chiefly from the Mediterranean area. Agricultural neglect during the war and growing labor costs had led to shortage of supplies and higher prices for citric acid than ever before. In the end the fermentation process was to destroy the citrus fruit industry as an economic source, though there were many headaches on the way.

Strains of *Aspergillus niger* were selected for the industrial process. One of the biggest problems was to isolate and maintain actively producing strains and to avoid mutation to nonproducers or strains that would convert sugar to anything but citric acid. Much of the laboratory technique employed for this purpose is still kept secret by firms engaged in the industry. There were immense problems with the medium, particularly with regard to trace elements that could exert an inhibiting effect. When in 1925 T. K. Walker set up the new department of Industrial Fermentation in the College of Technology, Manchester, his first field of research was the citric acid fermentation, particularly the critical pathways of the degradation of sugar to the free acid. There were so many attractive facets to this study that later Professor F. Challenger collaborated with him, and their many papers on the subject are listed in the chapter on the "Manchester School" in Volume 12 in this series.

In the present chapter it is the technique of fermentation that is of interest, because for the first time the method of surface culture was used on an industrial scale. It is hardly necessary to repeat the well-known fact that aspergilli, penicillia, and actinomyces are strongly aerobic and in still conditions will form a filamentous growth on the surface of a solid or liquid substrate containing suitable nutrients. They are susceptible to competing organisms, and many bacteria will multiply and consume nutrition more rapidly. In consequence pure culture is necessary if a desired end product is to be achieved. This was the knowledge that faced the pioneers in the field. They could culture and sporulate and select their organisms on solid media, but to study their metabolism in liquid media it was considered necessary to grow the organism on the surface of a very shallow layer of liquid. Only in this way could the required oxygen uptake be obtained. In the laboratory, conical flasks were the vessels of choice, enabling a large area of liquid usually not more than 1 cm in depth to be employed. For inoculum, spores were floated lightly onto the surface, and great care was taken never to disturb or submerge the mycelial growth since this was considered to be one of the main causes of changes in metabolic behavior. For industrial production of citric acid, considerable ingenuity was shown in designing suitable vessels that would permit the handling of larger volumes of liquid. These in principle consisted of rectangular or vertical cylindrical chambers that could be sterilized and contained a series of superimposed shallow trays filled by a suitable cascade system with the liquid medium, either presterilized or sterilized in the vessel itself. After inoculation the vessel was maintained at the correct incubation temperature, when a mat of mycelium spread across the whole surface of the liquid in each tray and the fermentation proceeded to completion. The contents of the trays were then bulked for citric acid recovery and the mycelial felt was discarded.

A great deal of citric acid has been made in this way, and all our needs have been supplied by it for more than a generation. It rapidly undercut the price of citric acid from natural sources, but this very success was a sad misfortune in another direction, because it colored the thinking of those who were to start on fermentation's greatest adventure, the production of penicillin. Not until modern techniques of antibiotic fermentation were fully established did the citric acid producers begin to realize that they had stumbled up a cul-de-sac, and had almost the whole fermentation problem to tackle again if they were to make real progress.

VI. The Entry of Antibiotics

It is of course not true to say that penicillin was the first antibiotic to be discovered. Many observations of different antibiotic substances had been reported in the literature of the previous 50 years, and were reviewed by Waksman (1941). Penicillin was, however, the first antibiotic to be brought into general use, and in Britain has always been regarded as the beginning of the story. Many writers have described the dramatic events that took place, so that there is no need to describe here the chance observation of Fleming (1929), the abortive attempts by Raistrick and his colleagues to capture this evanescent substance (Clutterbuck et al., 1932), or the careful library studies by Chain that selected penicillin as the target of research he was so successfully to pursue in conjunction with Florey and the team they gathered around them in Oxford. Microbiologists, biochemists, pathologists, they all made great contributions, from N. G. Heatley, E. P. Abraham, G. G. F. Newton, and Lady Florey herself, down to modest laboratory technicians like George Glister who extracted the first penicillin administered to a human patient.

The studies began in 1939, and first results were reported a year later (Chain et al., 1940). By 1941, in spite of the growing promise of the substance and appeals to the highest level in the government, Florey knew that there was no chance in Britain of a major industrial effort to make penicillin available for general clinical use. We were still inching away from the total defeat that had stared us in the face after Dunkirk, and everyone had a desperate and existing job to do.

What a wise and wonderful decision it was to let Florey go to the United States and appeal to scientific centers for help. His visit is well described by Herrell (1945). The fantastic range of studies that followed was soon coordinated by the Office of Scientific Research and Development to help the project gather momentum and to avoid unnecessary duplication. Information was circulated confidentially to all accredited investigators, and quite early in the scheme leading industrial fermentation companies were brought into the picture. The Office of Scientific Research and Development maintained liaison with the British Ministry of Health, and there was a regular exchange of information on all the factors relating to production.

Chain and Florey had from the first used the traditional method of surface culture for fermentation with Fleming's strain of *Penicillium notatum*. As their work progressed, the need for more and more penicillin for animal and clinical tests and for fundamental studies on the substance began to tax the resources of the Oxford laboratories to the limit. Bath tubs and milk churns were used as storage tanks in the extraction circuit. Every type of vessel that could hold a shallow layer of liquid was employed for the fermentation, even including bed pans. It was time for the Ministry of Health to provide further assistance.

It is sad but true that history shows our government to have blundered many times both in obtaining scientific and technical advice and in the direction of scientific effort under its immediate control. Even Churchill depended too much on a small handful of scientific pooh-bahs who true enough were brilliant in their own fields but as ignorant as the rest of us in others. In consequence the importance of many projects was not recognized as soon as it should have been, and quite a few projects were placed in the hands of people far less competent to undertake them than others who were available but undiscovered.

It was at last realized by our authorities that Florey and Chain needed help to make more penicillin for their work and that penicillin was a pharmaceutical substance. What more simple? Obviously the right thing to do was to bring the big pharmaceutical companies into their confidence and all would be well. No one in government circles had so far gathered that penicillin was made by a fermentation process. Let it be said at once that the pharmaceutical companies who were invited to help gave their assistance loyally and unsparingly, and are to be honored for what they did. Their staffs worked uncounted hours, inspired by an ideal, and from them came men who led the antibiotic industry with great success in later years. But what did they know? They were shown the technique used at Oxford, they knew at least vaguely of the wretchedly successful citric acid process, and the answer was clear. The surface culture technique must be expanded in scale as never before, so that acres could be covered with Erlenmeyer flasks and any similar vessels that would hold a shallow layer of liquid. Thus began the days of the "bottle plants," so named because one of the favorite receptables was the standard 1-pint milk bottle, that could be produced in large numbers without difficulty by existing glass factories.

Incubator rooms were built that would house an indoor sports arena, and more than one plant had at least 100,000 milk bottles incubating at any one time. It was impressive to see such an array, every bottle plugged with cotton and resting at a shallow angle on tiers of racks. The labor demand was heavy, but largely consisted of working housewives and girls straight from school. After fermentation came the operation of unplugging and emptying each bottle into bulk collecting tanks for recovery of the penicillin. Fortunately this was an area where the pharmaceutical houses had a variety of experience in the extraction of drugs. Chain had developed alternative methods of extraction that were simple to apply, and the only directives he had to give were that the process should be carried out quickly and at as low a temperature as possible. In spite of the fact that the titre in the flasks rarely reached 10 units/ml and was usually much lower, and many flasks were contaminated and produced nothing, a steady growth was maintained in the amount of penicillin produced. After Oxford had been furnished with a meager supply, there was still only sufficient to be used as a last resort in critical military cases and later for civilians for whom there was no other hope. And it worked, as Florey said it would. There was a story told in the last year of the war, after it had been demonstrated that a million units would cure one man of pneumonia or ten men of gonorrhea. Only one man could, and did, make the decision to get as many troops as possible back into the firing line. This may or may not be true, but it illustrates the desperation of the situation when penicillin had been shown to work its miracles and there was still not enough to go round.

Back in America, by 1943 the story had taken quite a different turning. Many useful discoveries had been made. New strains of *Penicillium notatum* and *P. chrysogenum* were isolated that gave substantially increased yields. At the Northern Regional Research Laboratory at Peoria it was shown that corn steep liquor was an outstanding nutrient that raised titres still higher. Best and most important of all, the technique of surface culture was proved to be a shibboleth, and deep fermentation came into being as the only route that would satisfy the world's demands. By the end of the year several industrial plants were on the drawing board, and the speed with which they were erected and brought into production in the following months was truly amazing. They made mistakes, and recognized them quickly, ripping out equipment that was unsatisfactory and replacing it with other units until the whole flowsheet ran smoothly.

By late 1943 the authorities in Britain knew that a deep fermentation process had been shown to work. The answer was not quite as simple as that. A scientific delegation had to be sent to the States to examine the situation and report back on the merits of the two processes. Happily there was at least one man in the delegation with seeing eyes and the ability to make an unambiguous decision. The British authorities were left in no doubt that deep fermentation was the method that would bring the results they wanted. The Glaxo company, which had made the bravest effort of all in surface culture, was also deeply impressed and at once began to make its own plans to follow suit.

If deep fermentation was the way, the government had first to create suitable plants. Here the penny dropped at last. This was a fermentation process, and industrial fermentation was already practiced in Britain. Surely they could find immediate help if they looked around. The Distillers Company undoubtedly ranked as leaders in the industry, and Graham Hayman, chairman of the company, was invited to meet high officials in Whitehall. The writer was fortunate to attend as his observer. Hayman had a fine technical understanding, and in his younger days had operated an industrial alcohol distillery himself. He was not a man to vacillate over decisions. When he was invited to lend one or more of his plants so that the pharmaceutical companies could struggle into deep fermentation, he refused point-blank, though he well knew that the government could requisition his plants if it so decided. His answer was brief. If this is a fermentation process, my company will tackle the job itself, and will make all its resources available. We have good friends among the industrial fermentation companies of America and will invite them to cooperate.

A short survey showed that none of the British fermentor plants was immediately suitable. The best was the solvents deep fermentation plant at Bromborough, which only needed aeration equipment to go into use, but it was already critically engaged on its own duties for the country. Ten days later a Distillers technical team was on its way to the States, and after clearance between Washington and Whitehall, moved to the Commercial Solvents Corporation penicillin plant at Terre Haute, Indiana. This plant had come into operation in 100 days from the sketching of the first line on a drawing board, and was already doing a fine job. Cooperation between the two companies was excellent, and as joint agents for the Ministry of Supply they proceeded with a program of designing, erecting, and operating a penicillin factory at Speke, on the outskirts of Liverpool, where the river Mersey was once again to bear the burden of accepting an effluent from a fermentation process. The factory was designed to be twice as big as the C.S.C. plant, which made it the biggest penicillin plant in the world at that time. It was planned to produce 2×10^{10} units of penicillin per week, or approximately fifteen hundredweight per annum. We thought big in those days.

Meanwhile Glaxo was proceeding with its own deep fermentation plant at Barnard Castle in County Durham. Distillers had Ministry backing and all the alleged priorities that this could bring in the supply of equipment, they could assign from their existing plants chemical and mechanical engineers, industrial microbiologists, biochemists, solvents experts, and operating personnel who had grown up among fermentation processes, and they had the best of American support. Yet a year later, both plants came into operation almost simultaneously. It says a great deal for Glaxo drive and enthusiasm that they could match their rivals in this way.

When the first batch of penicillin passed all its tests with flying colors at Speke, the bottle plants were closed down at once. In the deep fermentation plant, a single 300-gallon seed tank, using a new culture giving the then phenomenally high titre of 150 units/ml, was capable of making more penicillin than all the bottle plants put together, and the fourteen 10,000-gallon fermentors meant that Britain was really in the business. The penicillin so produced was recovered as the sodium salt in the form of an amorphous brown freeze-dried sterile powder, of approximately 40% purity. There were two basic methods of recovering penicillin from the fermented broth. In the first the penicillin was adsorbed from the filtered broth by activated carbon and eluted from the carbon with acetone. From this it was converted to a crude aqueous solution of the sodium salt, the latter then being acidified and the free penicillin extracted into butyl acetate. These last two steps were repeated at least once, giving an increase in purity and concentration each time. Finally the solution of sodium salt was Seitz-filtered and freeze-dried. The process was cumbersome on the large scale and threatened to monopolize the total British output of activated carbon. Understandably it was abandoned after 3 or 4 months in favor of Chain's alternative process, the basic steps of which have been generally applied ever since. In this the broth is acidified immediately after fermentation and the free penicillin acid extracted into butyl acetate or other suitable solvent, from which it can again be extracted with aqueous alkali as the appropriate salt. Steadily increasing fermentation titres gave a growing efficiency at this stage, and the development of centrifugal extractors showing almost theoretical performance was one of the features of the period.

Production of improved mutant cultures by X-rays, ultraviolet light, and chemical mutagenic agents has been widely reported and described. The world owes a great deal to the University of Wisconsin for the great advances it initiated. Not many years elapsed before descendants of some of the early deep culture strains were showing yields of 15,000 units/ml or more. There even came a time when the value of further mutation became much less significant, because the economic value of higher titres had to be weighed against higher medium costs and technical problems such as the aeration of broth containing a high density of mycelial growth and the extraction efficiency that could be obtained in such circumstances. Every producer has had to work out his own problems in this respect (Hastings, 1958).

Be that as it may, increasing titres meant that fermentor output was multiplied many times, and when Speke capacity had reached one hundred times its original design, extraction facilities had to be extended very considerably. The most fortunate result was that the increased concentration of penicillin submitted to the very start of the extraction process helped to eliminate a number of recovery stages. The development by C.S.C. of a method of crystallization of penicillin salts by azeotropic distillation under reduced pressure of a butanol-water solution was a most important step, leading as it did to essentially pure compounds that could be submitted to chemical analysis rather than to the vagaries of microbiological assay.

The crystallization process followed hard on the heels of the revelation by fundamental studies at Oxford that "natural" penicillin produced by all the known cultures contained at least four penicillins differing from one another only in the side chain attached to the 6-amino penicillanic acid nucleus. Of these the most widely active and stable form was shown to be benzylpenicillin or penicillin G. It was but a short step from this to attempt to produce a preponderance of penicillin G in the fermentation by adding nontoxic precursors that would make the appropriate radical available to the mold. Quite soon it was shown that phenylacetic acid or its derivatives would do this effectively, and all production units adopted the method. Extensive continuation of these studies by the Eli Lilly research laboratories at Indianapolis showed that many new semisynthetic penicillins could be obtained by adding other mono-substituted acetic acid derivatives to the medium, but it was some years before this work resulted in a useful practical result.

Following the establishment of the Glaxo and Ministry of Supply penicillin plants, Imperial Chemical Industries and Boots Pure Drug Company, who had both been participants in the bottle plant adventure, built their own plants for penicillin manufacture on the large scale. Imperial Chemical Industries had not been altogether convinced that deep culture was the right answer, and began by building what was probably the finest surface-culture plant in the world, with huge horizontal cylindrical incubator chambers capable of sterilization under pressure, fitted with an admirable cascade tray system and provided with every type of control device then known. As leading organic chemical as well as pharmaceutical manufacturers they were able to design and install an extraction process that compared favorably with any other. Moreover, as soon as the nature of penicillin G was known, their own research showed it could be precipitated as an insoluble salt in a high state of purity and quantitative yield from a fairly impure solution. The salt was the N-ethyl piperidine derivative, and it became for a time the basis for quantitative assay for penicillin G (British Pharmacopoeia, 1958). Although this salt was too toxic for clinical use, it formed a valuable intermediate in the purification process, as it could be regenerated to the free acid and thus to any clinically acceptable penicillin salts.

This step for a time helped I.C.I. to take a lead in the quality of penicillin produced in Britain, but they soon found that the output of their surface culture plant could not match their capacity to extract penicillin. Quite undaunted, their engineers coupled their horizontal cylindrical incubators in pairs face to face and erected them vertically as deep fermentors. This was one of the most rapid and ingenious improvizations in the British penicillin story. While the changes were made, they refused to go out of production, and persuaded the Ministry to supply primary penicillin extract from the Speke plant to keep their purification process going. By this time the fermentation yield at Speke was marching steadily ahead of its finishing capacity, so the combined operation made a very useful contribution to the national output.

By the end of 1947 the government had seen the supply of penicillin increase until it was available for all home needs and export could be permitted. The cost had fallen to a small fraction of what it was when deep fermentation began. At the time, a Socialist government was in power. Nye Bevan, a left-wing extremist who believed in total nationalization, was in office as Minister of Health, and the National Health scheme was the biggest consumer of pharmaceuticals in the country. The Ministry of Supply penicillin plant at Speke was a successful and viable concern, yet strangely enough, a decision was taken to release the factory to private enterprise. The Distillers Company was glad to accept the opportunity and formed the Biochemicals Division to operate the plant.

Plans were immediately made to expand by producing one or other of the new antibiotics then being reported. The first choice was streptomycin, as it was of special importance to Britain. Deaths from tuberculosis, chiefly pulmonary, were running at the rate of 50,000 per annum. It is interesting and heartening to read the official annual returns and to note how that figure has fallen to the relatively trivial numbers that occur today. Death on the roads is now a far more serious risk.

Glaxo first made streptomycin available as the calcium chloride complex. Distillers came to a cooperative arrangement with E. R. Squibb and Sons of New Jersey, and soon afterward issued streptomycin as the sulfate, a form which proved to be and still is more generally acceptable. The calcium chloride complex is now deleted from the British Pharmacopoeia. They also prepared dihydrostreptomycin by catalytic hydrogenation, but when accumulating medical evidence revealed the inherent dangers of the product it was withdrawn from the market.

Streptomycin is of course one of the basic antibiotics and cannot be recovered from the fermentation by the same solvent extraction technique as used for acids such as penicillin. It can be adsorbed onto activated carbon, but such a method is t⟨o⟩day no more than a laboratory tool for use with new and unknown antibiotics. For the first time in this country an ion exchange process was introduced for the large-scale recovery of drugs, first through static beds of resin in a series of columns, and later as fluidized beds through which the whole broth could be passed without removal of the mycelium. For a number of years the final solid streptomycin was produced as a freeze-dried powder, until crystallization methods became available using such solvents as methyl alcohol.

Following streptomycin, other antibiotics began to come into the picture. A number of American pharmaceutical houses had long established subsidiary companies in Britain, which formed a useful base for marketing in the sterling area and in Europe, and expansion into the field of antibiotics was a natural sequence. Much of this was done by cooperating with British manufacturers. Boots adapted their fermentor plant for the production of aureomycin, and Distillers commenced the manufacture of erythromycin. Pfizer, who had probably made a wider range of fermentation studies than any other company, decided to put up their own fermentor plant for the production of terramycin at Sandwich on the south coast.

By the end of the 1950s, competition became extremely keen from the rest of the world, particularly from Italy, who ignored all pharmaceutical patents, and from Eastern European countries desperate for western currency. Only manufacture on the largest scale and with the highest efficiency enabled the leading companies to remain in profitable business. Production of antibiotics is not an easy life. Capital investment is high, a qualified staff covering a wide range of scientific fields is required for control of operations, and continuous research is necessary even to keep abreast of the market. The reward per unit of production is far less than that obtained by the retail pharmacist who issues it over the counter.

Pharmaceutical houses that did not have manufacturing facilities of their own but were interested in some of the newer antibiotics with a limited field of application, found it quite inadvisable to enter into production. Instead, they made arrangements with existing fermentation companies to conduct the operation on their behalf. In this way Viomycin and polymyxin B were brought into use. Speke produced limited amounts of bacitracin for the home market, while I.C.I. itself developed and produced the important antifungal agent griseofulvin.

Within 15 years of the Second World War, Britain had created the facilities and developed the skills to produce any known antibiotic in quantity, and her efforts in this direction raised the total of British pharmaceutical exports to hitherto unknown levels. Many university departments adjusted their courses to provide men with the right technical training for this new industry. Throughout the field there were men from T. K. Walker's department in Manchester who had been present from the beginning of the battle. Now they were joined by a younger generation already well versed in the principles of the industry, so that scientific meetings on the subject often seemed like a family reunion. Walker's only contribution to the antibiotic era was as a teacher of industrial fermentation. His own researches lay elsewhere, but he will long be honored among those who made the antibiotic industry what it is today. He died in 1970 after a happy and active retirement.

The success of Britain in the antibiotic field attracted many students and postgraduate workers to this country. Many new antibiotic-producing organisms that were isolated by individual workers here and abroad were submitted either to our academic centers or to manufacturers for further study. Most of these turned out to produce antibiotics already known or too toxic for clinical use, but it was in this way that cephalosporin C was discovered. Professor Brotzu isolated a strain of *Cephalosporium* from a sewage outfall in Sardinia. He was able to grow it in pure culture and to demonstrate that it had antibiotic-producing powers, but after he had made and used crude extracts he recognized that a complete study was far beyond the resources of his laboratory. At this point he decided to appeal to Florey in Oxford, who put E. P. Abraham in charge of the investigation. The antibiotic had amphoteric properties and was rather difficult to isolate. It was named cephalosporin N and was later shown to be a penicillin (Abraham *et al.*, 1953). In order to obtain more of the substance for study, Florey asked Distillers to produce a modest quantity in its research pilot plant at Bromborough. Within a short time of commencing this work, Abraham made the discovery that in addition to cephalosporin N there were traces of another antibiotic that was not sensitive to penicillinase and could be presumed to have a different structure. An entirely new extraction procedure had to be worked out, and after a long struggle Bromborough succeeded in producing about 5 kg of freeze-dried extract containing not more than 5% of cephalosporin C. This in fact was sufficient for Abraham and his team to make extensive studies leading finally to a brilliant elaboration of the chemical structure. Later Glaxo gave most welcome help in further pilot studies. As an antibiotic the substance was rather disappointing. Even in the pure state it had a very low potency when compared with penicillin G against a number of test organisms, but it had an extremely low toxicity and since it was not inactivated by penicillinase it could well have some value. If a cephalosporinase existed, it was not commonly produced by random infecting organisms that frequently destroyed penicillin, such as the resistant strains of *Staphylococcus aureus*. Work continued at Oxford under the shelter of the National Research Development Corporation (N.R.D.C.), the government-sponsored body that had come into being after a largely ill-informed public outcry about the disappointing patent position in respect to penicillin. The National Research Development Corporation filed effective patents on every aspect of the cephalosporin field, and was later to show how wise it had been.

The history of government interest in scientific endeavor over the last 50 years is a sad and muddled story, but this is no place to hold a detailed inquest. The research institutions sponsored by the government have suffered from parsimony, retrenchment, or reorganization at the wrong time, redirection at the whims of successive administrations, and changes in overhead control as different ministries have been invented or eliminated or combined. It is fortunate that such organizations as the Medical Research Council and the Agricultural Research Council have achieved almost autonomous control under the strong personalities appointed from time to time to lead them. They have both given valuable help to the study of antibiotics, and many of the international standards adopted by the World Health Organization have originated as British standards in the M.R.C. laboratories at Mill Hill.

When industrial penicillin manufacture became soundly established in Britain, it came to the ears of the public that certain royalties were being paid to American companies. Was not this a British discovery, and where were the royalties that America should be paying to us. It is no simple matter to explain patent law to the average citizen. No patent law is perfect in any country, but at least the objective is to give an inventor some reward for his work. It is unfortunate but true that if Fleming had filed a patent on his discovery, his patent would essentially have expired before the world was ready and able to use it. This situation happens all too often. Chain was perhaps the most unfortunate of all, because patents on his basic extraction processes could have brought great financial reward, but Oxford gave this knowledge to the world, and honor has fallen where honor is due. Where the Americans obtained patents was in the invention of new manufacturing methods and new process steps that gave tremendous increases in yield or in quality of penicillin produced. They made these inventions by tremendous endeavor and sometimes great expenditure, and like all inventors in civilized countries throughout the world, they were entitled to their reward. When such inventions were of help to British production, their use was highly profitable. The royalties charged by American owners of the patents were in almost every case minimal and in some cases no more than a token intended to acknowledge the inventor's contribution to the art. The British public paid very much less, and not more, for its penicillin because these patents were used and the appropriate royalties paid.

A case in point is the development of penicillin V, or phenoxymethyl penicillin. The exhaustive studies by Eli Lilly workers on the production of novel penicillins by addition of precursors to the fermentation medium had lain on the shelf for several years, although phenoxymethyl penicillin was one of the products very clearly described therein. Biochemie G.m.b.H., the relatively small Austrian producers of antibiotics, had a bad spell for a period when their penicillin G fermentations were troubled by infection. This is no novelty among manufacturers, but the Biochemie research team attempted to cure the situation by adding different antibacterial agents to the fermentation in the hope that one would be discovered that would inhibit the infection without any adverse effect on the penicillin culture. One of the agents was a well-known phenoxymethyl acetic acid derivative that is sold to the public as a general antiseptic. When they tried to extract penicillin G from the test fermentation, a substantial white precipitate formed on addition of acid to the first crude penicillin salt solution. Much to their surprise, the precipitate was found to be crystalline and to have high antibiotic activity when redissolved in aqueous alkali. They had discovered the free acid form of phenoxymethyl penicillin, the first penicillin to be recovered as a stable solid in this state, and for the first time a penicillin that could be given orally without rapid decomposition in the stomach. They named it penicillin V.

Distillers were greatly impressed by the first reports, and quickly made arrangements for production in England on the large scale. Because of the patent position, they also had to negotiate with Eli Lilly in America, but the most amicable arrangements were made between the three companies, and before long penicillin V, competitive in cost of treatment with penicillin G and much simpler to administer, began to sweep injection forms aside except in specialized cases. Penicillin V from Britain became another valuable contribution to our pharmaceutical exports, and the additional markets helped to increase the scale of manufacture and to reduce the cost of our home production. Eli Lilly were so impressed with British performance in this respect that later they negotiated to acquire the Speke factory, which has since operated as a subsidiary of that company.

When N.R.D.C. was satisfied that cephalosporin C had been adequately protected by patent applications filed in Britain and abroad, it offered exploratory licenses to a limited number of competent antibiotic manufacturers in selected countries, with the promise of manufacturing licenses if anything useful should emerge. In England, Glaxo and Distillers, who had already given useful aid to the project, were both invited to participate in what was still a highly speculative venture from a commercial point of view.

At this time there was a sudden major advance in the penicillin story. Beecham research workers revealed that they had successfully split the side chain from the penicillin molecule, and had isolated and characterized the nucleus, 6-amino penicillanic acid (6-APA) (Chain et al., 1959; Batchelor et al., 1959). Earlier Japanese workers had claimed to have demonstrated the existence of this compound, but their work was not considered conclusive and was limited to the laboratory. Now there was a method to produce 6-APA in quantity, and Beecham's further work showed that it could be coupled with a number of acyl chlorides and similarly reactive groupings to give a vast range of semisynthetic penicillins that could not possibly be achieved by addition of precursors to the fermentation. Many laboratories were attracted to the study, with the prospect of new penicillins that might have a wider range of activity or specific action against infections for which there was still no completely satisfactory treatment, but Beecham kept well ahead of the field and did not neglect patent cover, so that their new contributions to the medical armory are now well known.

From the standpoint of the fermentation industry, these discoveries meant still further expansion. In order to make these novel penicillins, it was necessary to start with the cheapest penicillins produced by fermentation, either penicillin G or penicillin V, and remove the side chain in order to obtain 6-APA. One of the most interesting facts to emerge was that this could be done enzymatically, and a simple bacterial fermentation could be used to provide suitable enzymes that gave almost quantitative yields under the right conditions. Many microorganisms have since been discovered that will accomplish this feat.

Research workers on cephalosporin C had not been slow to note these events. Abraham had shown that the compound consisted of a side chain attached to a ring system that in spite of certain similarities was different from that of the penicillin molecule. It seemed obvious that if the side chain could be removed and replaced by other groupings, the possibility existed of obtaining new cephalosporins that would be much more active than the original compound but might still retain its known advantages of low toxicity and resistance to penicillinase. In spite of a vigorous search, no enzyme could be discovered that would effectively remove the side chain while leaving the ring system undisturbed. Abraham reported that he had done this chemically in very low yield, but it was not until Eli Lilly workers found a rather novel chemical approach that the idea became a practical proposition and 7-amino cephalosporanic acid was made available as an intermediate for new products. Since then, of course, the dream has come true, and many new semisynthetic cephalosporins have been elaborated, with high potency and great value in therapy. Their use is now worldwide. Lilly workers made the further novel discovery of converting the penicillin ring to the cephalosporin ring, so that either fermentation process could be used to provide the starting material according to the economics of the situation.

One result of these studies, of the greatest importance to the country, was that the patents so carefully filed by N.R.D.C. to protect British scientific endeavor have brought in very large sums indeed as royalties from overseas. This is a form of invisible export that we are not likely to neglect again. The problem of obtaining patent cover in foreign countries is complicated by differences in patent law, so that adequate protection is not always possible. The cost of obtaining patent cover in all significant countries of the world is very high indeed, and quite prohibitive to the individual inventor. For this reason, the establishment of N.R.D.C. as the guardian of such inventors has made a great difference in the possible rewards to this country.

It must be recognized however, that in the field of antibiotics as elsewhere, patents expire after a term of years. Success can then only be maintained by the greatest expertise both in manufacture and in marketing, and by continued research to bring forward new inventions. Unless this is done, this part of the fermentation industry will die in Britain as others have done before it.

VII. Enzymes

The use of an enzymatic process for the fission of penicillins was a notable addition to the small number of processes aimed at the production of industrial enzymes. The field of enzymology is far too vast to be reviewed here. From the beginning of the century the isolation of specific enzymes, at first from plant and animal tissues, and later from microbial sources, has increasingly occupied the attention of biochemists as a prelude to studies of the mechanism of enzyme action. Most of this work has consisted of fundamental studies carried out in our universities and medical schools, but almost inevitably information has been gleaned that holds the promise of industrial application. The importance of this area has been recognized in official quarters, and in 1970 the Science Research Council made financial grants on a scale never known before to a number of universities engaged on research in the field, including Liverpool, London, Oxford, Stirling, Strathclyde, and Swansea. Much of this support was allocated to possible technical applications of enzymes, including chemical engineering studies. It may well be that the next decade will show outstanding advances both in technology and in applications affecting the life of humanity at large.

The production of enzymes may be divided into the preparation of highly purified specific enzymes for use in biochemical research and analysis and in clinical diagnosis and treatment (Rogers, 1969); and the manufacture of crude enzyme preparations for industrial use. Until recently it was necessary to import many purified enzymes from abroad, but an interesting development was the establishment in 1968 of Whatman Biochemicals Ltd., with the support of N.R.D.C., for the production of pure enzymes, including many from microbial sources. A description of this plant has been given by Malby (1970). While most of these enzymes are produced in very small quantities and sold as fractions of a gram, commercial enzymes obtained from microbiological sources are manufactured on a large scale that permits their bulk use in industrial processes. Generally they are exocellular enzymes that permit a relatively simple recovery process without the necessity for cell rupture, although today this can be performed with ease in the modern homogenizers now available. The preparations are sold either as a liquid of defined activity under described conditions of use, or as a dry powder blended with a suitable inert diluent to give a standard product for marketing. The definition of potency has varied from one source to another, and is most usually agreed between producer and user to suit particular circumstances. The recommended definition of the International Union of Biochemistry that the standard unit of enzyme activity should be the amount that will catalyze the transformation of 1 μmole of substrate is clearly of value in relation to pure specific enzymes acting on a chemically defined substrate. In the case of nonspecific enzymes used in industry there are variations in conditions of use that make a single precise definition impossible to apply (Collier, 1970).

The best-known industrial application of enzymes on an industrial scale is the use of bacterial and fungal amylases for the conversion of starch to sugars, for the desizing of textiles, and more recently for destarching in sugar manufacturing processes. The conversion of starch is recognized as the oldest microbial enzyme process used in Britain, and dates from the very early part of the century, when it was shown that these could compete with the enzymes of malted grain in the conversion of starchy materials to fermentable sugars. It is obvious now that there were several earlier microbial enzyme processes in industrial use, of which flax retting is a typical example, but these ancient arts went unrecognized for what they were, and the use of fungal amylases represents the first deliberate commercial application of an enzyme to catalyze a reaction.

The use of amylases was followed in due course by the introduction of proteases, cellulases, and pectinases, all of bacterial or fungal origin, and there is hardly a section of the food-processing industry that does not make use of enzymes in one or more of these classes. They are all produced by fermentation techniques, though the number of British commercial producers is very small. Uses go far beyond the food industry, and Rogers (1969) listed a number of reviews of industrial applications, but it appears certain that the next few years will show a substantial expansion of such lists, and it is more than probable that some of these uses will be in directions not yet anticipated. The recent introduction of methods of insolubilization of enzymes, giving stability and prolonged high potency activity, seems bound to improve the economics of processes that are on the borderline of success at the moment.

The newer uses of enzymes include that already mentioned in the fission of penicillin to produce 6-APA, and the more recent application of proteolytic enzymes for detergent purposes. The latter has required fermentation capacity on the largest scale, and production of enzymes on a tonnage basis. In fact some fermentation companies have had to refuse production contracts because the output required would absorb far too large a proportion of their fermentor capacity. The right long-term answer is of course for detergent manufacturers to establish their own fermentation plants. The success of these products seems assured since they have now had a substantial period of use and sales are still expanding. The only doubt that has occurred has been the suggestion that such enzyme-containing products are responsible for a certain incidence of dermatitis. It is claimed by manufacturers that the incidence is quite trivial, and it has to be recognized that no biologically active substance exists that does not appear to induce dermatitis in a small fraction of the population.

In addition to the use of liquid or solid preparations of enzymes, the possibility of using living cells has attracted much attention, and the best known example of this is the production of molecular changes in steroids. In this case the original steroid is added to the culture medium in which the enzyme-producing organism is grown and becomes part of the substrate. This technique is eminently satisfactory in cases where the organism does not produce other enzymes that could interfere with the desired course of the reaction. Other examples of this technique are already known. They have the attraction of simplicity and cheapness, providing conversion can take place at a concentration and under conditions that do not involve a costly recovery process. The dream that chemical manufacturers will carry on their shelves microorganisms capable of performing every known reaction is not too far from practicality.

VIII. Gaps in the Program

There is little point in reviewing British accomplishment in the production of the wide range of sugar derivatives (Spencer and Gorin, 1968) that have been demonstrated in the literature to be yielded by fermentation methods, or the many amino acids that the Japanese have proved so enthusiastically to be capable of manufacture, because very little of industrial significance has so far taken place. Nor so far has the production of protein from petroleum had any major impact. The use of such protein for anything but animal feed faces the same serious hurdle of acceptability that brought torula yeast plants to a standstill a generation ago. It is possible to manipulate such material to forms acceptable to the human palate, just as a chef at the Waldorf Astoria can do with a number of prosaic foods, but few are willing or able to pay the price. Even in animal feeds, the protein seed meals appear able to put up the most determined opposition.

This is not to say that Britain is sadly lagging behind in such areas. The fact is that before any of the above products is brought into commercial manufacture, there must be a reasonable prospect of profitable operation to justify the necessary investment and allocation of production resources. Many processes have been studied in detail in the research departments of our fermentation companies, tested exhaustively on the pilot scale, carefully costed at different production levels, and subjected to market surveys. Most of these processes lie fully documented in company archives, waiting for the day when the changing pattern of man's behavior will bring them into profitable use.

IX. The Influence of Continuous Fermentation

This is in no sense a chemical engineering treatise. It is sufficient to say that British practice in the design of fermentation vessels and auxiliary equipment does not differ significantly from that of other countries. There has been steady progress in the design of fittings, of auxiliary equipment for addition of inocula, nutrients, and reagents, and in recording and controlling instruments, leading eventually to automation and computerization of operations. All these things are international, and instrument makers recognize no boundaries when business is to be obtained.

The only major change in comparatively recent times has been the development of continuous fermentation.. A great amount of research has taken place in this field, and the published literature is so extensive that it would be invidious to quote the studies of individual workers. Nevertheless, it is recognized in Britain that our leading center has been the government Microbiological Research Establishment at Porton, and many valuable papers on both the fundamental aspects and the operational requirements have been published by the team that started this work under the brilliant leadership of David Henderson.

It was in fact one of his concepts that Britain's defense against biological warfare should in part consist of mobile continuous fermentation units, small in size but with a large throughput capacity, that could be moved in a state of emergency to almost any part of the country for the production of antibiotics.

The mandate given to Porton to develop methods of defense against bacterial warfare has at times led to unfortunate publicity and attention from the lunatic fringe of the population. It appears clear that the work must at times have required the study of dangerous pathogenic organisms. The axiom "know thine enemy" is in this case perhaps more important and more difficult than in any other possible field of conflict. How anyone in his right mind could dispute such an approach is beyond belief. The major part of medical effort is devoted to the study and treatment of infections that chance puts in our way, and what lack of logic can forbid us to look for protection against infections not brought by chance? There have been demands that everything that goes on in such a defense establishment should be revealed to the general public, presumably so that our enemies should know us equally well. It is seldom acknowledged that what has been revealed of this research has been treasure indeed to the fermentation world.

Starting with the premise that every organism should be treated as carefully as a dangerous pathogen, fermentation equipment was devised at Porton that provided every precaution against the contamination of pure cultures. Such equipment has served as a model for industry, research centers, and plant manufacturers, not only in designing experimental units but on the full scale as well. The design of valves, joints, and internal fittings has improved considerably since studies at Porton began.

In early attempts at continuous fermentation, failure was frequent and nearly always due to accidental infection, either from the newly introduced medium or from the air supply or through vessel connections, and it was usually difficult to locate the source of infection. Only after lengthy investigation and elaboration of methods that would make such infections virtually impossible was it worthwhile proceeding with the fundamental studies that have helped so much to establish the basic principles of successful continuous fermentation.

In a batch fermentation a considerable part of the working cycle is dead time, when the fermentor is emptied at harvesting, cleaned, sterilized, cooled, and recharged with fresh medium. In addition, a substantial part of the actual fermentation time is usually taken up by the initial lag phase while the organism is multiplying and not producing at peak rate. In a continuous fermentation, as soon as the system is on balance the organism can produce at peak rate as long as satisfactory conditions are maintained, but if infection occurs the process must be shut down and recommenced with the same loss of effective operating time as in batch working. Unless a continuous fermentation can be maintained in operation for a very much longer time than the equivalent batch fermentation this key advantage is lost.

When fermentations at Porton began to achieve 1000 hours of continuous running and were only shut down to change the type of study, the staff members knew that the first great difficulty had been overcome. Their methods and their subsequent comprehensive studies have been freely given in published work and papers presented at scientific meetings in many countries, including Czechoslovakia, where Ivan Malek is rightly regarded as the European leader in this field. His efforts have led to a series of international symposia on continuous fermentation, on one occasion at Porton itself. The fifth symposium was held at Oxford in 1971.

One might imagine from the vigor with which these studies have been pursued and the advantages which are claimed that industrial fermentation in Britain would have largely changed over to continuous operation. This is not the case. Only in the production of yeast and in brewing of beer has there been real achievement. The Distillers Company yeast plant at Dovercourt is usually regarded as the finest example of continuous yeast production (Olsen, 1960). It must be recognized that this is a fermentation for cell production only, and is ideally suited to such a technique. The only important step in product recovery is the continuous removal of yeast from the spent fermentation liquor by means of centrifuges. Any fermentation process for the recovery of whole bacterial cells could be operated with comparable effectiveness.

In the brewing of beer, there are greater problems in the preparation of the fermentation medium than in the case of yeast manufacture, but here the product required is the whole fermented medium, which greatly simplifies the flow sheet compared with those fermentations aimed at recovery of a pure metabolic product by what may be a very complex extraction procedure.

It is rather difficult to persuade manufacturers to assemble their arguments against continuous fermentation. True, the major companies usually have a large battery of batch fermentors, generally installed at intervals over the years as the company has expanded its operations, so that they have a large capital investment that is never likely to be written off entirely at one time. With a large number of fermentors, they can supply extraction units continuously, a stage where continuous operation may well be considered more vital. If they manufacture several products, for example a range of antibiotics and vitamins, they seldom make them all at one time, but campaign their operations in a way to minimize their investment in recovery plant and the use of process labor. Although a variety of extraction procedures may be employed, a great deal of the equipment is suitable for the recovery of several products, and it is not too difficult to train modern process labor to operate a variety of procedures. If several different continuous fermentation units are in operation at the same time, they obviously need separate extraction units and separate crews to man them. If one continuous fermentation unit is used in turn for a variety of products, then it cannot be used to maximum economic advantage, since it must be shut down more frequently than necessary in order to avoid an excessive buildup of stocks of each product.

Such arguments will be a subject of debate for a long time to come, although change is likely to be in the direction of continuous fermentation. There are fields such as enzyme preparation and the manufacture of certain organic acids where the use of this technique appears very attractive indeed.

In the meantime it is a very pleasing thought that the wheel has turned full circle so that the most ancient fermentation in history has now become the leader in this new technique.

REFERENCES

Abraham, E. P., Newton, G. G. F., Crawford, K., and Burton, H. S. (1953). *Nature (London)* **171**, 343.

Adler, L., and Leurs, H. (1916). *J. Inst. Brew. London* **22**, 504.

Batchelor, F. R., Doyle, F. P., Nayler, J. H. C., and Rolinson, G. N. (1959). *Nature (London)* **183**, 257.

British Pharmacopoeia (1958). pp. 88–89.

Chain, E. B., Florey, H. W., Gardner, A. D., Heatley, N. G., Jennings, M. A., Orr-Ewing, J., and Sanders, A. G. (1940). *Lancet* **ii**, 226–228.

Chain, E. B., Ballio, A., Dentice di Accadia, F., Rolinson, G. N., and Batchelor, F. R. (1959). *Nature (London)* **183**, 180.

Chem. Ind. (London) (1959). p. 1242.

Clutterbuck, P. W., Lovell, R., and Raistrick, H. (1932). *Biochem. J.* **26**, 1907.

Collier, B. (1970). *Process Biochem.* **5**(Aug.), 39.

Fleming, A. (1929). *Brit. J. Exp. Pathol.* **10**, 226.

Hastings, J. J. H. (1958). *In* "Biochemical Engineering" (R. Steel, ed.), pp. 299–318. Heywood, London.

Herrell, W. E. (1945). "Penicillin and other Antibiotic Agents." Saunders, Philadelphia, Pennsylvania.

Lester Smith, E. (1948). *Nature (London)* **161**, 638.

Malby, P. G. (1970). *Process Biochem.* **5**(Aug.), 22.

Olsen, A. J. C. (1960). *Chem. Ind. (London)* p. 416.

Rogers, P. J. (1969). *Rep. Progr. Appl. Chem.* **54**, 391.

Ross, J. (1970). "Whisky." Routledge & Keegan Paul, London.

Spencer, J. F. T., and Gorin, P. A. J. (1968). *Progr. Ind. Microbiol.* **7**, 178–220.

Waksman, S. A. (1941). *Bacteriol. Rev.* **5**, 231–291.

Chemical Composition as a Criterion in the Classification of Actinomycetes

H. A. LECHEVALIER, MARY P. LECHEVALIER,
AND NANCY N. GERBER

Institute of Microbiology, Rutgers University–The State University of New Jersey, New Brunswick, New Jersey

I.	Introduction	47
	A. Chemical Classification of Higher Plants	48
	B. Chemical Classification of Algae and Protozoa	48
	C. Chemical Classification of Fungi and Lichens	49
	D. Chemical Classification of Viruses	50
	E. Chemical Classification of Bacteria	50
II.	Pigments in the Classification of Actinomycetes	50
	A. Introduction	50
	B. Phenazines and Phenoxazinones	51
	C. Prodiginines	53
III.	Antibiotics in the Classification of Actinomycetes	53
IV.	Deoxyribonucleic Acid Base Composition in the Classification of Actinomycetes	55
V.	Cell Wall and Whole-Cell Sugar Composition in the Classification of Actinomycetes	57
VI.	Lipid Composition in the Classification of Actinomycetes	60
VII.	Conclusion: A Tentative System of Classification of Actinomycetes	62
	References	69

I. Introduction

One could argue that most classifications are chemical. If we choose color as a taxonomic criterion, save for interference and diffraction phenomena, it is an expression of the pigments produced by the organisms in question. If we choose morphology, it is controlled by the chemical composition of the polymers which are formed. In this review, we plan to consider the classification of actinomycetes based on the presence or absence of specific compounds, be they secondary metabolites or constituents of fundamental molecules.

It is not our contention that actinomycetes may be classified on the basis of chemistry alone. It is, however, our feeling that actinomycetes may be classified conveniently into rather large groups, say at the generic level, by using a combination of morphological and chemical criteria (Lechevalier and Lechevalier, 1965, 1967). Since the state of taxonomy of a given group of organisms is but a reflection of the state of knowledge about these organisms at a given time, we have to be constantly ready to accept changes which will lead to improvements in classification systems.

Most taxonomists will agree that separation between taxa should not be based on a single criterion. The modern expression for this old adage is numerical taxonomy (Sokal and Sneath, 1963). The fact still remains that some criteria are more important than others. For example, the presence or absence of a true nucleus in a "cell" is full of implications (Stanier, 1964) and has more importance than the presence or absence of one specific enzyme, especially when the enzyme may be found in such diverse organisms as bacteria and man. However, here again, the state of knowledge at a given time has to be taken into consideration. For example, a bacterial enzyme and a mammalian enzyme may carry out the same overall reaction, but when enough is known about the actual structures of the proteins involved, differences of taxonomic significance may be found.

A. Chemical Classification of Higher Plants

Chemical criteria have long been used in the classification of most diverse organisms. In botany, early efforts at classifying plants were linked to their pharmaceutical properties which were often due to their content in alkaloids. This was demonstrated by the isolation of morphine by Sertürner and that of quinine by Pelletier and Caventou, during the second decade of the nineteenth century.

About 20% of all vascular plants contain alkaloids and more than a thousand of these compounds are known. Since they are easily detected, they are of great interest to taxonomists. However, the mere presence of a given alkaloid was soon found to be a criterion of dubious value as it became obvious that the same compound could be found in rather different plants. For example, anabasine can be found in members of both the Solanaceae and the Leguminoseae. A more detailed investigation revealed, however, that in the genus *Nicotiana*, anabasine is derived from one molecule of lysine and one of nicotinic acid whereas in the legumes, the alkaloid is derived from two molecules of lysine (Hegnauer, 1963). Thus, biosynthetic pathways of alkaloids may furnish important clues as to relationships among plants.

We would not like to give the impression, in this brief discussion of plant chemical taxonomy, that only biosynthetic pathways of rather complicated molecules are of some use to plant taxonomists. Simpler criteria such as the presence or absence of raphides (bundles of needle-shaped crystals of calcium oxalate) have a long and useful history, having probably been first used by Robert Brown around 1830.

B. Chemical Classification of Algae and Protozoa

Chemical criteria are used extensively in the classification of microorganisms. For example, algae are classified according to the pigments they contain, the composition of their cell walls, and the nature of the materials they accumulate as food (Lewin, 1962). In protozoa, the use of chemical criteria is less common, but comes into play at least in the case of those organisms having some sort of an envelope, such as a shell, a test, or a lorica. Such envelopes may be, depending on the group, formed of cellulose, chitin, silica, or calcium carbonate. The latter substance is also deposited in the sporangia of some Mycetozoa and not in that of others, a fact which has proved of great usefulness to taxonomists (Kudo, 1966).

C. CHEMICAL CLASSIFICATION OF FUNGI AND LICHENS

Mycologists also use chemical criteria in classification of fungi. However, the chemical mechanism of the reactions involved is often poorly understood (Bataille, 1948). One of the most useful chemical tests is the amyloid reaction given by starch, glycogen, and related substances with Melzer's reagent (a mixture of iodine, potassium iodide, and chloral hydrate). Combining morphology (presence of sphaerocysts) with chemistry (dark staining of basidiospore ornaments with Melzer's reagent) permits one to make a sure diagnosis of the *Lactarius-Russula* group. The skillful mycologist does not limit himself to this application of the amyloid reaction but uses subtle color changes of various parts of fungal "tissues" (Smith, 1966). Studies of cell wall composition of fungi have brought out (1) their lack of relationship to bacteria, (2) their kinship, though distant, to algae, (3) the presence of cellulose in many of the lower fungi, (4) the presence of chitin associated with glucan in most of the higher fungi with septate mycelia, and (5) the presence of mannan in yeast forms (Bartnicki-Garcia, 1968).

Chemical criteria are widely used in the classification of lichens. In 1866, W. Nylander, a Finnish botanist who lived most of his productive life in Paris, introduced color tests given by calcium hypochlorite (C) and potassium hydroxide (K) which he used either separately or in combination (KC). To this trio of reagents was later added *p*-phenylenediamine (P). The color tests with C, K, KC, and P, though most useful in taxonomy, enable one to detect only groups of substances within the "tissues" of lichens. Starting in 1936, the Japanese chemist Y. Asahina developed a battery of microchemical tests which permit the identification of many lichen chemicals with accuracy and speed by the formation of characteristic crystals on microscopic glass slides. These tests are widely used by lichenologists who are, in addition, exploring other techniques such as various forms of chromatography, for the detection of lichen chemicals (Hale, 1967).

D. CHEMICAL CLASSIFICATION OF VIRUSES

Chemical composition is basic in the classification of viruses. These are split into two broad groups, depending on the type of nucleic acid that they contain. In addition, the virion's content of protein and lipid furnishes a further chemical criterion for viral classification (Luria and Darnell, 1967).

E. CHEMICAL CLASSIFICATION OF BACTERIA

Bacteriologists have used a number of chemical criteria in taxonomy. Among the easiest chemicals to detect are various enzymes; among the most subtle are those detected by the humoral responses of vertebrates. Serological classification has achieved a high level of complexity in the case of streptococci and salmonellae. Streptococci are first separated biochemically into three groups depending on the production or lack of hemolytic enzymes. Further subdivisions are made following the method of Rebecca Lancefield, on the basis of group-specific antigens, which are carbohydrates forming a layer exterior to the murein of the bacterial cell (Krause, 1963). In the case of salmonellae, agglutination reactions have resulted in the recognition of more than 700 types. The antigens responsible for these serological reactions are either proteins (flagella) or carbohydrates (somatic and Vi antigens) (Kauffmann, 1966). In the case of the Actinomycetales we will consider pigments, antibiotics, deoxyribonucleic acid base composition, cell wall composition, whole cell sugar composition, and lipid composition as taxonomic criteria.

II. Pigments in the Classification of Actinomycetes

A. INTRODUCTION

It is a fact that similar actinomycetes tend to have a similar color. This is exploited mainly in the classification of streptomycetes, where one can recognize: (1) the production of diffusible pigments, (2) the color of the aerial mycelium, and (3) the color of substrate mycelium (Cross and MacIver, 1966; Shirling and Gottlieb, 1966, 1970; Gottlieb and Shirling, 1970). However, in actinomycetes, variants differing in color from their parents can be obtained rather easily both as the result of natural variation (Gordon and Mihm, 1962; Gordon and Rynearson, 1963; Gordon, 1966b; Gordon and Pang, 1970; Krassilnikov, 1970) and as the result of irradiation with ultraviolet light (Baldacci et al., 1962, 1963). Thus color is of little value as a basic criterion in the systematics of actinomycetes, although it is of great value in the presumptive identification of many actinomycetic species. Hence, we can say with confidence that strains of *Streptomyces griseus* are frequently greenish, that those of *S. fradiae*, and *Microbispora bispora* are usually pinkish and that those of *M. rhodochrous* are usually some shade of salmon. Great efforts have been made in standardizing the detection and the nomenclature of colors of actinomycetes (Tresner and Backus, 1963; Prauser, 1964; Pridham, 1965; Oliver, 1970); however, an internationally recognized system has yet to be devised.

It would seem that a rational approach to the use of pigments in the systematics of actinomycetes would involve the elucidation of their chemical nature (Preobrazhenskaya, 1970). The structures of actinomycetic pigments with interesting antibiotic activities, such as the red actinomycins and the yellow tetracyclines and polyenes, have been extensively investigated. Much less is known about the chemistry of substances which have been studied purely as pigments (Blinov and Khokhlov, 1970).

B. Phenazines and Phenoxazinones

In our laboratory, the pigments of strains of *Microbispora* were first investigated. Some members of this genus (*M. aerata, M. amethystogenes, M. parva*) deposit in the medium crystals having a metallic glitter. These were found to be iodinin (1,6-phenazinediol-5, 10-dioxide). Iodinin is a red pigment, highly insoluble in aqueous media, which had been isolated before the Second World War from cultures of nonfilamentous bacteria. Strains of *M. aerata* produce also brown and yellow pigments. Two brown to yellow pigments were isolated and were identified (1) as questiomycin A (2-aminophenoxazin-3-one), previously known to be produced by a species of *Streptomyces*, and (2) as 1,6-phenazinediol a reduction product of iodinin previously known to be elaborated by *Streptomyces thioluteus* and nonfilamentous bacteria. Another yellow pigment was also isolated from the same culture and was identified as 2-acetamidophenoxazine-3-one, a compound previously known but not as a natural product (Gerber and Lechevalier, 1964).

The cultures of *M. aerata* contained, in addition, an orange pigment which was isolated and characterized as 1,6-phenazinediol-5-oxide by Gerber and Lechevalier (1965). We then had a family of three phenazines (Fig. 1), one red, one orange and one yellow, all part of a biosynthetic path (Lechevalier, 1965). These three compounds were found not only in cultures of *M. aerata* but also in those of *Streptomyces thioluteus* (Gerber, 1967), *Brevibacterium iodinum* (Gerber and Lechevalier, 1965), *Actinomadura dassonvillei* (Gerber, 1966), *Streptosporangium amethystogenes* (Prauser and Eckardt, 1967, and

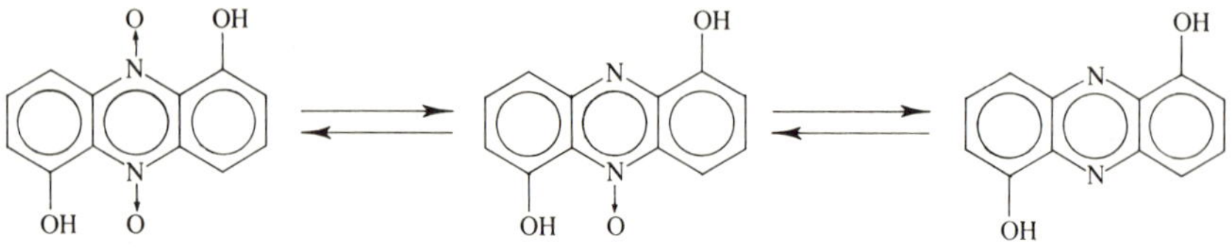

FIG. 1. A family of phenazines produced by strains of *Microbispora, Streptomyces, Actinomadura, Streptosporangium*, and some nonfilamentous bacteria. I = iodinin (1,6-phenazinediol-5, 10-dioxide) (red); II = 1,6-phenazinediol-5-oxide (orange); III = 1,6-phenazinediol (yellow). Disrupted cells of *Brevibacterium iodinum* can carry out the synthesis of iodinin from the phenazinediol, yeast cells can produce the phenazinediol from iodinin.

our unpublished observations), and an unidentified gram-negative, non-motile, rod-shaped bacterium (Gerber, 1969).

The occurence of these three phenazines did not seem to have much taxonomic value. In the case of *A. dassonvillei*, they were accompanied by another orange pigment, 1-phenazinol-10-oxide (Fig. 2) and a yellow phenoxazinone, 2-amino-1-carboxy-3H-phenoxazin-3-one, which is closely related to cinnabarin (Cambrie and LeQuesne,

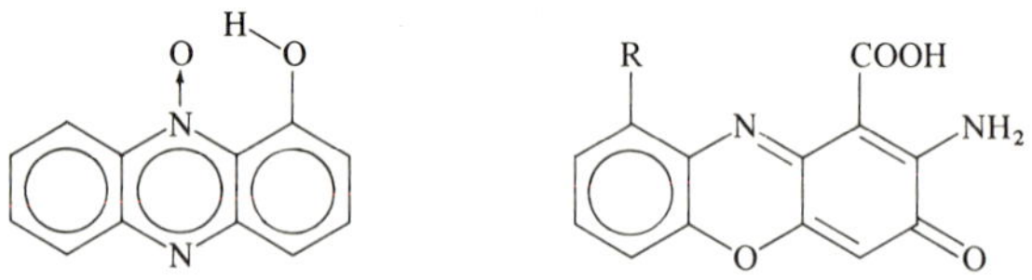

FIG. 2. Some phenazine and phenoxazinones from *Actinomadura dassonvillei* and *Pycnoporus cinnabarinus*. IV = 1-phenazinol-10-oxide (orange); V:R = H, 2-amino-1-carboxy-3H-phenoxazin-3-one (yellow); R = CH_2OH, cinnabarin (red).

1966) a product of two closely related species of *Polyporus, P. cinnabarinus* and *P. sanguineus*. These fungi have been assigned by different authors to various genera such as *Coriolus* and *Pycnoporus* (Nobles and Frew, 1962). Although cinnabarin has not been found, so far, in cultures of actinomycetes, the presence of related phenoxazinones would suggest that its isolation from actinomycetes may be only a question of time.

Thus, many different microorganisms can produce the same pigments. In the cases just discussed, a basic trio of phenazines is usually produced along with some phenoxazinones and additional phenazines. Not all phenazine-phenoxazinone-producers are equally prolific. A strain of *Streptomyces thioluteus* (Gerber, 1967) produces no less than 8 phenazines, together with 2 phenoxazinones and an assortment of pigments from different chemical families (aureothricins, dioxopiperazines). The previously mentioned gram-negative bacterium (Gerber, 1969) furnishes 11 phenazines but no pigments of other chemical types. One might ask whether all these iodinin producers elaborate the same collections of related compounds with only differences in yields, with the result that trace amounts of certain pigments pass through our hands undetected.

C. PRODIGININES

Prodigiosin (Fig. 3a) is the bright red pigment of *Serratia marcescens* (Williams and Hearn, 1967) and was probably responsible for many medieval "miracles" involving the appearance of blood stains on the Holy Host (Gaughran, 1969). Red pigments with the methoxytripyrrole nucleus of prodigiosin (prodiginine) have been isolated from *Streptomyces longisporus ruber* (Wasserman et al., 1969; Harashima et al., 1967) as well as from *Actinomadura madurae* and *A. pelletieri* (Gerber, 1969, 1970). In Fig. 3a, if $R = R_1 = H$, we have the formula of the methoxytripyrrole nucleus, prodiginine, which is not known to occur *per se* as a natural product.

In Fig. 3 are outlined the known, naturally occurring, substituted prodiginines. The nonfilamentous *Serratia marcescens* produces prodigiosin itself (Fig. 3a), *Streptomyces longisporus ruber* is the source of N-undecylprodiginine and metacycloprodigiosin (Fig. 3b). In our own survey we found that 5 strains of *A. madurae* produced two prodiginines with a 9-carbon aliphatic chain, one of which is cyclic (Fig. 3c) whereas 14 strains of *A. pelletieri* elaborated a mixture of two prodiginines each with an 11-carbon aliphatic chain. One of these is the N-undecylprodiginine previously mentioned; the other is cyclic, with only 10 carbons participating in the ring structure which is branched (Fig. 3d).

These patterns of production of prodiginines are not without exceptions. One strain of *Actinomadura madurae* gave a mixture of N-nonyl- and N-undecylprodiginines, another one produced only N-undecylprodiginine and one strain of *A. pelletieri* gave the two pigments usually associated with *A. madurae*. In all cases strain identification had been made previously (Gordon, 1966a).

III. Antibiotics in the Classification of Actinomycetes

Actinomycetes constitute the most prolific group of antibiotic producers (Waksman and Lechevalier, 1962; Waksman, 1963; Umezawa,

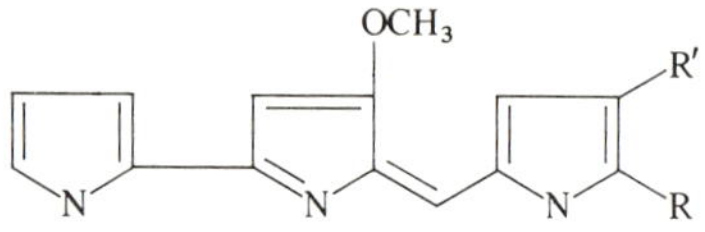

(VI)

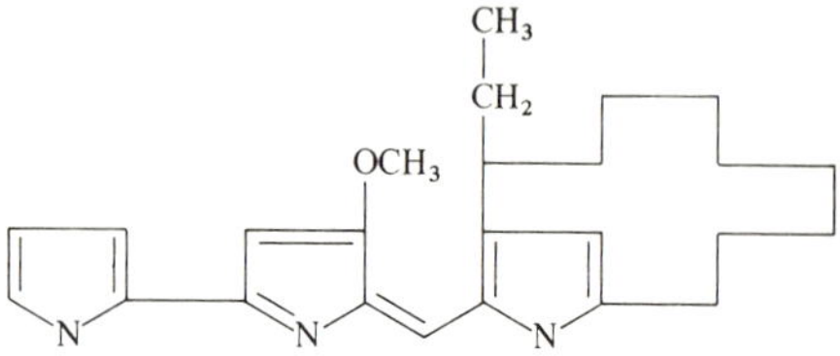

(VII)

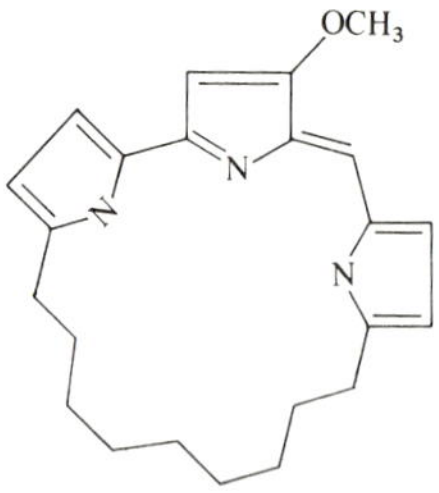

(VIII)

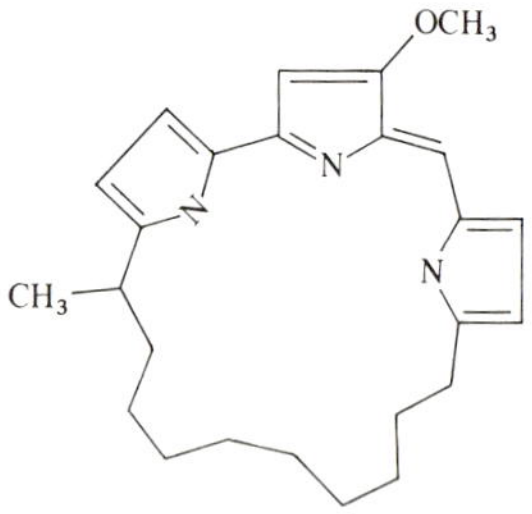

(IX)

FIG. 3. Prodigiosin and "prodigiosin-like" pigments. (a) VI: R= CH_3, R' = n-pentyl, prodigiosin, from *Serratia marcescens*. (b) VII = metacycloprodigiosin and VI:R = n-undecyl, R' = H, undecylprodiginine from *Streptomyces longisporus ruber*. (c) VIII = nonacycloprodiginine and VI:R = n-nonyl, R' = H, nonylprodiginine from *Actinomadura madurae*. (d) IX = methyldecacycloprodiginine and VI:R = n-undecyl, R' = H, undecyl prodiginine from *A. pelletieri*.

1967). Anyone who has screened actinomycetes for the detection of antibiotics has noted that similar actinomycetes tend to produce similar families of antibiotics. This led Krassilnikov (1959) to make the statement that the property of producing specific antibiotics is a very stable and specific characteristic of actinomycetic species and that antibiotic production can be used as one of the criteria in their classification.

Since actinomycetes produce several hundred antibiotics in various mixtures, the use of antibiotic production in taxonomy presents staggering problems (Baldacci, 1956). One easy way of determining whether one actinomycete produces a specific antibiotic is based on the assumption that an organism is always resistant to the antibiotic it produces. There are several difficulties (Teillon, 1953): (1) the producer is not necessarily the only organism resistant to the antibiotic it produces, (2) the producer may be distressingly sensitive to the lethal action of its own metabolites, and (3) cross-resistance phenomena come into play. We could thus conclude that antibiotic production is even less useful than pigment production in the speciation of actinomycetes. Indeed, in the case of speciation of streptomycetes, which are the most prolific antibiotic producers, pigment production seems universally used whereas antibiotic production is rarely mentioned (Cross and MacIver, 1966; Shirling and Gottlieb, 1970).

IV. Deoxyribonucleic Acid Base Composition in the Classification of Actinomycetes

Deoxyribonucleic acids (DNA) contain equimolar amounts of guanine (G) and cytosine (C) and of adenine (A) and thymine (T). One can express the composition of a DNA molecule as the percentage of G + C that it contains. Whereas species of highly organized organisms, such as various species of vertebrates, all contain DNA with about the same G + C percentage, bacteria vary from a low of 26 G + C % for some clostridia to a high of 75 G + C % for some of the actinomycetes (Sueoka, 1964). The most complete compilation of bacterial DNA G + C percentages is probably that of W. M. Normore and J. Brown published in the second edition of the CRC Handbook of Biochemistry. DNA base compositions of various actinomycetes have been compiled by DeLey (1970).

Yamaguchi (1967) analyzed DNA preparations from morphologically diverse actinomycetes. His results, which showed their DNA G + C percentages to range from 67.5 to 74.5 did not seem to have much taxonomic significance. At the most could he remark that DNA from organisms with a Type IV cell wall tended to have lower G + C percentages than those with a Type I cell wall (see Table II for cell wall types). Similar results were reported by Tewfik et al. (1968) and Tsyganov and co-workers (1966; Tsyganov and Krasikova, 1968). The determination of DNA base composition is, however, much more laborious than that of cell wall composition, especially since, as we will see later, cell wall composition can often be inferred on the basis of simple paper chromatography of whole cell hydrolyzates.

The most striking result obtained from the study of DNA base composition of various actinomycetes has been the low G + C percentage values observed in some thermophiles (Evreinova et al., 1965; Craveri and Manachini, 1965). In Table I is shown the relationship between the temperature requirement of actinomycetes and their base composition.

TABLE I
TEMPERATURE REQUIREMENTS OF SOME ACTINOMYCETES
AND THEIR DNA BASE CONTENT[a]

Temperature optimum (°C)	Organisms	G + C (%)
20 to 35°	Streptosporangium album	76.1
	Streptomyces argenteolus	76.6
	Streptoverticillum biverticillatum	79.8
	Microbispora rosea	73.7
	Micromonospora chalcea	79.8
	Nocardia blackwellii	79.1
40 to 50°	Pseudonocardia thermophila	79.3
	Thermomonospora viridis	73.9–74.9
	Streptomyces rectus	79.5
	Actinomadura sp.	77.4
	Micropolyspora sp.	74.1
	Streptomyces thermoviolaceus pingens	79.8
50 to 60°	Thermoactinomyces vulgaris	53.4–54.8
	Streptosporangium album thermophilum	53.7
	Thermomonospora citrina	43.8

[a] Modified from Manachini et al. (1967).

In summary, it seems that DNA base composition is of little value as a taxonomic criterion in the classification of actinomycetes. If, on the other hand, one does not rely exclusively on the DNA G + C percentages but rather on the order in which the bases are assembled along the DNA molecules, it is possible to determine the degree of reassociation of single-stranded DNA from various actinomycetes. In this way one should be able to gauge the degree of relatedness between organisms. Various factors can influence the results obtained in DNA hybridization determinations: one should measure not only the extent of DNA association but also the thermal stability of the duplexes formed (Walker, 1969). The early results published in this domain, which have been reviewed by Hütter (1970), showed that variations of DNA homologies observed within strains belonging to the same genus were as great as variations observed between DNA coming from strains obviously belonging to different genera. In recent studies, Farina and Bradley (1970) used reference DNA from *Actinoplanes philippinensis* and *Streptomyces venezuelae*. At least 59% of the single-stranded DNA from strains of *Actinoplanes, Dactylo-*

sporangium, and *Ampullariella* were bound in a thermally stable way to the *Actinoplanes* reference DNA. Only 29 to 33% of the single-stranded DNA from strains of *Planomonospora, Planobispora, Streptosporangium*, and *Spirillospora* were bound to the *Actinoplanes* reference DNA; in addition, the duplexes were less thermostable. Thus Actinoplanaceae can be separated into the same two groups by DNA homology and by cell wall composition (Table V). Use of reference DNA from S. *venezuelae* confirmed that *Microellobospora flavea* was more closely related to streptomycetes than to Actinoplanaceae.

The obvious conclusion is that the study of DNA homologies is bound to give us some inkling about the relationships between actinomycetes. The method is delicate, laborious, expensive, and thus does not lend itself to the rapid accumulation of large masses of data. So far, the results obtained tend to confirm those given by the study of cell wall composition.

V. Cell Wall and Whole-Cell Sugar Composition in the Classification of Actinomycetes

The cell wall of actinomycetes, like that of bacteria contains a murein (mucopeptide) layer (Avery and Blank, 1954; Romano and Sohler, 1956; Cummins and Harris, 1958). Cummins, summarizing what was known about cell wall composition of actinomycetes in 1962, noted that if one took into account only the major constituents of the "skeleton" of the cell wall, one could recognize at least two major groups of aerobic actinomycetes. He also noted that microaerophilic to anaerobic actinomycetes fell into two cell wall groups. Further studies have permitted us to distinguish some 9 cell wall groups (Table II). The recognition of cell wall Types I to IV, in which most of the aerobic actinomycetes fall, is based on the studies of Becker *et al.* (1965) and of Yamaguchi (1965). The properties of members of the genus *Actinomyces* have recently been reviewed by Slack and Gerencser (1970). *Oerskovia* and *Agromyces* are recently described genera (Prauser *et al.*, 1970; Gledhill and Casida, 1969) and *Mycoplana* is the generic name for certain gram-negative rudimentarily branched filamentous, motile bacteria (Sukapure *et al.*, 1970). L-2,6-diaminopimelic acid (DAP) may be easily distinguished from *meso*-DAP and D-DAP by simple one-way paper chromatography of whole cell hydrolyzates (Becker *et al.*, 1964). Thus actinomycetes with a Type I cell wall may easily be singled out. One should note that in our studies no attempt is made to differentiate between *meso*- and D-DAP.

TABLE II
MAJOR CONSTITUENTS OF CELL WALLS OF ACTINOMYCETES

Cell wall type	Major constituents[a]
Streptomyces or Type I	L-DAP, glycine[b]
Micromonospora or Type II	meso-DAP, glycine; hydroxy DAP may also be present
Actinomadura or Type III	meso-DAP
Nocardia or Type IV	meso-DAP, arabinose, galactose
Oerskovia	Lysine, aspartic acid, galactose
Actinomyces bovis	Lysine, aspartic acid
Actinomyces israeli	Lysine, ornithine
Agromyces	DAB, glycine[b]
Mycoplana	meso-DAP + numerous amino acids

[a] All cell wall preparations contain major amounts of alanine, glutamic acid, glucosamine, and muramic acid.
[b] DAP= 2,6-diaminopimelic acid; DAB= 2,4-diaminobutyric acid.

Boone and Pine (1968) proposed a rapid method for characterization of actinomycetes by cell wall composition. Further studies (Lechevalier, 1968; M. P. Lechevalier and Lechevalier, 1970) revealed that in most cases it is not necessary to prepare and analyze cell wall preparations from actinomycetes to be able to determine to which cell wall group they belong. This is especially useful in differentiating between cell wall Types II, III, IV which are characteristic of most of the aerobic actinomycetes not of the *Streptomyces*-type (Type I). This presumptive cell wall identification may be done by determining the whole-cell sugar pattern of the actinomycete. Correspondences between cell wall types and whole-cell sugar patterns are given in Table III. One will note that actinomycetes with cell walls of Type III fall into two groups: those containing madurose and those in which madurose is lacking. Madurose has been identified as 3-0-methyl-D-galactose by Lechevalier and Gerber (1970).

We can thus conclude that cell wall types and whole-cell sugar patterns are not difficult to determine. In addition, these characters are stable characteristics of a given strain (Šuput *et al.*, 1967).

The study of cell wall composition and whole-cell sugar patterns has permitted us to solve some taxonomic riddles. First, it has furnished a convenient way of drawing a line between softish, baldish, nonsporulating streptomycetes and morphologically similar nocardiae. The streptomycetes, which contain the L-form of DAP, are easily distinguished by the method of Becker *et al.* (1964). The method is, of course, equally useful in separating from streptomycetes, hard,

TABLE III
CELL WALL TYPES AND WHOLE-CELL SUGAR PATTERNS OF AEROBIC
ACTINOMYCETES CONTAINING Meso-DIAMINOPIMELIC ACID[a]

Cell wall		Whole-cell sugar pattern	
Type	Distinguishing major constituents[b]	Type	Diagnostic sugars
II	Glycine	D	Xylose, arabinose
III	None	B	Madurose[c]
		C	None
IV	Arabinose, galactose	A	Arabinose, galactose

[a]No differentiation is made between meso-DAP and D-DAP.
[b]All cell wall preparations contain major amounts of alanine, glutamic acid, glucosamine, and muramic acid.
[c]Madurose = 3-O-methyl-D-galactose.

filamentous, and conidia-bearing strains of *Nocardia*. Second, it has lent strong support to the concept that there are indeed three types of single spore-forming actinomycetes. This fact was suspected on the basis of morphology but chemical data pointed up the fact that *Micromonospora* (cell wall Type II) is different from *Thermoactinomyces* (cell wall Type III) which in turn could be distinguished from *Thermomonospora* (cell wall Type IV).

Medical mycologists have wondered for some time whether the mycetoma-forming species *madurae* and *pelletieri* (Type III cell wall) should be placed in the genus *Nocardia* or in the genus *Streptomyces*. Neither *Streptomyces* (Type I cell wall) nor *Nocardia* (Type IV cell wall) seems suitable. The genus *Actinomadura* (H. A. Lechevalier and Lechevalier, 1970) was thus proposed for organisms with a Type III cell wall which form chains of conidia. This group seems related chemically and morphologically to *Microbispora*.

At first glance, some groupings made on the basis of cell wall composition may seem morphologically unsound. For example, it would seem logical to separate the sporangia-forming actinomycetes into two groups: those with motile and those with nonmotile sporangiospores. Cell wall composition does indeed reveal two groups; some Actinoplanaceae have a Type II cell wall while others have a Type III cell wall. However, motile forms are found in both groups. As mentioned previously, the studies of Farina and Bradley (1970), using DNA hybridization have confirmed the value of the grouping on the basis of cell wall type. Indeed, motility appears to be an unstable property which, in addition, may be apparent only under stringent nutritional conditions. This was true in the case of *Amorphosporangium* first described as a genus harboring organisms with irregularly shaped sporangia containing nonmotile spores (Couch, 1963). Subsequently, the spores of this group were found to be motile (Hanton, 1968).

Cell wall composition seems to be a practical taxonomic criterion only if major constituents are taken into account. We favor grouping into a given genus only strains which have similar cell wall composition. However, we recognize that different organisms may have the same major constituents in their cell walls. For example, the very different *Actinomadura madurae* and *Dermatophilus congolensis* have cell walls of Type III and contain major amounts of madurose.

VI. Lipid Composition in the Classification of Actinomycetes

Lipid patterns can be used to separate streptomycetes from nocardiae (Mordarska, 1968) confirming the separation made on the basis of the isomer of DAP present in their cell walls. But, more important, lipid composition is of assistance in the characterization of the genus *Mycobacterium*. Nocardiae, mycobacteria, and many corynebacteria all have a Type IV cell wall and a Type A whole-cell sugar pattern. Their morphology presents an uninterrupted spectrum from mucoid bacteria with a shade of branching to highly filamentous, highly branched organisms bearing chains of conidia. Clearly, other criteria are needed to separate these organisms into workable groups. As summarized by Asselineau (1966), bacteria of the *Nocardia-Mycobacterium-Corynebacterium* complex contain α-branched, β-hydroxylated fatty acids, called collectively mycolic acids, which are closely associated with the cell wall. Studies reported so far (Lanéelle-Carrieu, 1969) seem to confirm that there are three rather distinct types of mycolic acids (Fig. 4): (a) those with carbon skeletons of about 80 carbon atoms associated with strains of *Mycobacterium* (mycolic acids *stricto sensu*), (b) those with skeletons of about 50 carbons found in strains of *Nocardia* (nocardomycolic acids) and, (c) those

$$CH_3-R-\underset{|}{CH}-CH-COOH$$
$$\overset{OH}{\underset{(CH_2)_n}{|}}$$
$$\underset{CH_3}{|}$$

$R = (CH_2)_m$ or (C_mH_{2m-2})
where $m = 14$ or 16
$n = 11, 13, 15$

(Corynomycolic acids)

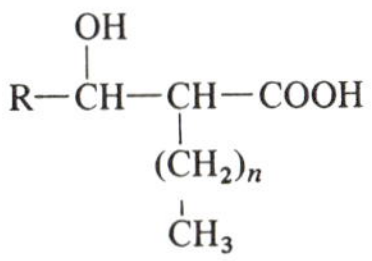

$$R-\underset{\underset{CH_3}{|}}{\underset{(CH_2)_n}{|}}\overset{OH}{\underset{|}{CH}}-CH-COOH \qquad R = C_{32}H_{39\text{-}61} \pm 7CH_2$$
$$n = 9, 11, 13, 15$$

(Nocardomycolic acids)

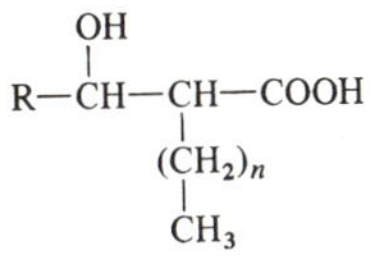

$$R-\underset{\underset{CH_3}{|}}{\underset{(CH_2)_n}{|}}\overset{OH}{\underset{|}{CH}}-CH-COOH \qquad R = C_{43\text{-}61}H_x$$
$$n = 19, 21, 23$$

(Mycolic acids)

FIG. 4 General structures of mycolic acids.

with smaller skeletons of about 30 atoms of carbon which seem to be associated with certain strains of *Corynebacterium* (corynomycolic acids).

In our own studies (M. P. Lechevalier et al., 1971), we have shortened the classical procedures for isolation and purification of these compounds. Fractions containing methylated mycolates are injected into a gas chromatograph where they are pyrolyzed in the injector into a fatty ester moiety and an aldehyde moiety (Fig. 5). True mycolates

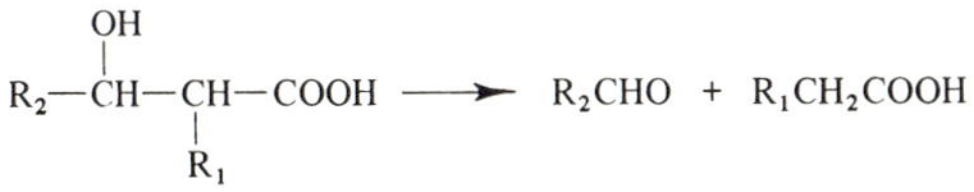

$$R_2-\underset{\underset{R_1}{|}}{\overset{\overset{OH}{|}}{CH}}-CH-COOH \longrightarrow R_2CHO + R_1CH_2COOH$$

(mycolic acid)   (aldehyde)   (fatty acid)

FIG. 5. Pyrolytic cleavage of a mycolic acid.

are easily singled out since they release unbranched saturated fatty esters having 22 to 26 carbons. From the other two types of mycolates, shorter fatty esters (C_{12} to C_{18}) are released. A distinction may be made between corynomycolates and nocardomycolates on the basis of the aldehyde moiety formed during pyrolysis. The aldehydes from nocardomycolates being rather large (C_{31} to C_{38}), are eluted much less readily from the gas chromatograph than those from corynomycolates which are smaller (C^{16} to C^{18}). These compounds may, therefore, be distinguished by their retention times.

Mycolic acid analyses of 96 strains of actinomycetes considered to belong either to the genus *Nocardia* or to the genus *Mycobacterium* revealed that they indeed fell into two groups depending on the size of the fatty acid released by pyrolysis (M. P. Lechevalier et al., 1971). Those with the large mycolic acids may conveniently be placed in the genus *Mycobacterium;* those with the smaller ones previously described, in *Nocardia.*

Organisms having quite dissimilar types of cell walls are currently placed in the genus *Corynebacterium.* An even larger and more diverse group has been called, more generally, "coryneforms" (Keddie et al., 1966; Sukapure et al., 1970; Veldkamp, 1970). This might mean that corynebacteria are a center of evolution from which many different types of bacteria, including the actinomycetes, have evolved, or, more prosaically, it may simply be a reflection of how little we know about the organisms we so casually place together in the coryneform group.

As Canhan et al. (1970) recently said: "Chemical taxonomy is a two-edged sword. It can serve to expose errors either of chemistry or of taxonomy. It does this by drawing attention to misfits." The problem becomes especially critical when the "misfit" is the type species of a genus! *Nocardia farcinica* is technically the type species of the genus *Nocardia.* It was isolated by Nocard from the pus of a diseased steer from the Guadeloupe. Nocard described this organism in 1888 and Trevisan coined the name *Nocardia* the next year. The original cultures of Nocard have been lost. Recently, organisms which contain true mycolic acids rather than nocardomycolic acids have been isolated from cases of bovine farcy (Chamoiseau, 1969; Chamoiseau and Asselineau, 1970). It thus seems sensible to consider *N. farcinica* a *nomen dubium* (Horan et al., 1970) and to accept *N. asteroides* as the type species of the genus *Nocardia* as suggested by Gordon and Mihm (1962).

VII. Conclusion: A Tentative System of Classification of Actinomycetes

In 1965, we proposed that actinomycetes should be classified on the basis of morphology and chemical composition (Lechevalier and Lechevalier, 1965). This concept, which was further developed in later publications (H. A. Lechevalier and Lechevalier, 1967, 1970), was well received by some (Küster, 1968; Williams et al., 1968; Hütter, 1970; Prauser, 1970) but was considered too chemical by others (Baldacci and Locci, 1970; Thirumalachar, 1970).

H. A. Lechevalier and Lechevalier (1970) and Prauser (1970) recently proposed two tentative schemes of classification of actinomycetes which are based on morphology and chemical composition. The main difference between the two systems is that Prauser is willing to base his first division of the actinomycetes into two large groups on the basis of fragmentation of the primary mycelium, a characteristic which seems unstable and which is often subject to misunderstanding (see article "Chemical Characteristics and Classification of Nocardiae," by H. A. Lechevalier *et al.*, 1971).

It is quite difficult to separate actinomycetes into meaningful groups on the basis of morphology alone. This has been attempted in Table IV and we can note that the result is not always satisfactory: members of the same genus can appear in more than one group.

Taking the various morphological groups in Table IV in order, we

TABLE IV
MORPHOLOGIC CLASSIFICATION OF THE MOST IMPORTANT ACTINOMYCETES

Morphology	Possible generic assignment
I. No sporangia formed	
1) Mycelium dividing in more than one plane	*Dermatophilus, Geodermatophilus*
2) Mycelium dividing only perpendicularly to the main axis of the hyphae	
a) Only aerial mycelium formed	*Sporichthya*
b) Primary (substrate) mycelium only	
No conidia, no motile elements	*Mycobacterium, Actinomyces, Agromyces, Nocardioides, Nocardia,* or an imperfect form of everything that follows under b and c
No conidia, motile elements present	*Oerskovia, Mycoplana*
Single conidia	*Micromonospora*
c) Both mycelia formed	
No spores	*Mycobacterium, Nocardioides, Nocardia,* or an imperfect form of everything that follows under c
Single spores on aerial mycelium	*Thermomonospora*
Single spores on both mycelia	*Thermoactinomyces*
Longitudinal pairs of spores on aerial mycelium	*Microbispora*
Short chains of conidia on aerial mycelium	*Actinomadura, Streptomyces, Nocardia*
Short chains of conidia on both mycelia	*Micropolyspora, Nocardia*
Long chains of conidia on aerial mycelium	*Streptomyces, Streptoverticillium,*

(Continued)

TABLE IV (continued)

I. (Continued)	Nocardia, Actinomadura, Pseudonocardia
Long chains of conidia on both mycelia	Streptomyces, Micropolyspora
II. Sporangia formed	
Sporangia each with one single spore	Planomonospora
Sporangia each with a longitudinal pair of spores	Planobispora
Sporangia with one single chain of spores	
Sporangiospores nonmotile	Microellobosporia
Sporangiospores motile	Dactylosporangium
Sporangia with many spores	
Sporangia globose to lageniform, spores motile	Actinoplanes, Amorphosporangium, Ampullariella, Spirillospora
Sporangia globose, spores nonmotile	Streptosporangium

find *Dermatophilus* (animal pathogens) and *Geodermatophilus* (soil forms) grouped together. It would be quite difficult to explain in simple terms the difference between the morphology of these two groups of organisms. The most accurate statement that could be made is that most strains of *Dermatophilus* have a better developed mycelial phase than those of *Geodermatophilus* (Ishiguro and Wolfe, 1970). Both types of organisms have Type III cell walls. However, madurose is present in strains of *Dermatophilus* and absent in those of *Geodermatophilus*.

Sporichthya presents no special problem: the three strains so far isolated have Type I cell walls. Our morphological observations (Lechevalier *et al.*, 1968) have been confirmed by Williams (1970).

Strains of *Agromyces* and of *Actinomyces* may be separated from each other and from strains of *Mycobacterium* and *Nocardia* by cell wall composition (Table III). *Nocardioides* (see article "Phage Typing of Nocardioform Organisms," by Prauser, 1971) is characterized by a cell wall of Type I. Strains of *Mycobacterium* can be separated from those of *Nocardia* on the basis of the mycolic acids they contain (M. P. Lechevalier *et al.*, 1971). We do not believe that one can separate mycobacteria from nocardiae solely on the basis of morphology. Our feeling has been well expressed by Bönicke and Juhasz (1965): "It is ... very difficult to select any specific morphological property present in one genus and absent in the other."

Oerskovia may be easily separated from *Mycoplana* on the basis of gram stain and thus, cell wall composition.

Strains of *Micromonospora* present no special problems. Their morphology is unique and, quite alone among the nonsporangiate actinomycetes, they have a Type II cell wall.

Strains of *Thermomonospora* (Type IV cell wall) can be separated from those of *Thermoactinomyces* (Type III cell wall). Some strains of organisms from both groups of organisms may bear dichotomously branched sporophores and are placed by some authors in the genus *Actinobifida* (Locci et al., 1967; Henssen, 1970). The conidia formed by all these single spore formers may be true endospores (Cross, 1970).

Strains of *Microbispora* form longitudinal pairs of spores on the aerial mycelium and have a Type III cell wall as do members of the genera *Actinomadura* and *Microtetraspora* (Thiemann et al., 1968) which form longer chains of conidia on the aerial mycelium. These latter organisms are easy to distinguish from strains of *Streptomyces* (Type I cell wall) and those of *Nocardia* (Type IV cell wall). The creation of the genus *Actinomadura* (H. A. Lechevalier and Lechevalier, 1970) has received the endorsement of the studies of Goodfellow (see article "Numerical Taxonomy of the Genus Nocardia and Allied Genera," by Goodfellow, 1971).

In the genus *Micropolyspora* the cell wall is of Type IV, as in *Nocardia*. The spores, borne on both the substrate and the aerial mycelia are very spherical and are formed basipetally (Lechevalier et al., 1961; Dorokhova et al., 1969).

The most important group of organisms forming long chains of conidia on their aerial hyphae fall into the *Streptomyces* complex. These organisms are easily characterized by the determination of their cell wall type (I). Species of *Actinomadura* have a Type III cell wall and those of *Nocardia* and *Pseudonocardia* have a Type IV cell wall.

The streptomycetes present a formidable group of widely distributed species and subgenera. Attempts at separating these organisms into workable, related groups have not always been too successful. The names *Actinopycnidium* and *Actinosporangium* have been proposed for streptomycetes forming, respectively, pycnidia or cephalosporal accumulations of spores. The studies of Williams (1970) have shown that the production of pycnidia by actinomycetes is not easy to demonstrate. Likewise, the use of the generic name *Chainia* for streptomycetes forming sclerotia is warmly supported by Thirumalachar (1970) but seems to have few other fans. On the contrary, the use of the generic name *Streptoverticillium* for streptomycetes forming verticillate sporophores (Locci et al., 1969) seems to grow in popularity.

Since it is entirely a matter of personal judgment whether a chain of conidia is long or short, one might possibly confuse certain *Streptomyces*, such as strains of *S. griseus* which form chains of conidia in the agar with strains of *Micropolyspora*. Cell wall composition permits us to separate these two types since streptomycetes have a Type I cell wall and micropolysporae have a Type IV cell wall.

Morphology is quite satisfactory for the separation of sporangia-forming actinomycetes into groups. However, the morphologically very similar group formed by *Actinoplanes* (globose to lageniform sporangia, globose spores with one polar tuft of flagella), *Amorphosporangium* (same but sporangia irregularly shaped and spores often nonmotile), and *Ampullariella* (same as *Actinoplanes* but rod-shaped spores with one polar tuft of flagella) all have a Type II cell wall and might well all be grouped in the same genus as proposed by Szaniszlo and Gooder (1967). The genus *Spirillospora*, also with motile spores, can be separated from the rest of this group since its members have a cell wall of Type III as do species of *Streptosporangium*. Also with Type III cell walls are the morphologically fascinating members of the genera *Planomonospora* and *Planobispora* (Thiemann, 1970). The sporangia of strains of *Planomonospora* contain one single spore; those of *Planobispora* contain two spores usually separated by a transverse septum.

If we keep all these considerations in mind, we can locate most actinomycetes among one of the genera listed in the scheme presented in Table V. No attempt at grouping the genera into families has been attempted in this key. Some family groupings are obvious, others less so:

Dermatophilaceae
   Cell wall Type III
   Unusual division

Dermatophilus
Geodermatophilus

Actinoplanaceae
   Cell wall Types II, III, or even I
   Sporangia

Actinoplanes
Dactylosporangium
Planomonospora
Planobispora
Spirillospora
Streptosporangium
Microellobosporia ?

Streptomycetaceae
   Cell wall Type I
   No sporangia

Streptomyces
Streptoverticillium
Microellobosporia ?
Sporichthya ?
Nocardioides ?

CHEMICAL COMPOSITION AND CLASSIFICATION OF ACTINOMYCETES

Micromonosporaceae
  Cell wall Type II
  No sporangia

Micromonospora

Thermoactinomycetaceae
  Cell wall Type III
  No sporangia

Thermoactinomyces
Microbispora
Actinomadura

Nocardiaceae
  Cell wall Type IV
  No mycolic acids, but nocardomycolic
  acids may be present

Thermomonospora
Micropolyspora
Pseudonocardia
Nocardia

Mycobacteriaceae
  Cell wall Type IV
  Mycolic acids

Mycobacterium

Actinomycetaceae
  Cell walls not of Types I to IV

Actinomyces
Agromyces
Oerskovia

Substrate mycelium only

Mycoplana

To improve classification of actinomycetes, the most pressing need at the moment is to find ways and means of simplifying the differentiation between the true mycolic acids, present in mycobacteria, and the nocardomycolic acids, present in nocardiae. Not even listed in Tables IV and V are the coryneform bacteria which still present formidable taxonomic problems (Veldkamp, 1970). A study of their corynomycolic acids should yield fruitful results.

TABLE V
KEY TO THE GENERA OF ACTINOMYCETES

A. Mycelium dividing in all planes. Cell wall of Type III.
  a) Madurose present. Animal pathogens.
       Dermatophilus
  b) Madurose absent. Soil forms.
       Geodermatophilus
B. Mycelium dividing only perpendicularly to the main axis of the hyphae.
  AA. No substrate mycelium formed. Cell wall of Type I.
       Sporichthya
  BB. Substrate mycelium formed.
    a. Cell wall with major amounts of 2,6-diaminopimelic acid (DAP)

(Continued)

TABLE V *(Continued)*

aa. Sporangia formed, cell wall of Type II
    1. Globose to lageniform sporangia with numerous motile spores.
       Actinoplanes
    2. Sporangia with one single row of motile spores.
       Dactylosporangium
bb. Sporangia formed. Cell wall of Type III.
    1. One single motile spore per sporangium.
       Planomonospora
    2. Two motile spores per sporangium.
       Planobispora
    3. Numerous motile sporangiospores.
       Spirillospora
    4. Numerous nonmotile sporangiospores.
       Streptosporangium
cc. Sporangia formed. Cell wall of Type I. One single row of nonmotile spores per sporangium.
       Microellobosporia
dd. No sporangia. Cell wall of Type I.
    1. Usually abundant aerial mycelium bearing chains of conidia. Substrate mycelium usually nonfragmenting.
       11. Conidiophores not verticillate.
           Streptomyces
       22. Conidiophores verticillate.
           Streptoverticillium
    2. Aerial mycelium formed sparingly. Primary mycelium usually fragmenting.
       Nocardioides
ee. No sporangia. Cell wall of Type II. Single conidia formed on the substrate mycelium only.
       Micromonospora
ff. No sporangia. Gram-negative cell wall with numerous amino acids. Mycelium rudimentary, breaking into motile elements.
       Mycoplana
gg. No sporangia. Cell wall of Type III.
    1. Single spores formed. No madurose.
       Thermoactinomyces
    2. Longitudinal pairs of conidia formed. Madurose present.
       Microbispora
    3. Chains of conidia longer than two.
       11. Chains of conidia short. Madurose present.
           Actinomadura, madurae-type
       22. Chains of conidia long. Madurose absent.
           Actinomadura, dassonvillei-type
hh. No sporangia. Cell wall of Type IV.
    1. Single spores formed on aerial mycelium only.
       Thermomonospora
    2. Short chains of conidia formed both on the substrate and on the aerial mycelium.
       Micropolyspora

TABLE V *(Continued)*

3. Chains of cylindrical conidia formed both on the substrate and the aerial mycelia.
   Pseudonocardia
4. Morphology very variable. Aerial mycelium may be absent. Sporulation, when it occurs, by formation of chains of conidia. Substrate mycelium may fragment. Nocardomycolic acids produced.
   Nocardia
5. Substrate mycelium usually fragmenting. Aerial mycelium usually not formed. Spores not formed. Mycolic acids formed.
   Mycobacterium

b. Cell wall with major amounts of 2,4-diaminobutyric acid. No aerial mycelium formed. No motile elements formed but fragmentation of substrate mycelium.
   Agromyces
c. Cell wall with major amounts of lysine.
   aa. Primary mycelium breaking into motile elements. No aerial mycelium formed.
      Oerskovia
   bb. No aerial mycelium formed. No spores formed. Usually associated with man and animals. Usually microaerophilic to anaerobic.
      Actinomyces

REFERENCES

Asselineau, J. (1966). "The Bacterial Lipids." Hermann, Paris.
Avery, R. J., and Blank, F. (1954). *Can. J. Microbiol.* **1**, 140–143.
Baldacci, E. (1956). *G. Microbiol.* **2**, 50–62.
Baldacci, E., and Locci, R. (1970). *In* "The Actinomycetes" (H. Prauser, ed.), pp. 419–424. Fischer, Jena.
Baldacci, E., Locci, R., Farina, G., and Vegetti, G. (1962). *G. Microbiol.* **10**, 213–235.
Baldacci, E., Locci, R., Farina, G., and Vegetti, G. (1963). *G. Microbiol.* **11**, 137–166.
Bartnicki-Garcia, S. (1968). *Annu. Rev. Microbiol.* **22**, 87–108.
Bataille, F. (1948). "Les Réactions Macrochimiques chez les Champignons." Reprint (1969) Cramer, Lehre, Germany.
Becker, B., Lechevalier, M. P., Gordon, R. E., and Lechevalier, H. A. (1964). *Appl. Microbiol.* **12**, 421–423.
Becker, B., Lechevalier, M. P., and Lechevalier, H. A. (1965). *Appl. Microbiol.* **13**, 236–243.
Blinov, N. O., and Khokhlov, A. S. (1970). *In* "The Actinomycetales" (H. Prauser, ed.), pp. 145–153. Fischer, Jena.
Bönicke, R., and Juhasz, S. E. (1965). *Bull. Int. Union Against Tuberc.* **36**, 14–22.
Boone, C. J., and Pine, L. (1968). *Appl. Microbiol.* **16**, 279–284.
Cambie, R. C., and LeQuesne, P. W. (1966). *J. Chem. Soc. C.* pp. 72–74.
Canhan, S. C., Anchel, M., and Bistis, G. N. (1970). *Mycologia* **62**, 599–603.
Chamoiseau, G. (1969). *Rev. Elevage Med. Vet. Pays Trop.* **22**, 195–204.
Chamoiseau, G., and Asselineau, J. (1970). *C. R. Acad. Sci., Ser. D* **270**, 2603–2604.
Couch, J. N. (1963). *J. Elisha Mitchell Sci. Soc.* **79**, 53–70.
Craveri, R., and Manachini, P. L. (1965). *Atti Congr. Naz. Microbiol., 13th* pp. 101–109.

Cross, T. (1970). *J. Appl. Bacteriol* **33**, 95-102.
Cross, T., and MacIver, A. M. (1966). *In* "Identification Methods for Microbiologists" (B. M. Gibbs and F. A. Skinner, eds.), Part A, pp. 103-109. Academic Press, New York.
Cummins, C. S. (1962). *Ann. Inst. Pasteur* **103**, 385-391.
Cummins, C. S., and Harris, H. (1958). *J. Gen. Microbiol.* **18**, 173-189.
DeLey, J. (1970). *In* "The Actinomycetales" (H. Prauser, ed.), pp. 317-327. Fischer, Jena.
Dorokhova, L. A., Agre, N. S., Kalakoutskii, L. V., and Krassilnikov, N. A. (1969). *J. Microsc. (Paris)* **8**, 845-854.
Evreinova, T. N., Tsaplina, I. A., Agre, N. S., and Davydova, I. M. (1965). *Mikrobiologiya* **34**, 411-417.
Farina, G., and Bradley, S. G. (1970). *J. Bacteriol.* **102**, 30-35.
Gaughran, R. L. (1969). *Trans. N.Y. Acad. Sci.* **31**, 3-24.
Gerber, N. N. (1966). *Biochemistry* **5**, 3824-3829.
Gerber, N. N. (1967). *J. Org. Chem.* **32**, 4055-4057.
Gerber, N. N. (1969). *J. Heterocycl. Chem.* **6**, 297-300.
Gerber, N. N. (1970). *Tetrahedron Lett.* pp. 809-812.
Gerber, N. N., and Lechevalier, M. P. (1964). *Biochemistry* **3**, 598-602.
Gerber, N. N., and Lechevalier, M. P. (1965). *Biochemistry* **4**, 176-180.
Gledhill, W. E., and Casida, L. E. (1969). *Appl. Microbiol.* **18**, 340-349.
Goodfellow, M. (1971). *Recent Progr. Microbiol., Symp. Int. Congr. Microbiol., 10th, Mexico, 1970* (in press).
Gordon, R. E. (1966a). *J. Gen. Microbiol.* **45**, 355-364.
Gordon, R. E. (1966b). *J. Gen. Microbiol.* **43**, 329-343.
Gordon, R. E., and Mihm, J. M. (1962). *J. Gen. Microbiol.* **27**, 1-10.
Gordon, R. E., and Pang, H.-N. (1970). *Appl. Microbiol.* **19**, 862-864.
Gordon, R. E., and Rynearson, T. R. (1963). *Can. J. Microbiol.* **9**, 737-739.
Gottlieb, D., and Shirling, E. B. (1970). *In* "The Actinomycetales" (H. Prauser, ed.), pp. 67-77. Fischer, Jena.
Hale, M. E. (1967). "The Biology of Lichens." Arnold, London.
Hanton, W. K. (1968). *J. Gen. Microbiol.* **53**, 317-320.
Harashima, K., Tsuchida, N., Tanaka, T., and Nagatsu, J. (1967). *Agr. Biol. Chem.* **31**, 481-489.
Hegnauer, R. (1963). *In* "Chemical Plant Taxonomy" (T. Swain, ed.), pp. 389-427. Academic Press, New York.
Henssen, A. (1970). *In* "The Actinomycetales" (H. Prauser, ed.), pp. 205-210. Fischer, Jena.
Horan, A. C., Lechevalier, M. P., and Lechevalier, H. (1970). *Bacteriol. Proc.* p. 20.
Hütter, R. (1970). *In* "The Actinomycetales" (H. Prauser, ed.), pp. 55-66. Fischer, Jena.
Ishiguro, E. E., and Wolfe, R. S. (1970). *J. Bacteriol.* **104**, 566-580.
Kauffman, F. (1966). "The Bacteriology of Enterobacteriaceae." Munksgaard, Copenhagen.
Keddie, R. M., Leask, B. G. S., and Grainger, J. M. (1966). *J. Appl. Bacteriol* **29**, 17-43.
Krassilnikov, N. A. (1959). *Ann. Inst. Pasteur* **96**, 434-447.
Krassilnikov, N. A. (1970). *In* "The Actinomycetales" (H. Prauser, ed.), pp. 123-131. Fischer, Jena.
Krause, R. M. (1963). *Bacteriol. Rev.* **27**, 369-380.
Kudo, R. R. (1966). "Protozoology," 5th Ed. Thomas, Springfield, Illinois.
Küster, E. (1968). *In* "The Ecology of Soil Bacteria" (T. R. G. Gray and D. Parkinson, eds.), pp. 322-336. Univ. of Toronto Press, Toronto, Canada.

Lanéelle-Carrieu, M-A. (1969). Étude d'acides hydroxylés d'origine bactérienne et de quelques formes liées de ces acides. Thesis, Toulouse Univ., Toulouse.
Lechevalier, H. A. (1965). In "Biogenesis of Antibiotic Substances" (Z. Vaněk and Z. Hošťálek, eds.), pp. 227–232. Czech. Acad. Sci., Prague.
Lechevalier, H. A., and Lechevalier, M. P. (1965). Ann. Inst. Pasteur 108, 662–673.
Lechevalier, H. A., and Lechevalier, M. P. (1967). Annu. Rev. Microbiol. 21, 71–100.
Lechevalier, H. A., and Lechevalier, M. P. (1970). In "The Actinomycetales" (H. Prauser, ed.), pp. 393–405. Fischer, Jena.
Lechevalier, H. A., Solotorovsky, M., and McDurmont, C. I. (1961). J. Gen. Microbiol. 26, 11–18.
Lechevalier, H. A., Lechevalier, M. P., and Horan, A. C. (1971). Recent Progr. Microbiol., Symp. Int. Congr. Microbiol., 10th, Mexico, 1970 (in press).
Lechevalier, M. P. (1968). J. Lab. Clin. Med. 71, 934–944.
Lechevalier, M. P., and Gerber, N. N. (1970). Carbohyd. Res. 13, 451–454.
Lechevalier, M. P., and Lechevalier, H. A. (1970). In "The Actinomycetales" (H. Prauser, ed.), pp. 311–316. Fischer, Jena.
Lechevalier, M. P., Lechevalier, H. A., and Holbert, P. E. (1968). Ann. Inst. Pasteur 114, 277–286.
Lechevalier, M. P., Horan, A. C., and Lechevalier, H. A. (1971). J. Bacteriol. 105, 313–318.
Lewin, R. A. (1962). "Physiology and Biochemistry of Algae." Academic Press, New York.
Locci, R., Baldacci, E., and Petrolini, B. (1967). G. Microbiol. 15, 79–91.
Locci, R., Baldacci, E., and Baldan, B. (1969). G. Microbiol. 17, 1–60.
Luria, S. E., and Darnell, J. E. (1967). "General Virology," 2nd Ed. Wiley, New York.
Manachini, P. L., Craveri, A., and Craveri, R. (1967). Atti Congr. Soc. Ital. Microbiol., 14th pp. 13–22.
Mordarska, H. (1968). Arch. Immunol. Ther. Exp. 16, 45–50.
Nobles, M. K., and Frew, B. P. (1962). Can. J. Bot. 40, 987–1016.
Nocard, E. (1888). Ann. Inst. Pasteur 2, 293–302.
Oliver, T. J. (1970). In "The Actinomycetales" (H. Prauser, ed.), pp. 155–162. Fischer, Jena.
Prauser, H. (1964). Z. Allg. Mikrobiol. 4, 95–98.
Prauser, H. (1970). In "The Actinomycetales" (H. Prauser, ed.), pp. 407–418. Fischer, Jena.
Prauser, H. (1971). Recent Progr. Microbiol., Symp. Int. Congr. Microbiol., 10th, Mexico, 1970 (in press).
Prauser, H., and Eckardt, K. (1967). Z. Allg. Mikrobiol. 7, 409–410.
Prauser, H., Lechevalier, M. P., and Lechevalier, H. A., (1970). Appl. Microbiol. 19, 534.
Preobrazhenskaya, T. P. (1970). In "The Actinomycetales" (H. Prauser, ed.), pp. 91–96. Fischer, Jena.
Pridham, T. G. (1965). Appl. Microbiol. 13, 43–61.
Romano, A. H., and Sohler, A. (1956). J. Bacteriol. 72, 865–868.
Shirling, E. B., and Gottlieb, D. (1966). Int. J. Syst. Bacteriol. 16, 313–340.
Shirling, E. B., and Gottlieb, D. (1970). In "The Actinomycetales" (H. Prauser, ed.), pp. 79–89. Fischer, Jena.
Slack, J. M., and Gerencser, M. A. (1970). In "The Actinomycetales" (H. Prauser, ed.), pp. 19–27. Fischer, Jena.
Smith, A. H. (1966). In "The Fungi. Vol. 2: The Fungal Organism" (G. C. Ainsworth and A. S. Sussman, eds.), Vol. 2, pp. 151–177. Academic Press, New York.

Sokal, R. R., and Sneath, P. H. A. (1963). "Principles of Numerical Taxonomy." Freeman, San Francisco, California.
Stanier, R. Y. (1964). In "The Bacteria. Vol. 5: Heredity" (I. C. Gunsalus and R. Y. Stanier, eds.), pp. 445–464. Academic Press, New York.
Sueoka, N. (1964). In "The Bacteria. Vol. 5: Heredity" (I. C. Gunsalus and R. Y. Stanier, eds.), pp. 419–443. Academic Press, New York.
Sukapure, R. S., Lechevalier, M. P., Reber, H., Higgins, M. L., Lechevalier, H. A., and Prauser, H. (1970). *Appl. Microbiol.* **19**, 527–533.
Šuput, J., Lechevalier, M. P., and Lechevalier, H. A. (1967). *Appl. Microbiol.* **15**, 1356–1361.
Szaniszlo, P. J., and Gooder, H. (1967). *J. Bacteriol.* **94**, 2037–2047.
Teillon, J. (1953). "Essai de différenciation biologique des Streptomyces antibiotiques." Librairie Générale de l'Enseignment, Paris.
Tewfik, E., Bradley, S. G., Kuroda, S., and Wu, R. Y. (1968). *Develop. Ind. Microbiol.* **9**, 242–249.
Thiemann, J. E. (1970). In "The Actinomycetales" (H. Prauser, ed.), pp. 245–257. Fischer, Jena.
Thiemann, J. E., Pagani, H., and Beretta, G. (1968). *J. Gen. Microbiol.* **50**, 295–303.
Thirumalachar, M. J. (1970). In "The Actinomycetales" (H. Prauser, ed.), pp. 424–429. Fischer, Jena.
Tresner, H. D., and Backus, E. J. (1963). *Appl. Microbiol.* **11**, 335–338.
Tsyganov, V. A., and Krasikova, N. V. (1968). *Mikrobiologiya* **37**, 969–971.
Tsyganov, V. A., Namestnikova, V. P., and Krasikova, N. V. (1966). *Mikrobiologiya* **35**, 92–95.
Umezawa, H., ed. (1967). "Index of Antibiotics from Actinomycetes." Penn. State Univ. Press, University Park, Pennsylvania.
Veldkamp, H. (1970). *Annu. Rev. Microbiol.* **24**, 209–240.
Waksman, S. A. (1963). *Advan. Appl. Microbiol.* **5**, 235–315.
Waksman, S. A., and Lechevalier, H. A. (1962). "Antibiotics of Actinomycetes." Williams & Wilkins, Baltimore, Maryland.
Walker, P. M. B. (1969). *Progr. Nucl. Acid Res. Mol. Biol.* **9**, 301–326.
Wasserman, H. H., Rodgers, G. C., Keith, D. D., and Nadelson, J. (1969). *J. Amer. Chem. Soc.* **91**, 1263–1269.
Williams, R. P., and Hearn, W. R. (1967). In "Antibiotics. Vol. III: Biosynthesis" (D. Gottlieb and P. D. Shaw, eds.), pp. 410–432. Springer-Verlag, Berlin and New York.
Williams, S. T. (1970). *J. Gen. Microbiol.* **62**, 67–73.
Williams, S. T., Davies, F. L., and Cross, T. (1968). In "Identification Methods for Microbiologists" (B. M. Gibbs and D. A. Shapton, eds.), Part B, pp. 111–124. Academic Press, New York.
Yamaguchi, T. (1965). *J. Bacteriol.* **89**, 444–453.
Yamaguchi, T. (1967). *J. Gen. Appl. Microbiol.* **13**, 1994–2000.

Prevalence and Distribution of Antibiotic-Producing Actinomycetes

JOHN N. PORTER

Lederle Laboratories Division, American Cyanamid Company, Pearl River, New York

I.	Introduction	73
II.	Isolation and Enumeration of Actinomycetes	74
	A. Selective Isolation	74
	B. Prevalence in Soils	76
	C. Distribution in Aqueous Habitats	77
	D. Actinomycetes from Air	78
	E. Thermophilic Actinomycetes	78
	F. Microbial Changes during Storage	79
	G. Effect of Soil Amendments	80
III.	Classification of Actinomycetes Isolated from Soils	81
IV.	Antibiotic-Producing Actinomycetes	83
	A. Antibiotic Screening	83
	B. Distribution of Antibiotic-Producing Actinomycetes	84
	C. Variety of Microorganisms Capable of Producing the Same Antibiotic	86
V.	Identification and Classification of Antibiotics Produced by Actinomycetes	86
	A. Antimicrobial Spectra	86
	B. Physical and Chemical Properties	87
VI.	In Retrospect	89
	References	90

I. Introduction

The actinomycetes comprise an ubiquitous order of bacteria which exhibit wide physiological and morphological diversity. The majority of species are aerobic, saprophytic, mesophilic forms whose natural habitat is soil. In addition, actinomycetes may also be isolated from a wide variety of terrestrial and aquatic habitats and can even be recovered from the air, where they temporarily exist as spores or mycelial fragments. These microorganisms are particularly abundant in slightly alkaline soils rich in organic matter.

Antibiotics, because of their industrial importance, are the best known products of actinomycete (particularly *Streptomyces*) metabolism but other metabolites such as extracellular enzymes, pigments, and growth-promoting factors (probiotics) may be produced as well (Rainbow and Rose, 1963; Lilly and Stillwell, 1965). It is quite likely that with continued and widening interest in the field of microbial metabolites the actinomycetes will be shown to be capable of producing additional products of scientific and perhaps commercial interest.

Many of these microbial products, including the antibiotics, are considered to be "secondary metabolites" because they seem to have no direct role in those aspects of metabolism which support necessary functions in the cell, namely, energy, growth, and reproduction. There is great structural variety among secondary metabolites but organisms also have the ability to produce very closely related metabolites, some antibiotically active and some not (Bu'Lock, 1961; Pramer, 1968).

Without entering into the question of why a microorganism produces an antibiotic it should be remembered that the soil contains populations of bacteria and other organisms all interacting with one another. As discussed by Brock (1966), there is an ecological approach versus a screening approach to finding new antibiotics. The former is a rational approach based on microbial interrelationships and the latter is an empirical approach based on the theory that antibiotics are produced at random by microorganisms. Pochon and Tardieux (1968), in following this general line of thinking, pointed out that many soil organisms are not present in very large numbers and may not be recognized by standard isolation techniques. Therefore, enrichments may stimulate species or functional groups of species, representatives of which can then be isolated under appropriate medium and other culture conditions.

Waksman (1969), in an essay written after three decades of experience with antibiotics and their discovery, put it another way, "Any new method for the isolation of antibiotic-producing organisms will tend to yield new types of antibiotics."

In this discussion it is not my intent to review the actinomycetes and their antibiotics. Several excellent reviews already exist on these subjects (Waksman, 1959, 1961, 1963; Waksman and Lechevalier, 1962; Sevcik, 1963; Lechevalier and Lechevalier, 1967). Rather, I wish to call the attention of the reader to the variety of actinomycetes literally lying at his feet and to the fact that an intensive study of relatively few soil samples can produce substantial yields in terms of these microorganisms and their metabolites. Consequently, an overview is presented of the isolation and classification of actinomycetes and of the detection and classification of antibiotics with emphasis on the more recent contributions to the literature on these subjects.

II. Isolation and Enumeration of Actinomycetes

A. Selective Isolation

In order to isolate successfully a wide variety of actinomycetes from soil samples it is necessary to eliminate or greatly curtail fungal and bacterial spreaders in the isolation medium without producing an

adverse effect on actinomycetes. This can be accomplished in one, or a combination of more than one, of the following ways:
1. Control of the medium constituents
2. Addition of inhibitors to the medium
3. Prior treatment of the soil sample

Because actinomycetes are in general inhibited by antibacterial agents the most effective method of discouraging unwanted bacterial spreaders has been through manipulation of the composition of the isolation medium. Streptomycetes, in contrast to most other bacteria, grow well on a glycerol medium containing L(+) arginine as the sole source of nitrogen (Benedict et al., 1955). Starch-potassium nitrate agar was preferred for the same reason by Flaig and Kutzner (1960) in enumerating *Streptomyces* in an ecological study in which they also employed a glucose-peptone medium to obtain the total microbial count.

In contrast to most bacteria and fungi, many strains of *Streptomyces*, as well as other soil-inhabiting actinomycetes, produce chitinase. Lingappa and Lockwood (1962) took advantage of this to develop a simple medium consisting only of 0.2% colloidal chitin and 20% agar. Williams and Davies (1965), however, found this medium to be less satisfactory than starch-casein agar because of the small size and lack of distinguishing features of colonies and they also noted that some fungi with large radial spread occasionally developed. The starch-casein agar, in which glycerol may be substituted for starch, was subsequently used to isolate large numbers of streptomycetes from a composted soil (Küster and Williams, 1964).

Fungal contaminants can be virtually eliminated by adding antifungal agents to the isolation medium. Antibiotics have proven particularly successful. Nystatin, cycloheximide, or pimaricin added at a level of approximately 50 μg/ml to selective media such as those noted above have effectively eliminated most undesired contaminants in isolation plates (Corke and Chase, 1956; Porter et al., 1960; Williams and Davies, 1965). In my experience pimaricin has been the most effective antifungal antibiotic of the three.

A reduction in numbers of bacteria and fungi can also be achieved by the careful use of heat or the centrifugation of soil suspensions. In attempting to isolate actinomycetes to be studied for pectinolytic properties Agate and Bhat (1963) dried plates of glucose-yeast extract agar at 37°C overnight. The plates were then inoculated with soil suspensions and heated in the hot air oven at 110°C for 10 minutes. It was necessary to restrict the pretreatment to precisely 10 minutes to avoid melting the medium. Although this procedure eliminated contaminants it also decreased the total count of actinomycetes.

Rehacek (1959) used centrifugation of soil suspensions at a relative centrifugal force of 904 g at the surface and 1609 g at the bottom of the cuvette to eliminate contaminants and favor actinomycetes. He recommended that soil samples be ground prior to dilution to prevent entrapment of spores by soil particles during centrifugation.

Procedures have also been developed to isolate a greater variety of actinomycetes from individual soil samples. For example, the predominance of *Streptomyces* strains on isolation plates can be altered by incorporating a series of carbon sources which can be metabolized by some strains but not others in a synthetic basal medium (Porter and Wilhelm, 1961a).

Five species and two varieties of *Microbispora* and *Streptosporangium* were isolated by placing particles of soil on soil-extract agar and incubating for a few weeks at 30°C (Nonomura and Ohara, 1960a,b). Considerable modification was later made of this method (Nonomura and Ohara, 1969a). Soil samples were dried at room temperature, ground in a mortar, and heated in a hot air oven at 120°C for 1 hour. Suspensions were plated out on a synthetic medium containing B vitamins and in which polymyxin (4 mg/liter) and penicillin (0.8 mg/liter) were sometimes incorporated. Plates were then incubated at 30°C for 40 days or at 50°C for 20 days. Eubacteria and streptomycetes were substantially reduced which facilitated the preferential isolation of *Microbispora* and *Streptosporangium*.

B. Prevalence in Soils

There have been numerous surveys designed to show the numbers of actinomycetes present in various types of soils and the portion they comprise of the total microbial population (Waksman, 1959). An early representative study was that of Danish soils, in which a dextrose-casein medium was used for isolation. The numbers of actinomycetes varied from none to 13 million/gm and the percent of the total microflora from 0 to 73. Fewest actinomycetes were found in soils below pH 5 and most in soils of pH 6.8–8 (Jensen, 1930). In a subsequent survey of Australian soils Jensen (1934) concluded that other conditions favoring actinomycetes over bacteria and fungi were low moisture and high temperature.

Flaig and Kutzner (1960) isolated 1492 streptomycetes from various habitats which included tilled fields, grasslands, and forests. The highest streptomycete count was 4.8 million/gm in field soil and the highest percentage was 21.5 under grassy vegetation. As did Jensen they also found a low count in acid soils but several soils with pH values near neutrality were not much better. The greatest variety occurred in grassland soils and the least in forest soils, with field soils intermediate.

Using a glycerine-glycine agar for isolation Jagnow (1956) noted that microbial counts reached a peak in the summer months during warm weather and when there was considerable precipitation. He, too, found the fewest actinomycetes in soils below pH 5. Actinomycetes constituted up to 60% of the microbial population of dry chalk sods and dry forest habitats. These soils also contained the greatest proportion of antagonistic streptomycetes.

Counts and percentages of actinomycetes in soil samples as obtained in various surveys must be regarded as approximations, and interpretations must be made with caution. Different media and methods of isolation are used from investigator to investigator. Moreover, Jagnow found that four parallel samples from the same habitat could vary as much as 100% in microbial and actinomycete counts. Perhaps this is because the methods we use only skim the surface and repeated isolations would turn up more and more organisms (Porter *et al.*, 1964).

Partly because of a widening search for new antibiotics among the more obscure microorganisms there has been an increasing interest in soil-inhabiting actinomycetes other than *Streptomyces*. Some recent surveys include the proportion of nonstreptomycetes to streptomycetes. Species of *Nocardia, Micromonospora*, and *Streptomyces* were present at various levels of seven soils representing the four major soil types of South India (Rangaswami *et al.*, 1967). Total microbial counts varied from 0.5 million to 52 million/gm, while actinomycete counts varied from 0.2 to 10.81 million/gm and the percent from 5.02 to 41.02 of the total. *Streptomyces* isolates accounted for 90–95% of the total actinomycetes in the various soil types.

In a review, Lechevalier and Lechevalier (1967) referred to one of their studies in which 16 soil samples were plated out on three different media and plates incubated at 28 and 37°C. More than 5000 cultures were isolated, approximately 95% being *Streptomyces*. The number of nonstreptomycete strains was as follows: *Nocardia asteroides* types and species, 99; *Nocardia madurae* types, 5; *Micromonospora*, 70; *Thermoactinomyces*, 7; *Microbispora*, 9; *Thermomonospora*, 11; *Microellobosporia*, 2; *Actinoplanes*, 10; *Streptosporangium*, 5; *Pseudonocardia*, 3; *Micropolyspora*, 5; *Mycobacterium*, 7.

C. Distribution in Aqueous Habitats

Although actinomycetes occur most commonly in soils and composts they may also be found in freshwater and marine habitats. Micromonosporae have been isolated from lakes and lake bottoms where they may comprise 10–50% of the total microbial population (Umbreit and McCoy, 1941). Species of *Actinoplanes* and *Streptosporangium* have also been isolated from fresh water (Couch, 1955).

There appears to be some difference of opinion as to whether actinomycetes are normal inhabitants of marine waters. From a collection of approximately 100 samples from five geographically different inshore locations, Grein and Meyers (1958) accumulated over 200 isolates, of which *Streptomyces* was the most prevalent, while *Nocardia* and *Micromonospora* made up about one-fifth of the isolates. They concluded that "the actinomycetes collected, especially the streptomycetes, do not represent an autochthonous marine flora. More likely, they may be terrestrial forms that have become adapted to the salinity of sea water and sediments."

On the other hand, Weyland (1969) decided that there is a natural actinomycete flora in marine sediments after isolating 1348 strains well out to sea away from littoral zone localities. His isolates represented four genera, *Nocardia, Micromonospora, Microbispora,* and *Streptomyces.*

D. Actinomycetes from Air

By the use of suitable entrapment and sampling methods actinomycetes may be demonstrated in the air (Andersen, 1958). Air-borne dissemination is primarily related to the amount of dust to which spores and mycelial fragments cling. Any action that disturbs the dry soil of fallow land, such as first raindrops, launches soil particles into the air and produces an increase in the numbers of air-borne streptomycetes (Lloyd, 1969b).

When samples of hay were shaken in a perforated drum in a wind tunnel and air samples taken with the Andersen sampler, a high incidence of actinomycete spores (over 60 million/gm) was found with moldy hays in contrast to other hays. Thermophiles were very prominent (Gregory and Lacey, 1963). A direct relation has been established between the prevalence of actinomycetes in moldy hays and the respiratory disease, farmer's lung, the primary cause of which has now been identified as *Micropolyspora faeni* (Cross et al., 1968).

E. Thermophilic Actinomycetes

Actinomycetes have been isolated at 40–60°C from a wide variety of natural substrates, including most soils, composts, hays, dung, and manure (Waksman *et al.,* 1939; Cross, 1968). There seems to be no geographical limitation and thermophiles have been isolated from both tropical and subpolar regions (Kosmachev, 1956).

In general, isolations of thermophilic actinomycetes have been carried out on media employed to isolate mesophiles. Tendler and Burkholder (1961), however, found that many such media were not particularly useful, especially in isolating *Thermoactinomyces*. Working with 1000 isolates of *Streptomyces* and *Thermoactinomyces* they developed seven preferred media for culture isolation, maintenance, culture rejuvenation, and certain physiological tests. Difficulties of transfer and maintenance on agar at 60°C led Uridil and Tetrault (1959) to devise a high protein medium combined with a commercial liquid form of colloidal silica.

Because many actinomycetes isolated at elevated temperatures are thermotolerant rather than truly thermophilic the erection of taxa based on this character alone is open to question. There will be less confusion if other criteria can be correlated with thermophily.

Thermophiles isolated from stable manure have been classified as species belonging to the genera *Streptomyces, Thermomonospora, Thermopolyspora, Thermoactinomyces, Pseudonocardia, Nocardia,* and *Microbispora* (Henssen, 1957a,b; Henssen and Schnepf, 1967).

Most cultures isolated from peat on nutrient agar at 45 and 55°C proved to be *Thermoactinomyces vulgaris* (Küster and Locci, 1963). The number of thermophiles isolated from moldy hay may be large but few species are represented and these belong to *Thermopolyspora (Micropolyspora), Streptomyces,* and *Micromonospora* (Corbaz et al., 1963).

F. Microbial Changes during Storage

In order to obtain the maximum variety of microorganisms from soil samples, culture isolations should be made as soon after collection as possible. In many instances months or even years pass between the time of collection and the time of isolation. There appears to be a continual decline in the microbial population the rate of which is influenced by such storage conditions as temperature, moisture, and organic content. Changes can probably be kept to a minimum by holding samples in a freezer.

Jagnow (1956) collected samples from a wide variety of soils and found that storage of dry soil for several months produced a sharp decrease in numbers over the first 5 months because of the death of mycelial fragments. The proportion of actinomycetes to the total microflora did not noticeably change, however. In a subsequent study he concluded that there were characteristic differences in survival which could be correlated with the nature of the strains and their origins (Jagnow, 1957).

Marked changes in the microbial content of soils collected in Honduras were observed as a result of isolations made 16 hours after collection and again 3 months later after shipment to Massachusetts (Stotzky et al., 1962). Actinomycetes increased in numbers whereas all other groups showed a pronounced decrease. The soils were not dried prior to shipping and it was pointed out that increased actinomycete sporulation probably occurred during the storage period.

Experiments have also been carried out to determine the fate of streptomycetes introduced into sterile soils. Valyi-Nagy and Kulcsar (1963) observed a loss of viability of over 50% in a 1-year period. Cultures stored in soils from which they were isolated survived better than they did in washed clay or washed quartz sand.

There are conflicting opinions as to whether streptomycetes occur in soils mainly as vegetative mycelia or as spores. Lloyd (1969a) undertook to add conidia and germlings to natural soil. Transient hyphae were produced, sporulated, and disappeared when germinated conidia were added to soil. Conidia added to unsterilized soil mostly failed to germinate although some spores apparently can germinate and produce new populations of spores. There is a continual decline, however, because these are insufficient to overcome the death rate. He concluded that the primary means for *Streptomyces* survival was the partial resistance of spores to destruction and an intermittent replenishment of the original spore population by mass germination when new organic substances become available.

G. Effect of Soil Amendments

Enrichment techniques have often been used in soil bacteriology to obtain greater populations of desired organisms. It is also known that in the presence of sufficient moisture organic soil amendments can have a profound effect on microbial populations. It was pointed out in the last section that mass germination of streptomycetes takes place upon the addition of fresh organic substances.

It appears that actinomycetes are stimulated more by some agents than by others. Both Waksman and Starkey (1924) and Porter and Wilhelm (1961b) noted large increases in actinomycete numbers when dried blood was added to moistened soil samples. The latter investigators observed similar results upon the addition of certain plant meals such as cottonseed meal. They reported a pronounced shift in the predominant streptomycete strains on isolation plates but that there was less variety than on plates prepared from nontreated aliquots.

Apparently because of a change of pH favorable to actinomycetes in contrast to fungi, the addition of $CaCO_3$ to soil causes a substantial increase in the ratio of actinomycetes to fungi. Such an effect was obtained by Tsao et al. (1960) who added $CaCO_3$ at a rate of 100 mg/gm of soil in petri dishes. The soil was moistened and incubated at 26°C for 7–9 days. The same observation was made by El-Nakeeb and Lechevalier (1963) who incubated three different soil samples with 1 gm $CaCO_3$ per 1 gm of soil at high relative humidity for 10 days at 28°C. The highest plate counts of actinomycetes were obtained by combining this technique with isolation on arginine-glycerol-salts medium.

III. Classification of Actinomycetes Isolated from Soils

Whether his purpose is to study soil ecology or to accumulate microorganisms for a screening program, the investigator who isolates cultures from soil may initially be faced with an unassorted assemblage of cultures many of which will be morphologically and physiologically synonymous. Failure to recognize identical isolates can lead to inaccurate conclusions and failure to eliminate them as soon as possible can waste time and effort in a screening program. For the opposite reasons it is just as important to recognize differences, subtle as some of these may be.

With some exceptions, physiological reactions have proven less reliable than morphology in classifying streptomycetes. Culture media, pH, incubation temperature, and other conditions of growth must be carefully controlled in order to make valid and meaningful comparisons. An international cooperative effort to define acceptable criteria and methods of studying *Streptomyces* for taxonomic purposes has been carried out and the results and recommendations published (Shirling and Gottlieb, 1966).

The primary criteria for studying *Streptomyces* soil isolates that have most frequently been used are the nature of the spore chain, cultural characteristics on various media, and spore color (Jensen, 1930; Lindenbein, 1952; Jagnow, 1956; Mayama and Tawara, 1959; Nomi, 1960; Porter *et al.*, 1962).

Pridham *et al.* (1958) grouped streptomycetes in sections based on spore-chain morphology and within the sections based on spore colore *en masse*. Using only the nature of the spore-bearing hyphae and spore chains Nishimura *et al.* (1960) classified 848 strains of *Streptomyces* and placed them in 14 series comprising 5 sections. Cross and MacIver (1966) devised a key which employed melanin reaction on peptone-iron agar, nature of the spore surface, morphology of the aerial mycelium, color of the substrate mycelium, and carbon utilization (limited to eight carbon sources).

The electron microscope has made possible the realization that spore morphology is a valuable taxonomic criterion in *Streptomyces* speciation. Tresner *et al.* (1961) studied 600 cultures representing 120 species or varieties. Spores were smooth, warty, spiny, or hairy. An association between spore types and spore colors was also noted. For example, all isolates bearing blue to blue-green spores had spiny spores and those which produced white, yellow, cream, or buff colors had smooth-walled spores.

In a more recent application, spore morphology was used to distinguish between *Streptomyces hygroscopicus* and *S. platensis*. Both species are members of the *S. hygroscopicus* complex characterized by tightly spiraled spore chains, brownish-gray spore masses, hygroscopic patches on the aerial mycelium, and lack of chromogenicity. *Streptomyces hygroscopicus*, however, has short-cylindrical phalangiform spores whereas *S. platensis* has elliptical spores (Tresner *et al.*, 1967).

Continuing their investigation of potentially useful criteria for taxonomic purposes, Tresner *et al.* (1968) surveyed approximately 1300 strains of *Streptomyces* for tolerance to NaCl in the growth medium. The uniformity of response between strains belonging to different species suggested that the degree of tolerance to NaCl in concentrations up to 13% would be of assistance in classifying actinomycetes.

In summary, the following are considered today to be the most reliable criteria for classifying streptomycetes:
1. Spore-chain morphology
2. Spore morphology
3. Spore color *en masse*
4. Chromogenicity (melanin production)
5. Carbon source utilization patterns
6. Color of the nonsporulated aerial mycelium
7. Color of the substrate medium
8. Level of NaCl tolerance
9. Soluble pigments

As pointed out in a previous section, about 5% of the actinomycete soil population can be expected to be nonstreptomycetes. A number of genera and species have been recently described and there is uncertainty as to some of their relationships and appropriate classification. Important criteria used in their study include number and position of conidia, fragmentation of the vegetative mycelium, formation or lack of sporangia, presence or absence of aerial mycelium, and cell wall composition (Lechevalier and Lechevalier, 1965, 1967; Küster,

1968; Luedemann, 1969). Scanning electron microscopy promises to provide additional information useful in elucidating generic relationships (Williams and Davies, 1967; Williams, 1970).

Thermophilic actinomycetes have been classified using similar criteria. An additional, but not too reliable, characteristic is the temperature growth range (Tendler and Burkholder, 1961; Corbaz *et al.*, 1963; Nonomura and Ohara, 1969b).

IV. Antibiotic-Producing Actinomycetes

A. ANTIBIOTIC SCREENING

With numerous variations of each, there are two methods of screening for antibiosis. One involves the direct screening of cultures growing on agar and the other the determination of activity in liquid media. A full discussion has been presented in reviews by Waksman (1959) and by Waksman and Lechevalier (1962) but it may be pertinent here to mention some of the techniques available.

For screening on agar, cultures may be examined *in situ* on isolation plates or grown individually and tested for activity against selected bacteria and fungi. One of the earliest methods was the "crowded plate" technique in which cultures were isolated when zones of inhibition were noted in crowded soil isolation plates. Since there was no control over what was being inhibited this technique was not very useful.

To overcome this objection, Kelner (1948) devised a four-layer plate. (1) A foundation layer of sterile actinomycete agar. (2) A seed layer of 0.5 ml of soft agar containing a soil dilution designed to give 30–50 colonies per plate. These colonies will grow as surface colonies suitable for replication and are more readily distinguishable than deep colonies. (3) A layer composed of 5–10 ml of sterile agar added after growth of the colonies on layer two. (4) A top layer of 3 ml or more containing a suspension of bacteria.

Kelner's method and methods involving direct contact of test bacteria with actinomycetes colonies have the disadvantage of making it difficult to isolate contaminant-free cultures. Furthermore, only one test culture can be used on each isolation plate. These disadvantages can be overcome by adapting the Lederberg replica plate technique to antibiotic screening by stamping a series of plates from the isolation plate and applying a different test organism to each (Lechevalier and Corke, 1953). This method has the added advantage that more than one agar medium can be employed, thereby increasing the possibility of detecting medium-related zones of inhibition.

Screens on agar may also be carried out with individual isolates.

The well-known cross-streak plate has been much employed but both the producing actinomycete and test bacteria or fungi must be grown on the same medium. To overcome this handicap, cultures can be grown over the surface of agar plates and plugs or blocks cut out. These plugs may then be placed on plates seeded with desired test organisms (Ball et al., 1957; Vanek et al., 1958).

Properly devised screens on agar have the advantage that numerous organisms can be screened quite soon after isolation and before there is a loss of antibiotic-producing capacity on subculture. Screens on liquid media are more time-consuming, laborious, and require more expensive equipment but they can provide larger amounts of antibiotic for identification purposes and useful information for volume scale-up in larger vessels.

Variations in the composition of liquid media seem nearly infinite. Some idea of the abundance of media recommended for producing individual antibiotics can be gained from a perusal of the literature, both in patents and journals. One gets the impression that the choice of media is often empirical and is made after successful use with other antibiotics. As examples, Warren et al. (1955) recommended ten nonsynthetic media which had proven particularly useful in antibiotic studies. Horvath et al. (1964) screened streptomycetes on a standard set of four media. After employing a relatively small number of primary screening media, modifications can be made for additional fermentation studies if antibiotic yields are lower than desired.

B. Distribution of Antibiotic-Producing Actinomycetes

Several studies have shown that about 30 to 40% of *Streptomyces* soil isolates are of sufficient antibiotic interest when screened to be retained for further investigation (Waksman et al., 1942; Chun, 1956; Jagnow, 1956; Flaig and Kutzner, 1960; Routien, 1961; Rehm, 1961; Whaley and Boyle, 1967). Obviously, the results depend on the screening methods and media used and on the number and variety of test organisms. It is probably safe to say that nearly all streptomycetes may be found to possess some antimicrobial properties if enough conditions are employed.

Some surveys have also attempted to show that there is a relation between the percentage of antibiotic-producing strains and the sites of collection of soil samples. There may be such a relation but many of the data are unconvincing because repetitive strains were not recognized and eliminated. Furthermore, most surveys have been relatively superficial, insufficient isolation procedures having been used to isolate a maximum variety of strains from each soil sample. To see if this might be the case, my co-workers and I carried out an intensive study of five soil samples using several soil amendments and repeated isolations on glycerol-arginine agar over a 6-week period. The total number of actinomycetes isolated was 2817 but this was reduced to 647 following a comparative study of the isolates. Each of the 647 strains was grown in three liquid media in shake flasks and 162 were found to produce sufficient activity for antibiotic identification studies. No relation was established between the five soils and the types of antibiotics produced (Porter et al., 1964).

In reviewing the literature on the distribution of antibiotic producers Calam (1964) observed that investigators tend to choose soils with the richest flora and that these would contain a wide variety of antibiotic-producing actinomycetes. Certain strains may be the normal inhabitants of poorer soils and would be overlooked if these soils were passed over. The ability to find soils with numerous active strains may in part be related to the location of the investigator's laboratory. He concluded his review by saying, "It would seem that the best soils come from places in warm climates where the soil is neutral or alkaline, fertile and more or less cultivated, and containing a suitable amount of moisture." How well greenhouse soils fit these requirements!

Thermophilic antibiotic-producing actinomycetes have also been the object of several surveys. Of 289 strains isolated from Indian Pamir soils at 55°C, 36% inhibited *Staphylococcus aureus* and a total of 48% had some sort of antibiotic activity. *Micromonospora vulgaris* was the most prevalent organism (Agre and Orleanskii, 1962). Kudrina and Maksimova (1963) isolated 1100 cultures under thermophilic conditions from 385 Chinese soil samples. The percent of antagonists varied from 20 to 63 in the various samples but an average of only 17% had a broad antibiotic spectrum. Cross (1968) listed eight antibiotics produced by truly thermophilic actinomycetes. He pointed out, however, that none of these are active against gram-negative bacteria and none are of industrial importance.

In their review on the biology of actinomycetes, Lechevalier and Lechevalier (1967) noted that relatively few actinomycete-produced antibiotics have been produced by nonstreptomycetes and put forward two possible reasons: nonstreptomycetes are not frequently encountered on isolation plates and they often possess more complex nutritional requirements and grow more slowly than streptomycetes. Their infrequency of isolation may therefore be related to the methods and media used. Of the antibiotics produced by nonstreptomycete actinomycetes, the gentamicin complex is the only one of genuine commercial interest at the present time.

C. Variety of Microorganisms Capable of Producing the Same Antibiotic

It has been recognized for some time that different strains of the same species of *Streptomyces* may be capable of producing different antibiotics, which makes antibiotic production a characteristic of questionable taxonomic value. Certain antibiotics may also be produced by more than one well-defined species of *Streptomyces* and some interesting examples exist where an antibiotic is not only produced by a *Streptomyces* but by nonstreptomycetes and even fungi as well.

Iodinin has been isolated from *Streptomyces thioluteus Brevibacterium iodinum, Microbispora aerata, M. amethystogenes, M. parva, Streptosporangium amethystogenes* var. *nonreducens,* and several Nocardia spp. (Prauser and Eckardt, 1967).

In addition to being produced by species of *Streptomyces,* prodigiosin is produced by *Serratia marcescens* (Perry, 1961); cephalosporin N by species of *Emericellopis* and *Paecilomyces* (Miller *et al.,* 1962; Grosklags and Swift, 1957; Pisano *et al.,* 1961); helvolic acid by *Aspergillus* and *Emericellopsis* (Waksman *et al.,* 1943; Okuda *et al.,* 1964; Umezawa, 1964); and nebularine by *Agaricus nebularis* (Nakamura, 1961).

V. Identification and Classification of Antibiotics Produced by Actinomycetes

A. Antimicrobial Spectra

Some idea of antibiotic relationships and a very tentative classification can be obtained from antimicrobial spectra on agar cross streaks or from culture broths and cell extracts. A representative spectrum should contain selected gram-positive and gram-negative bacteria, at least one *Mycobacterium,* molds, and yeasts. It should include a number of strains both sensitive and resistant to the more commonly encountered antibiotics or families of antibiotics. It is also helpful to determine the effect of pH on antibiotic diffusibility or activity (Ogata and Igarashi, 1954; Stapley, 1958; Horvath *et al.,* 1964; Waksman and Lechevalier, 1962).

Antibiotic spectra are important and useful but it is imperative to substantiate antibiotic relationships by other means. Crude culture fluids containing antibiotics are usually mixtures, and the presence of dissimilar components may prevent a clear interpretation of antimicrobial spectrum data. Also, rather different antibiotics, such as certain polyenes, may exhibit much the same antimicrobial activity.

B. Physical and Chemical Properties

Waksman and Lechevalier (1962) list the following properties and techniques as useful in characterizing antibiotics: solubility, stability, color reactions and fluorescence, light absorption (ultraviolet, visible, infrared), paper chromatography of the whole antibiotic and of decomposition products, electrophoresis, countercurrent distribution, elementary analysis, and physical constants. Use has more recently been made of preparative thin-layer chromatography (Betina and Baráth, 1964) and mass spectrometry.

Ideally, to obtain complete characterization, an antibiotic should be isolated in pure form as a single component. This, of course, is impractical in a screening program, nor does the investigator initially need complete information since most of his leads will fall into rather well-defined categories of known antibiotics. Antibiotic isolation can then be confined to those leads which are presumptively novel.

An antibiotic classification and tentative identification can be achieved on the basis of antimicrobial spectrum, ionic character, and behavior on paper chromatograms. A determination of the ultraviolet spectrum is useful and necessary in dealing with antifungal antibiotics. These data can usually be obtained with small amounts of relatively impure substances, further purification and characterization being confined to those antibiotics of the greatest interest.

It is even possible to obtain useful information without growing organisms in liquid media. In an extension of their method of obtaining antimicrobial spectra, Vanek *et al.* (1958) placed agar blocks on strips of chromatographic paper and then developed the strips in a system of ten solvents. Agar blocks were also used to study the ionic character of the antibiotics and, after extraction with ethanol, to detect polyenes by obtaining ultraviolet absorption spectra. On the basis of the data collected, 386 active actinomycetes were classified in five antibiotic groups but actual antibiotic identification of each activity was not attempted. Of the active cultures, 16.3% produced only polyene antibiotics while 34.2% produced mixtures of polyene and nonpolyene antibiotics.

Most preliminary identification procedures have been carried out on shake-flask broths or crude preparations derived from them. In our study of 162 antibiotically active actinomycetes isolated from five soil samples we used paper chromatography of concentrated culture broths as the primary method of analysis. This revealed that 104 cultures produced 82 different essentially single components which were then classified in 21 antibiotic categories. The other 58 isolates produced mixtures or activities too weak to analyze (Porter *et al.*, 1964).

Snell et al. (1956) used ascending one-directional paper chromatography to differentiate gram-positive polypeptide antibiotics. The most useful solvent system was t-butyl alcohol, acetic acid, water—74:3:25. Separation of families but not mixtures of closely related antibiotics within families was achieved in this study.

Yajima (1955) and Ammann and Gottlieb (1955) employed seven solvent systems (not completely identical) as an aid in classifying antifungal antibiotics.

A method of circular paper chromatography with eight basic solvent systems, supplemented by paper electrophoresis, was used by Doskocilova and Vondracek (1961) to identify antibiotics in shake-flask broths. The method was first worked out on known antibiotics and then applied to unknowns.

Workers in Czechoslovakia have devised chromatographic systems for use with fungally produced antibiotics. Balan et al. (1962) studied 60 active cultures by means of ten solvent systems and reported observations on 44 different antibiotics derived from these cultures. Betina (1964) analyzed antibiotics in four solvent systems:

1. Distilled water
2. n-Butanol saturated with water
3. Ethyl acetate saturated with water
4. Benzene saturated with water

As a result of the data obtained from these four systems, antibiotics were divided into 5 classes and 14 subclasses. Additional solvent systems were used to further analyze antibiotics in each class. The scheme was applied to 62 known antibiotics from actinomycetes and lichens as well as fungi.

Variations on a theme include the use of paper strips impregnated with buffers at different pH levels to produce "pH chromatograms" and solvent systems which contain increasing concentrations of ammonium chloride to produce "salting-out chromatograms." The various chromatographic methods and their application to identification of antibiotics of differing degrees of purity have been reviewed by Betina (1965, 1967), who also included in the earlier publication an extensive bibliography on the subject.

A more recent development in identification procedures has been the use of thin-layer chromatography (TLC) on silica gel. The method was applied to antibiotics by Nicolaus et al. (1961) who carried out bioautography using sensitive test organisms plus triphenyltetrazolium salt. Antibiotic areas become light yellow while the remainder of the plate is red brown.

Ikekawa et al. (1963) studied 50 samples of known antibiotics by

TLC. They recommended a standard combination of seven solvent systems for antibiotic identification.

Betina and Baráth (1964) described a related method with aluminum oxide as the stationary phase. By this procedure 85 antibiotics were found and classified in concentrates from 50 strains of fungi isolated from Indonesian soils (Baráth et al., 1964).

VI. In Retrospect

During the past 30 years notable changes have taken place in the approach to finding new antibiotics. The first decade was a time of realization that products of microbial metabolism had vast medicinal — ergo, commercial — value and if one searched hard enough there was plenty of gold to be found in the hills. The fifties constituted a time of feverish screening programs. The rush was on and whoever got there first would surely come in a winner. During the sixties the emphasis shifted toward more exploration at the original site and to a search for related substances with greater intrinsic value than the original.

If there is a theme to my essay it is that traditional, superficial methods of seeking microorganisms capable of producing novel antibiotics are no longer likely to be very successful because too many people have already been there. But the several millions of microorganisms that inhabit a gram of soil rich in organic matter comprise a microcosm which has never been completely invaded by man. It is unlikely that anyone has ever isolated every strain and species present in a single gram of such a soil.

Investigators more than 20 years ago had the advantage of working in a newly opened field where additional discoveries were relatively easy to come by. But today we have the advantage of sophisticated techniques and knowledge which they did not possess. As I have tried to indicate, we know now more about how to isolate microorganisms, more about their ecological and taxonomic relationships, and more about how to elicit, evaluate, and characterize the products of their metabolism. In short, a microbial product approach to chemotherapy should continue to be scientifically and financially rewarding.

One final word on the nomenclature of antibiotics and producers of antibiotics. Most antibiotic-producing microorganisms are so ubiquitous that they can be found in almost anyone's garden. Also, as we have seen, some antibiotics are produced by several species or even genera of microorganisms. Therefore, it would seem that the least applicable choice of a name would be the geographical site of isolation of the culture. Since antibiotics are chemicals, they should preferably be named with their chemical structure or family relationships in mind. The organisms which produce them should be named according to distinguishing morphological or physiological features.

REFERENCES

Agate, A. D., and Bhat, J. V. (1963). *Antonie van Leeuwenhoek J. Microbiol. Serol.* **29**, 297-304.
Agre, N. S., and Orleanskii, V. K. (1962). *Mikrobiologiya* **31**, 95-102.
Ammann, A., and Gottlieb, D. (1955). *Appl. Microbiol.* **3**, 181-186.
Andersen, A. A. (1958). *J. Bacteriol.* **76**, 471-484.
Balan, J., Nemec, P., and Betina, V. (1962). *Folia Microbiol. (Prague)* **7**, 243-246.
Ball, S., Bessell, C. J., and Mortimer, A. (1957). *J. Gen. Microbiol.* **17**, 96-103.
Baráth, Z , Betina, V., and Nemec, P. (1964). *J. Antibiot., Ser. A* **17**, 144-147.
Benedict, R. G., Pridham, T. G., Lindenfelser, L. A., Hall, H. H., and Jackson, R. W. (1955). *Appl. Microbiol.* **3**, 1-6.
Betina, V. (1964). *J. Chromatogr.* **15**, 379-392.
Betina, V. (1965). *Chromatogr. Rev.* **7**, 119-178.
Betina, V. (1967). *Antimicrob. Ag. Chemother.* 1966, pp. 637-643.
Betina, V., and Baráth, Z. (1964). *J. Antibiot., Ser. A* **17**, 127-128.
Brock, T. D. (1966). *Advan. Appl. Microbiol.* **8**, 61-75.
Bu'Lock, J. D. (1961). *Advan. Appl. Microbiol.* **3**, 293-342.
Calam, C. T. (1964). *Progr. Ind. Microbiol.* **5**, 3-53.
Chun, D. (1956). *Antibiot. Chemother. (Washington, D.C.)* **6**, 324-329.
Corbaz, R., Gregory, P. H., and Lacey, M. E. (1963). *J. Gen. Microbiol.* **32**, 449-455.
Corke, C. T., and Chase, F. E. (1956). *Can. J. Microbiol.* **1**, 12-16.
Couch, J. N. (1955). *J. Elisha Mitchell Sci. Soc.* **71**, 148-155.
Cross, T. (1968). *J. Appl. Bacteriol.* **31**, 36-53.
Cross, T., and MacIver, A. M. (1966). *In* "Identification Methods for Microbiologists" (B. M. Gibbs and F. A. Skinner, eds.), Part A, pp. 103-110. Academic Press, New York.
Cross, T., MacIver, A. M., and Lacey, J. (1968). *J. Gen. Microbiol.* **50**, 351-359.
Doskocilova, D., and Vondracek, M. (1961). *Antibiotiki* **6**, 649-659.
El-Nakeeb, M. A., and Lechevalier, H. A. (1963). *Appl. Microbiol.* **11**, 75-77.
Flaig, W., and Kutzner, H. J. (1960). *Arch. Mikrobiol.* **35**, 207-228.
Gregory, P. H., and Lacey, M. E. (1963). *J. Gen. Microbiol.* **30**, 75-88.
Grein, A., and Meyers, S. P. (1958). *J. Bacteriol.* **76**, 457-463.
Grosklags, J. H., and Swift, M. E. (1957). *Mycologia* **49**, 305-317.
Henssen, A. (1957a). *Arch. Mikrobiol.* **26**, 373-414.
Henssen, A. (1957b). *Arch. Mikrobiol.* **27**, 63-81.
Henssen, A., and Schnepf, E. (1967). *Arch. Mikrobiol.* **57**, 214-231.
Horvath, I., Lovrekovich, I., and Magyar, K. (1964). *Z. Allg. Mikrobiol.* **4**, 225-231.
Ikekawa, T., Iwami, F., Akita, E., and Umezawa, H. (1963). *J. Antibiot., Ser. A* **16**, 56-57.
Jagnow, G. (1956). *Arch. Mikrobiol.* **25**, 274-296.
Jagnow, G. (1957). *Arch. Mikrobiol.* **26**, 175-191.
Jensen, H. L. (1930). *Soil Sci.* **30**, 59-77.
Jensen, H. L. (1934). *Proc. Linn. Soc. N. S. W.* **59**, 101-117.
Kelner, A. (1948). *J. Bacteriol.* **56**, 157-162.
Kosmachev, A. E. (1956). *Mikrobiologiya* **25**, 546-552.
Kudrina, E. S., and Maksimova, T. S. (1963). *Mikrobiologiya* **32**, 623-631.

Küster, E. (1968). *In* "The Ecology of Soil Bacteria" (T. R. G. Gray and D. Parkinson, eds.), pp. 322–336. Univ. of Toronto Press, Toronto.
Küster, E., and Locci, R. (1963). *Arch. Mikrobiol.* **45**, 188–197.
Küster, E., and Williams, S. T. (1964). *Nature (London)* **202**, 928–929.
Lechevalier, H. A., and Corke, C. T. (1953). *Appl. Microbiol.* **1**, 110–112.
Lechevalier, H. A., and Lechevalier, M. P. (1965). *Ann. Inst. Pasteur* **108**, 662–673.
Lechevalier, H. A., and Lechevalier, M. P. (1967). *Annu. Rev. Microbiol.* **21**, 71–100.
Lilly, D. M., and Stillwell, R. H. (1965) *Science* **147**, 747–748.
Lindenbein, W. (1952). *Arch. Mikrobiol.* **17**, 361–383.
Lingappa, Y., and Lockwood, J. L. (1962). *Phytopathology* **52**, 317–323.
Lloyd, A. B. (1969a). *J. Gen. Microbiol.* **56**, 165–170.
Lloyd, A. B. (1969b). *J. Gen. Microbiol.* **57**, 35–40.
Luedemann, G. (1969). *Advan. Appl. Microbiol.* **11**, 101–133.
Mayama, M., and Tawara, K. (1959). *Shionogi Kenkyusho Nempo* **9**, 1179–1184.
Miller, I. M., Stapley, E. O., and Chaiet, L. (1962). *Bacteriol. Proc.* p. 32.
Nakamura, G. (1961). *J. Antibiot., Ser. A* **14**, 94–97.
Nicolaus, B. J. R., Coronelli, C., and Binaghi, A. (1961). *Farmaco, Ed. Prat.* **16**, 349–370.
Nishimura, H., Mayama, M., and Tawara, K. (1960). *J. Antibiot., Ser. A* **13**, 228–342.
Nomi, R. (1960). *J. Antibiot., Ser. A* **13**, 236–342.
Nonomura, H., and Ohara, Y. (1960a). *Hakko Kogaku Zasshi* **38**, 401–405.
Nonomura, H., and Ohara, Y. (1960b). *Hakko Kogaku Zasshi* **38**, 405–409.
Nonomura, H., and Ohara, Y. (1969a). *Hakko Kogaku Zasshi* **47**, 463–469.
Nonomura, H., and Ohara, Y. (1969b). *Hakko Kogaku Zasshi* **47**, 701–709.
Ogata, K., and Igarashi, S. (1954). *Takeda Kenkyusho Nempo* **13**, 78–91.
Okuda, S., Iwasaki, S., Tsuda, K., Sano, Y., Hata, T., Udagawa, S., Nakayama, Y., and Yamaguchi, H. (1964). *Chem. Pharm. Bull.* **12**, 121–124.
Perry, J. J. (1961). *Nature (London)* **191**, 77–78.
Pisano, M. A., Fleischman, A. I., Littman, M. L., Dutcher, J. D., and Pansy, F. E. (1961). *Antimicrob. Ag. Annu.* 1960, pp. 41–47.
Pochon, J., and Tardieux, P. (1968). *In* "The Ecology of Soil Bacteria" (T. R. G. Gray and D. Parkinson, eds.), pp. 123–137. Univ. of Toronto Press, Toronto.
Porter, J. N., and Wilhelm, J. J. (1961a). *Develop. Ind. Microbiol.* **2**, 247–252.
Porter, J. N., and Wilhelm, J. J. (1961b). *Develop. Ind. Microbiol.* **2**, 253–259.
Porter, J. N., Wilhelm, J. J., and Tresner, H. D. (1960). *Appl. Microbiol.* **8**, 174–178.
Porter, J. N., Wilhelm, J. J., and Simpson, R. B. (1962). *Develop. Ind. Microbiol.* **3**, 240–244.
Porter, J. N., Wilhelm, J. J., Kirby, J. P., and Hausmann, W. K. (1964). *Antimicrob. Ag. Chemother.* 1963, pp. 100–103.
Pramer, D. (1968). *In* "The Ecology of Soil Bacteria" (T. R. G. Gray and D. Parkinson, eds.), pp. 220–233. Univ. of Toronto Press, Toronto.
Prauser, H., and Eckardt, K. (1967). *Z. Allg. Mikrobiol.* **7**, 409–410.
Pridham, T. G., Hesseltine, C. W., and Benedict, R. G. (1958). *Appl. Microbiol.* **6**, 52–79.
Rainbow, C., and Rose, A. H., eds. (1963). "Biochemistry of Industrial Microorganisms," 708 pp. Academic Press, New York.
Rangaswami, G., Oblisami, G., and Swaminathan, R. (1967). "Antagonistic Actinomycetes in the Soils of South India," 156 pp. Phoenix Press, Visveswarapuram, Bangalore, India.
Rehacek, Z. (1959). *Mikrobiologiya* **28**, 236–241.
Rehm, H. J. (1961). *Zentralbl. Bakteriol., Parasitenk., Infektionskr. Hyg., Abt. 2* **114**, 345–355.

Routien, J. B. (1961). *J. Bacteriol.* **81,** 218–225.
Sevcik, V. (1963). "Antibiotica aus Actinomyceten," 647 pp. Fischer, Jena.
Shirling, E. B., and Gottlieb, D. (1966). *Int. J. Syst. Bacteriol.* **16,** 313–340.
Snell, N., Ijichi, K., and Lewis, J. C. (1956). *Appl. Microbiol.* **4,** 13–17.
Stapley, E. O. (1958). *Appl. Microbiol.* **6,** 392–398.
Stotzky, G., Goos, R. D., and Timonin, M. I. (1962). *Plant Soil* **16,** 1–18.
Tendler, M. D., and Burkholder, P. R. (1961). *Appl. Microbiol.* **9,** 394–399.
Tresner, H. D., Davies, M. C., and Backus, E. J. (1961). *J. Bacteriol.* **81,** 70–80.
Tresner, H. D., Backus, E. J., and Hayes, J. A. (1967). *Appl. Microbiol.* **15,** 637–639.
Tresner, H. D., Hayes, J. A., and Backus, E. J. (1968). *Appl. Microbiol.* **16,** 1134–1136.
Tsao, P. H., Leben, C., and Keitt, G. W. (1960). *Phytopathology* **50,** 88–89.
Umbreit, W. W., and McCoy, E. (1941). "Symposium on Hydrobiology," pp. 106–114. Univ. of Wisconsin Press, Madison, Wisconsin.
Umezawa, H. (1964). "Recent Advances in Chemistry and Biochemistry of Antibiotics," 266 pp. Microb. Chem. Res. Found., Tokyo.
Uridil, J. E., and Tetrault, P. A. (1959). *J. Bacteriol.* **78,** 243–246.
Valyi-Nagy, T., and Kulczar, G. (1963). *Acta Biol. (Budapest)* **14,** 25–31.
Vanek, Z., Dolezilova, L., and Rehacek, Z. (1958). *J. Gen. Microbiol.* **18,** 649–657.
Waksman, S. A. (1959). "The Actinomycetes. Vol. I: Nature, Occurence and Activities," 327 pp. Williams & Wilkins, Baltimore, Maryland.
Waksman, S. A. (1961). "The Actinomycetes. Vol. II: Classification, Identification and Descriptions of Genera and Species," 363 pp. Williams & Wilkins, Baltimore, Maryland.
Waksman, S. A. (1963). *Advan. Appl. Microbiol.* **5,** 235–315.
Waksman, S. A. (1969). *Advan. Appl. Microbiol.* **11,** 1–16.
Waksman, S. A., and Lechevalier, H. A. (1962). "The Actinomycetes. Vol. III: Antibiotics of Actinomycetes," 430 pp. Williams & Wilkins, Baltimore, Maryland.
Waksman, S. A., and Starkey, R. L. (1924). *Soil Sci.* **17,** 373–378.
Waksman, S. A., Umbreit, W. W., and Cordon, T. C. (1939). *Soil Sci.* **47,** 37–61.
Waksman, S. A., Horning, E. S., and Spencer, E. L (1943). *J. Bacteriol.* **45,** 233–248.
Waksman, S. A., Horning, E. S., Welsch, M., and Woodruff, H. B. (1942). *Soil Sci.* **54,** 281–296.
Warren, H. B., Jr., Prokop, J. F., and Grundy, W. E. (1955). *Antibiot. Chemother.* **5,** 6–12.
Weyland, H. (1969). *Nature (London)* **223,** 858.
Whaley, J. W., and Boyle, A. M. (1967). *Phytopathology* **57,** 347–351.
Williams, S. T. (1970). *J. Gen. Microbiol.* **62,** 67–73.
Williams, S. T., and Davies, F. L. (1965). *J. Gen. Microbiol.* **38,** 251–261.
Williams, S. T., and Davies, F. L. (1967). *J. Gen. Microbiol.* **48,** 171–177.
Yajima, T. (1955). *J. Antibiot., Ser. A* **8,** 189–195.

Biochemical Activities of *Nocardia*

R. L. RAYMOND AND V. W. JAMISON

Sun Oil Company,
Marcus Hook, Pennsylvania

I.	Introduction	93
II.	Hydrocarbon Oxidation	94
	A. Biosynthesis	94
	B. Degradation	96
	C. Cooxidation	99
III.	Nonhydrocarbon Oxidation	105
IV.	Transformation of Steroids and Sterols	111
	A. Transformation of Steroids	112
	B. Transformation of Sterols	118
V.	Summary	119
	References	120

I. Introduction

It seems almost inevitable that the microbiologist searching for new reactions and products will become addicted to a specific genus. Such is our lot, and this particular essay is an attempt to bring attention to a group of microorganisms which we believe may have considerably more potential in microbial transformations than has previously been recognized. We have been intrigued with members of the genus *Nocardia* because of their ability to carry out hydroxylations and cleave aromatic rings, reactions which in many instances involve the mechanism of direct addition of molecular oxygen to a substrate. In this regard they seem to have an advantage over the somewhat more destructive pseudomonads in the accumulation of products. On the other hand, the isolation and characterization of oxygenases and mixed function oxidases which are responsible for molecular oxygen incorporation, have been almost exclusively confined to the pseudomonads, and for the moment, their presence in nocardia is by implication only.

Before proceeding it seems necessary to deal with the very sticky problem of what kind of nocardia and biochemistry will be covered in this essay. Our interests are directed at the saprophytes which have had their origin in the soil. Our definition is well delineated in the seventh edition of Bergey's *Manual of Determinative Bacteriology*. This group of nocardia are generally nonacid fast and form a mycelium which usually breaks up into short rods or cocci. They are aerobic, gram-positive, and the gross appearance of colonies resembles the genus *Mycobacterium*. Many more additional characteristics could be described but since each of the above are controversial it would appear to be a needless exercise to compound the problems of *Nocardia* taxonomy. For a discussion of the problems of classification of this genus the reader is referred to the following papers: Gordon and Mihm (1957, 1959), Schneidau and Shaffer (1957), McClung and Adams (1962), Gordon (1966), and Waksman (1967). By limiting ourselves to those cultures which have the specific *Nocardia* label, we are very much aware that a number of corynebacteria, mycobacteria, and streptomyces papers have not been included which may be nocardia by our definition.

Our interest in the biochemistry of nocardia is restricted for the most part to reactions which lie outside the conventional metabolic fate of proteins, sugars, etc. In so doing we have attempted to review the current status of utilization by members of the genus *Nocardia* of hydrocarbons, nonhydrocarbon aromatics, herbicides, and steroids. Where possible the order of magnitude of product accumulation and the type of system employed have been described.

II. Hydrocarbon Oxidation

The ability of the genus *Nocardia* to metabolize hydrocarbons has been recognized for many years, and indeed provides the basis for division of the genus into two groups in Bergey's, those that utilize paraffins and those that do not. Throughout the description of species in the hydrocarbon-utilizers group, the only specific hydrocarbon mentioned in the older literature is naphthalene. Usually "paraffin" was the hydrocarbon of choice for "baiting" and determining hydrocarbon utilization. Since "paraffin" is a very complex mixture of hydrocarbons, the nature of which will vary widely, depending upon the crude oil from which it was manufactured, and the manufacturing methods, it would not be too surprising to see great variations in the reported utilization of these hydrocarbon mixtures. In recent years several species have been examined for their ability to metabolize a wide variety of very pure hydrocarbons. We have chosen to categorize these reactions into oxidations which result in (1) biosynthesis of cell constituents, (2) degradation, (3) cooxidation reactions resulting in products with the initial hydrocarbon structure still intact.

A. Biosynthesis

Little evidence exists in the literature to suggest that oxidation of hydrocarbons by nocardia results in significant extracellular accumulation of industrially important biosynthetic products such as amino acids, antibiotics, vitamins, and organic acids. For the most part the oxidized hydrocarbons enter the usual metabolic pathways which results in cell biosynthesis. The paraffinic-type hydrocarbons have been most extensively studied in this regard and marked changes in cell composition have been noted depending to a great extent on the hydrocarbon-type used for growth.

Under the proper conditions nocardia are capable of very rapid growth on n-paraffinic hydrocarbons. Raymond and Davis (1960) found that a nocardia culture isolated from the soil using ethane as the sole carbon source, could attain cell concentrations of 12–15 gm (dry weight) per liter in 48 hours when n-octadecane was the growth substrate. These rates were accomplished in fermentors with impellers operated at very high speeds to keep the hydrocarbon emulsified. It was found that when the growth substrates were n-hexadecane or n-octadecane the nocardial cells contained high concentrations of lipids (> 70%) if a high C-N ratio was maintained in the medium. These lipids were made up of glycerides and waxes (60:40) with the waxes having a composition directly related to the starting substrate as illustrated (1).

$$CH_3(CH_2)_{14}CH_3 \longrightarrow CH_3(CH_2)_{14}CH_2O\overset{O}{\overset{\|}{C}}(CH_2)_{14}CH_3 \quad (1)$$

$\underline{n}$-HEXADECANE   CETYL PALMITATE

Waxes were not present when the cells were grown on glucose, n-hexane, or n-tridecane. Davis (1964a) found that the triglycerides of this *Nocardia* grown on n-alkanes (C_{13}–C_{20}), contained fatty acids of chain length with an odd or even characteristic, resembling the starting substrates, i.e., triglycerides from tridecane-grown cells had fatty acids of C_{13}, C_{15}, and C_{17} chain lengths while n-hexadecane-grown cells contained C_{12}, C_{14}, C_{16}, and C_{18} fatty acids.

When this same *Nocardia* culture was grown on propane or butane (Davis, 1964b), the lipid content of the cells was of the order of 50%, but poly-β-hydroxybutyrate was now a significant component. Rate of cell synthesis from propane and butane was of the order of 1 gm per liter per day, considerably lower than for the longer chain alkanes.

Further evidence of high lipid concentrations in other *Nocardia* species is absent from the literature. Linday and Donald (1961) examined *N. opaca* and *N. petrofila* for fatty acid production when grown on liquid paraffins, but their data indicate little if any accumulation. Following the discovery of extracellular ester formation from alkanes by *Micrococcus cerificans* (Stewart and Kallio, 1959) Finnerty *et al.* (1962) were unable to find similar production by *N. opaca, N. rubra, N. erythropolis,* or *N. polychromogenes* when *n*-hexadecane was used as a growth substrate. Wagner *et al.* (1969) studied three cultures, *Nocardia* sp. (NBZ 23), *Nocardia* sp. (NBS 28) and *N. opaca* for utilization of *n*-alkane mixtures (C_{10}–C_{20}) in 14-liter glass fermentors. In a mineral medium without added growth factors, cell concentrations of 14 gm/liter were obtained in 40 hours. Generation times ranged from 0.5 to 7.0 hours. Lipid concentration for the *Nocardia* varied, depending upon nitrogen source, and did not exceed 24%. They did find that in *Nocardia* sp. NBZ 23, the fatty acid profiles of the major fatty acids of the lipid, reflected the length of the parent alkane growth substrate. They did not report any evidence of wax esters. Recently, Ratledge (1970) reported lipid concentrations of 8.9, 2.4, and 11.2% for *N. hydrocarboxydans, N. opaca,* and *N. petrophila,* respectively. However, their cell yields on the C_{11-15} *n*-alkane fraction in 15-liter fermentor experiments were very low (0.03, 0.46, and 0.9 gm per liter per 3 days) and probably reflect unsatisfactory cultivation conditions for *Nocardia*.

There appears to be some evidence that growth on hydrocarbons may increase the concentration of certain vitamins in some *Nocardia*. Fukui *et al.* (1967) reported that hexadecane-grown cultures of *N. gardneri* and *N. lutea* contained significantly more vitamin B_{12} than the same cultures grown on glucose. Experiments were carried out in 500-ml. shaken flasks and the total B_{12} concentration ranged from 32.3 to 67.7 µg/liter. Morikawa and Kamikubo (1969) found similar results for *N. gardneri* on several other *n*-alkane substrates.

Davis (1967) noted that several nocardial cultures grown on *n*-alkanes, dearomatized kerosenes, or gas oils produced carotenoids. Highest concentrations were observed in butane-grown cells in illuminated systems.

B. Degradation

Although there is considerable overlap between cell biosynthesis from hydrocarbons and degradation it seems worthwhile to separate that body of literature which deals with growth of cultures on hydrocarbons from that in which the composition of the cells has been changed by the hydrocarbon substrate.

In a study of some 300 strains of *Proactinomyces (Nocardia)* isolated from English soils, Erikson (1949) found that the predominate strains, *Nocardia opaca, N. salmonicolor,* and *N. paraffinae* could grow on liquid or wax paraffins as the sole source of carbon. Webley and DeKock (1952) using *N. opaca* strain T_{16} (isolated by Erikson), found that dodecane and tetradecane supported very good growth when used as the sole source of carbon. In studying oxygen uptake by washed cells they found that the longer chain hydrocarbons, i.e., hexadecane, were oxidized more rapidly than the shorter chain compounds, i.e., decane. They believed the oxgen uptake represented true oxidation and not just stimulation. Methylene-blue reduction by cell extracts in the presence of dodecane and hexadecane led to the conclusion that a dehydrogenase was present. Treccani *et al.* (1955) isolated a *Nocardia* sp. P_2 which could grow on *n*-alkanes from C_5 to C_{28} and concluded, after oxygen uptake studies had been carried out, that the metabolism of hydrocarbons was similar to fatty acid metabolism and probably involved the loss of only one carbon at a time. Subsequently, Webley *et al.* (1956) presented evidence that a β-oxidation mechanism was operable in the metabolism of the saturated aliphatic hydrocarbons by *N. opaca* T_{16} and another culture designated *Nocardia* sp. (P_2). When phenyldodecane was incubated with washed cells of strain P_2 almost complete oxidation resulted (**2**).

$$\underset{\substack{\text{1-PHENYLDODECANE} \\ \text{1g / LITER}}}{\text{C}_6\text{H}_5\text{-CH}_2(\text{CH}_2)_{10}\text{CH}_3} \xrightarrow[\text{CONVERSION}]{85\%} \underset{\text{PHENYLACETIC ACID}}{\text{C}_6\text{H}_5\text{-CH}_2\text{COOH}}$$

(2)

Analogous products from 3-phenyleicosane resulted when *N. opaca* was employed, leading to the conclusion (partially based on earlier work with the ω-phenyl-substituted fatty acids) that an initial ω-oxidation was followed by β-oxidation of the resulting fatty acid. It is significant that the shorter chain ethyl-, *n*-propyl- and *n*-butylbenzenes were not attacked by the methods employed in these studies.

A very extensive study of the effect of substitution on alkane metabolism by nocardia was carried out by McKenna (1966). Species studied included *Nocardia opaca*, *N. rubra*, *N. erythropolis*, and *N. polychromogenes*. Her growth data was obtained in 50-ml. shaken flasks at a hydrocarbon concentration of 0.25%. Generally, it was found that any substitution (methyl, ethyl, phenyl, etc.) decreased the availability of the alkane for growth. For the alkylalkanes and phenylalkanes this substitution effect on growth is shown by the examples below (**3**) in which pristane and 1-phenylundecane supported good growth and the other compounds would not.

ALKYLALKANES

$$\underset{\substack{\text{HEPTAMETHYLNONANE}}}{CH_3\underset{\underset{CH_3}{|}}{\overset{\overset{CH_3}{|}}{C}}CH_2\underset{\underset{CH_3}{|}}{\overset{\overset{CH_3}{|}}{C}}CH_2CHCH_2\underset{\underset{}{}}{\overset{\overset{CH_3}{|}}{C}}CH_3}$$

$$\underset{\text{PRISTANE}}{CH_3\overset{\overset{CH_3}{|}}{C}H(CH_2)_3CH(CH_2)_3\underset{\underset{CH_3}{|}}{C}H(CH_2)_3\underset{\underset{CH_3}{|}}{C}HCH_3}$$

PHENYLALKANES (3)

$$\underset{\text{1-PHENYLUNDECANE}}{\overset{CH_2(CH_2)_9CH_3}{\underset{|}{\bigcirc}}}$$

$$\underset{\text{6-PHENYLUNDECANE}}{\overset{CH_3(CH_2)_4CH(CH_2)_4CH_3}{\underset{|}{\bigcirc}}}$$

With certain exceptions a parallelism with fatty acid metabolism was noted for branched alkane utilization. It was concluded that the presence of branched-chain fatty acids (nocardic acids) in the structural makeup of the cellular lipids might reflect the ease with which these types of compounds are utilized by the *Nocardia*.

Intermediates in the degradation of alkanes by nocardia other than those noted in the section on biosynthesis, are limited to monoalkenes. Wagner *et al.* (1967) claim to have isolated and purified 1-hexadecene from a hexadecane-grown *Nocardia* sp. Abbott and Casida (1968) obtained appreciable conversions of pentadecane, hexadecane, and octadecane to mixed internal monoalkenes by a culture of *N. salmonicolor*. Accumulation of these alkenes was accomplished by growing the cells on glucose, followed by incubation on the alkane. Yields of the monoalkenes from 5.0 ml of hexadecane incubated with 20 ml of a cell suspension appear to be of the order of 250 mg. The distribution of the mixture was calculated to be as shown (**4**).

The degradation of aromatic hydrocarbons by nocardia has been limited to a relatively few compounds. Treccani (1953) described a *Nocardia* sp. isolated by the enrichment technique from an oil-soaked soil which could grow on naphthalene and accumulate salicylic acid. Treccani *et al.* (1954) were able to detect naphthalene diol, salicylic acid, and catechol from this culture when it was grown on naphthalene in a simple salts medium. Wegner (1969) has described a *Nocardia* sp. (NRRL 3385) which grows on naphthalene and accumulates up to 5 gm/liter of 1,2-dihydro-1,2-dihydroxynaphthalene and some *o*-hydroxybenzalpyruvic acid.

$$CH_3(CH_2)_{14}CH_3 \longrightarrow CH_3(CH_2)_5CH=CH(CH_2)_7CH_3$$

n-HEXADECANE   7-HEXADECENE 80%

$$CH_3(CH_2)_6CH=CH(CH_2)_6CH_3$$

8-HEXADECENE 18%   (4)

$$CH_3(CH_2)_4CH=CH(CH_2)_8CH_3$$

6-HEXADECENE 2%

The degradation of benzene and toluene have been observed for only a very limited number of *Nocardia*. Wieland *et al.* (1958) isolated a culture growing on benzene that accumulated muconic acid and was reported to be similar to a nocardia or mycobacterium. Treccani and Bianchi (1959) reported extracting a benzene-oxidase from benzene-grown *Nocardia* cells which converted benzene to β-ketoadipic acid. Significant quantities of either the muconic or β-ketoadipic acids were not reported to have been isolated in either of these studies.

Wodzinski and Johnson (1968) isolated a *Nocardia* species, by an enrichment technique, which could grow on benzene. Studies in a 500-ml gas-tight Erlenmeyer shaken flask gave a cell yield of 79% (grams of cells produced/grams hydrocarbon used).

Forro (1965) studied the metabolism of a soil isolate of *Nocardia corallina* which could grow on toluene. Cell yields were best at a concentration of 0.04% toluene in 2-liter stoppered flasks. Products isolated but not conclusively identified from the oxidation included 3-methyl-catechol, α-hydroxy-ϵ-keto-2,4-heptadienoic acid and pyruvate.

C. Cooxidation

As originally defined by Foster (1962), "cooxidation" described a fermentation system in which nongrowth hydrocarbons were oxidized when present as cosubstrates in a medium in which one or more different hydrocarbons were furnished for growth. In recent years this concept has been broadened somewhat, to include substrates for growth which are nonhydrocarbon. This method of utilizing one substrate for providing carbon and energy for cell synthesis while at the same time carrying out a transformation on a second substrate, in this instance a hydrocarbon, has widened considerably the spectrum of hydrocarbons that can be metabolized by nocardia.

Davis and Raymond (1961) during studies on the utilization of white mineral oil by soil isolates of nocardia, noted the accumulation of organic acids in the culture fluid. These were shown to be naphthenic acids, and it was postulated that the nocardia were growing on the alkanes and oxidizing the naphthenes present to acids. Subsequently, strains designated *Nocardia* 107–332, *Nocardia* M.O., and *Nocardia* (Bough) were shown to bring about the oxidation of a number of cyclic hydrocarbons in 4-liter stirred fermentors. If the cyclic hydrocarbons had short alkyl substituents and a normal paraffin, such as, n-hexadecane or n-octadecane was provided for growth, a number of products were isolated in good yields, a few of which are illustrated below (5).

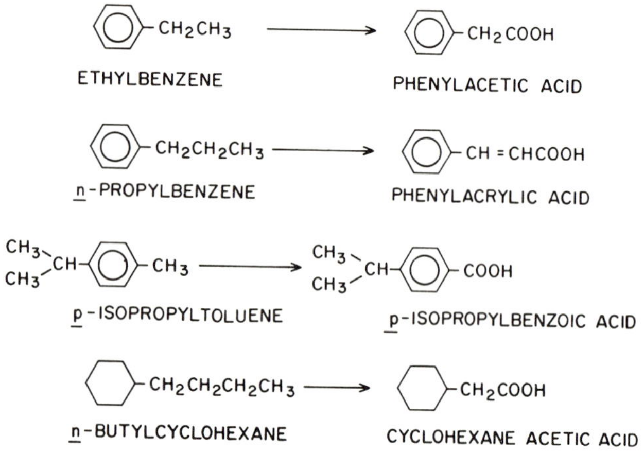

$$
\begin{array}{ccc}
\text{Ph-CH}_2\text{CH}_3 & \longrightarrow & \text{Ph-CH}_2\text{COOH} \\
\text{ETHYLBENZENE} & & \text{PHENYLACETIC ACID} \\
\\
\text{Ph-CH}_2\text{CH}_2\text{CH}_3 & \longrightarrow & \text{Ph-CH=CHCOOH} \\
\underline{n}\text{-PROPYLBENZENE} & & \text{PHENYLACRYLIC ACID} \\
\\
(\text{CH}_3)_2\text{CH-Ph-CH}_3 & \longrightarrow & (\text{CH}_3)_2\text{CH-Ph-COOH} \\
\underline{p}\text{-ISOPROPYLTOLUENE} & & \underline{p}\text{-ISOPROPYLBENZOIC ACID} \\
\\
\text{Cy-CH}_2\text{CH}_2\text{CH}_2\text{CH}_3 & \longrightarrow & \text{Cy-CH}_2\text{COOH} \\
\underline{n}\text{-BUTYLCYCLOHEXANE} & & \text{CYCLOHEXANE ACETIC ACID}
\end{array}
\tag{5}
$$

An important observation made in these studies was the necessity of keeping concentrations of many of the hydrocarbons at very low levels in the medium. Metabolism of aromatic hydrocarbons with long alkyl chains, i.e., phenyldodecane, gave similar results except that it was not necessary to have a cosubstrate present for growth, confirming the earlier observations of Webley et al. (1956) with *N. opaca*.

Dihydroxylation of aromatic hydrocarbons by nocardia under cooxidation conditions was first reported by Raymond et al. (1967). The cultures used in these studies were isolated from the soil using n-paraffinic hydrocarbons as growth substrates. In addition to their dihydroxylating ability, these strains of *Nocardia corallina*, *N. minima*, and *N. salmonicolor* were shown to oxidize methyl substituents on both mono- and dicyclic aromatics. When the cultures were grown on n-hexadecane in shaken flasks, representative products from aromatic hydrocarbons which could not support growth are shown (6).

Concentration of products was in the 1–2 gm/liter range. No evidence of further degradation of these compounds was found. Of the mono-

Reaction (6)

p-XYLENE → p-TOLUIC ACID + 2,3-DIHYDROXY-p-TOLUIC ACID   (6)

2,6-DIMETHYLNAPHTHALENE → 6-METHYL-2-NAPHTHOIC ACID

cyclic aromatics tested (benzene, toluene, o-xylene, m-xylene, p-xylene) significant accumulation of products was observed for o- and p-xylene. Ten methyl-substituted naphthalenes (1-methyl-; 2-methyl-; 1,3-dimethyl-; 1,4-dimethyl-; 1,5-dimethyl-; 1,6-dimethyl-; 1,8-dimethyl-; 2,3-dimethyl-; 2,6-dimethyl-; 2,7-dimethyl) were subjected to cooxidation conditions and it was found that those compounds having a β-methyl group were oxidized at this position and the corresponding naphthoic acid accumulated. Optimum pH for the mono- and dicyclic oxidations appeared to be pH 8.0 or slightly above. It was again noted that concentration of both the growth and non-growth substrates had to be carefully controlled, and usually incremental addition of substrate was advantageous.

Jamison et al. (1969) using a cosubstrate isolation technique, found a species (V-49) of *Nocardia corallina* with increased capability for aromatic hydrocarbon oxidation. Employing shaken flasks and 7.5-liter fermentors containing anion-exchange resins, the products shown (7) resulted from the cooxidation of p-xylene.

p-XYLENE → 3,6-DIMETHYLPYROCATECHOL + α,α'-DIMETHYL-CIS,CIS-MUCONIC ACID   (7)

The major product was dimethylmuconic acid (DMMA) with concentrations reaching 5 gm/liter in 96 hours. These studies of p-xylene cooxidation were extended to larger scale equipment by Hosler and Eltz (1969). In 520-liter equipment (a schematic of the fermentor with monitoring and control devices is shown in Fig. 1) they found that it was very important to control the level of the xylene and growth substrate for maximum acid production. A typical

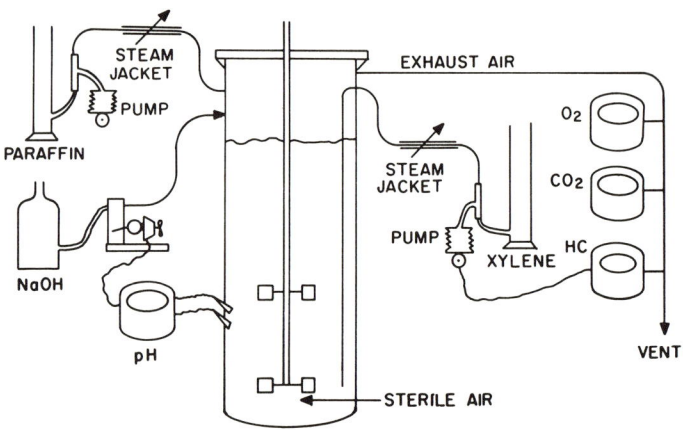

FIG. 1. Schematic drawing of equipment used in co-oxidation of *p*-xylene. After Hosler and Eltz (1969).

fermentation with *N. corallina* V-49 in which the *p*-xylene concentration was held between 3 and 20 mg/liter is shown in Fig. 2. The pH optimum for this oxidation was between 6.5 and 7.0. In similar studies with *N. salmonicolor* A-100 the products were *p*-toluic acid (PTA) and dihydroxy-*p*-toluic acid (DHPT). A typical time course for this fermentation is shown in Fig. 3. In both the *N. corallina* V-49 and *N.*

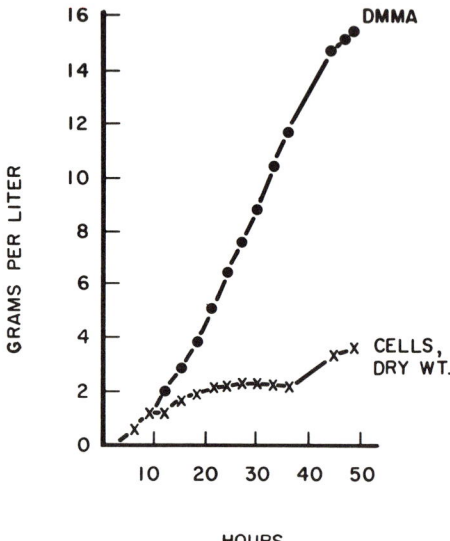

FIG. 2. Typical DMMA fermentation. After Hosler and Eltz (1969).

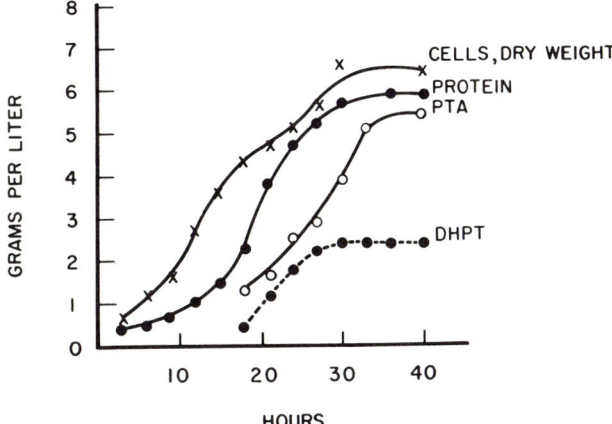

FIG. 3. Typical DHPT fermentation. After Hosler and Eltz (1969).

salmonicolor A-100 systems, the oxygenase systems had to be induced with *p*-xylene for ≈2 hours before product formation was initiated. In addition to *p*-xylene it was noted that naphthalene, tetralin, or biphenyl could serve as inducers for the DMMA fermentation and *p*-chlorobenzoic acid for the DHPT fermentation.

In addition to *p*-xylene cooxidation, *N. corallina* V-49 has been shown to oxidize a number of other hydrocarbons. Jamison *et al.* (1969, 1971) examined cycloparaffins, mono-, di-, and polycyclic aromatics for product formation. Cycloparaffins were not cooxidized in the resin-agar plate systems used. Of all the possible methyl-substituted monoaromatics, only four compounds gave sufficient product for isolation and identification. These included *m*-xylene which gave *m*-toluic acid and those shown in (8) from the other three hydrocarbons.

The oxidation of tetramethylbenzene proved to be the more facile of these reactions. Products identified from monoaromatics with larger substituents included those shown in (9).

Although cycloparaffins were not oxidized, tetralin oxidation was observed (10).

Twelve methyl-substituted naphthalenes were tested and naphthoic acids were characterized from 2,7-dimethyl-, 2,6-dimethyl-, 1,2-dimethyl-, 1,7-dimethyl-, and 2-methylnaphthalenes. The compounds are listed in decreasing order of concentration of product found in the cooxidation systems. Anthracene, phenanthrene, and several methyl-substituted analogs were not oxidized.

With the exception of the long-chain alkyl-substituted benzenes, the hydrocarbon substrates in the systems just described could not provide carbon and energy for growth. Recently, Raymond *et al.*

1,2,4-TRIMETHYL-BENZENE → 3,4-DIMETHYL-BENZOIC ACID + 2,3-DIHYDROXY-4,6-DIMETHYL-BENZOIC ACID

1,2,3-TRIMETHYL-BENZENE → 2,6-DIMETHYL-BENZYL ALCOHOL + 2,3-DIMETHYL-BENZOIC ACID   (8)

1,2,3,4-TETRAMETHYL-BENZENE → 2,3,4-TRIMETHYL-BENZOIC ACID

p-DIETHYLBENZENE → p-DIACETYL BENZENE + 4-(1-HYDROXYETHYL)-ACETOPHENONE   (9)

BIPHENYL → 4-BENZOYL-BUTYRIC ACID

TETRALIN → 4-PHENYL-(2'-HYDROXY) BUTYRIC ACID   (10)

(1971) have described two new soil isolates, *Nocardia coeliaca* 95-5, and *Nocardia* sp. 87-5 which could grow slowly on naphthalene but only accumulate product in significant amounts when a second substrate is present. When Cerelose or *n*-hexadecane was the growth substrate, the major product isolated from naphthalene cooxidation was 4-(2-hydroxyphenyl)-2-ketobutyric acid (**11**).

$$\underset{\text{NAPHTHALENE}}{\bigcirc\!\bigcirc} \longrightarrow \underset{\substack{\text{4-(2-HYDROXYPHENYL)-}\\\text{2-KETOBUTYRIC ACID}}}{\bigcirc\!\!\!-\!\!\overset{\text{OH}}{}\overset{\text{O}}{\underset{\|}{\text{C}}}\!\!-\!\text{COOH}} \tag{11}$$

Methyl-substituted naphthalenes were also oxidized if one ring remained unsubstituted. Biphenyl was oxidized to 4-benzoyl butyric acid. A limited capacity to oxidize *p*-xylene to DMMA was also noted, presenting the interesting possibility that both ortho- and meta-ring cleavage mechanisms were present in this culture.

III. Nonhydrocarbon Oxidation

There are several compounds, that have been used as substrates for nocardia, that are important and have not been discussed under the hydrocarbon section. The enzymatic mechanisms in the degradation or oxidation of the substrates mentioned in the hydrocarbon section are probably very similar, if not the same, as those systems that will be discussed in the nonhydrocarbon section. However, since the compounds that will be discussed here are not strictly hydrocarbons (compounds that contain hydrogen and carbon atoms only), we have chosen to treat them separately.

Webley *et al.* (1955, 1957, 1958, 1959) used *Nocardia opaca* strain T_{16} (Webley and DeKock, 1952) to study the breakdown of substituted fatty acids. When odd numbered ω-phenyl-substituted fatty acids were employed as substrates, two products resulted. If phenylpropionic, phenylvaleric, and phenylheptylic acids (odd number of carbons on the side chain) were used, the resulting product was benzoic acid with cinnamic acid as an intermediate product (**12**).

$$\underset{\text{PHENYLVALERIC ACID}}{\bigcirc\!\!-\!\text{CH}_2\text{CH}_2\text{CH}_2\text{CH}_2\text{COOH}} \longrightarrow \underset{\text{BENZOIC ACID}}{\bigcirc\!\!-\!\text{COOH}} \tag{12}$$

When phenylacetic, phenylbutyric, phenylcaproic, and phenylcaprylic acids (even number of carbon atoms in the side chain) were utilized, *o*-hydroxyphenylacetic acid was recovered (**13**).

PhCH₂CH₂CH₂COOH ⟶ o-HO-C₆H₄-CH₂COOH (13)

PHENYLBUTYRIC ACID → o-HYDROXYPHENYLACETIC ACID

In the first experiment 10 μmoles of phenylvaleric acid was converted to 10.7 μmoles of benzoic acid by 250 mg (dry weight) of cells in 144 hours with no cinnamic acid in the final fermentation broth. The yield of o-hydroxyphenylacetic acid in all cases was very low ($<15\%$). This work supports the belief that the mechanism for the breakdown of phenyl-substituted fatty acids by *N. opaca* takes place by β-oxidation by constitutive enzymes present in the cells.

Later using γ-(3-chlorophenoxy)-butyric acid they showed that the intermediate to 3-chlorophenoxyacetic acid was β-hydroxy-γ-(3-chlorophenoxy)butyric acid (**14**).

(3-Cl-C₆H₄-O)-CH₂CH₂CH₂COOH ⟶ (3-Cl-C₆H₄-O)-CH₂CH(OH)CH₂COOH + (3-Cl-C₆H₄-O)-CH₂COOH   (14)

γ-(3-CHLOROPHENOXY)-BUTYRIC ACID   β-HYDROXY-γ-(3-CHLOROPHENOXY)BUTYRIC ACID   3-CHLOROPHENOXY-ACETIC ACID

In more complete experiments on the breakdown of substituted fatty acids, rates of oxidation were studied and it was found that many factors affected the β-oxidation rate when using washed cell suspensions of *Nocardia opaca* strain T_{16} and strain P_2. An oxygen bridge between the fatty acid and the ring slowed down the rate of oxygen uptake, i.e., ω-phenylbutyric acid was oxidized more rapidly than ω-phenoxybutyric acid. The type of substitution and position on the phenyl ring affected the rate of oxidation, however, the longer the side chain the less effect the substitution had. Compounds that were chloro-substituted were oxidized much slower than those compounds that had a methyl substitution on the phenyl ring. The rate of oxidation was affected by the position of substitution ($p > m > o$). The ortho-position virtually inhibited the breakdown of the fatty acid side chain. The type of ring substitution was also important. Phenyl substituents were more readily oxidized than indolyl substituents, which were more readily oxidized than naphthyl substituents. As would be expected, the authors found that substitution on the fatty acid side chain greatly reduced the rate of breakdown of the fatty acid to β-hydroxyphenylacetic acid.

Taylor and Wain (1962) using *Nocardia coeliaca* carried out studies to show β-oxidation of ω-phenoxyalkanecarboxylic acid where the alkane fraction ranged from C_1 to C_{10}. The authors found that when one to eight methylene groups were used, the β-oxidation mechanism was predominant, however, when nine or ten methylene groups were used, evidence of more than one type of oxidation occurred. The oxidation of phenoxyundecanoic acid with whole cells gave several products after 24 hours. The propionic derivative was seen in addition to the butyric and acetic derivatives. The latter two acid derivatives would not be expected if the fatty acid side-chain breakdown was by β-oxidation only. The authors postulated that with the longer fatty acid side chains α-oxidation would occur until the methylene groups had been reduced to nine and then the β-oxidation mechanism would take over. At least two enzymes operate in the α-oxidation system, (1) a peroxidase and (2) a dehydrogenase. When using ω-phenoxyalkanecarboxylic acids that had a chloro-substitution on the ring, the authors found the same results as Webley *et al.* (1957).

Hirsch and Alexander (1960) made a study of the degradation of halogenated fatty acids by several soil isolates of *Nocardia*. The organisms were capable of utilizing α-halogenated compounds, however, when the corresponding β-derivatives were employed, the enzymatic attack was much slower, if at all. If 2,2-dichloropropionate was the substrate used for *Nocardia* sp. strain #398, #399, #400, #401, and #414, 90–100% of the halogen was released after a period of 3 weeks. Most of the halogenated fatty acids were attacked with the release of the halide. The β-substituted propionates were an exception. The organisms were unable to degrade 2,3-dichloro- and 2,2,3-trichloropropionate which suggests that β-substituted forms are less subject to degradation (**15**).

$$CH_3-\underset{\underset{CL}{|}}{\overset{\overset{CL}{|}}{C}}-COOH \longrightarrow CL^- + CELLS + CO_2$$
2,2-DICHLOROPROPIONIC ACID

$$CH_3-\underset{}{\overset{\overset{BR}{|}}{C}H}-COOH \longrightarrow BR^- + CELLS + CO_2 \qquad (15)$$
2-BROMOPROPIONIC ACID

$$\underset{}{\overset{\overset{BR}{|}}{C}H_2}-CH_2-COOH \longrightarrow NO\ REACTION$$
3-BROMOPROPIONIC ACID

Kaufman (1964) made a study of the degradation of dalapon (2,2-dichloropropionic acid) in five different soils. In two of the soils (Celeryville muck and Lakeland sandy loam) he was able to isolate species of *Nocardia* that were capable of degrading 24% of the dalapon in 8 days. If yeast extract was added to the medium, the amount of dalapon degraded was 42%.

Cain and Cartwright (1960) induced cells of *Nocardia* sp. with nitrobenzoic acid to open the ring of protocatechuic acid and catechol with oxidase enzymes. A partially purified suspension of oxidase obtained from *N. erythropolis* produced β-carboxymuconic acid when exposed to protocatechuic acid. In contrast to the enzymes of *Neurospora* and *Pseudomonas*, no cofactors or metal requirements were needed to catalyze the reaction of the protocatechuic oxidase (**16**).

$$\underset{\text{PROTOCATECHUIC ACID}}{\text{C}_6\text{H}_3(\text{COOH})(\text{OH})_2} \longrightarrow \underset{\beta\text{-CARBOXYMUCONIC ACID}}{\text{HOOC-CH=CH-C(COOH)=CH-COOH}} \qquad (16)$$

Crude cell-free extracts of *N. erythropolis* (clarified by ultracentrifuging) would attack the β-carboxymuconic acid and reduce it to CO_2 and β-oxoadipic acid in almost 100% yields. If a purified decarboxylase complex was used, the authors found that its activity was enhanced by the addition of Co^{++}, Mg^{++}, or Mn^{++} which is similar to the decarboxylase complex of *Pseudomonas*. When crude extracts of *N. opaca* were exposed to catechol, the oxidase activities caused the accumulation of *cis,cis*-muconic acid.

In an earlier paper Cartwright and Cain (1959) described the metabolism of several nitrobenzoate compounds by species of the genus *Nocardia*. Oxidation of these compounds and hydroxy-aromatic compounds was demonstrated by whole cells, freeze-dried cells, and crude extracts of *N. erythropolis*, *N. opaca*, and *Nocardia* M1. The oxidation of *p*-nitrobenzoate was shown to follow the pathway indicated in (**17**).

If *m*-nitrobenzoate was used as a substrate, the pathway was identical, however, if *o*-nitrobenzoate was the substrate, the NO_2 group was not replaced by a hydroxy group but protocatechuic acid was formed after the removal of the NO_2 group.

In two papers Cain (1966a,b) used *Nocardia opaca* to study the oxidation of *o*-nitrobenzoic acid and anthranilic acid. Growth of *N. opaca* on anthranilate induces the system of transport for anthranilate uptake

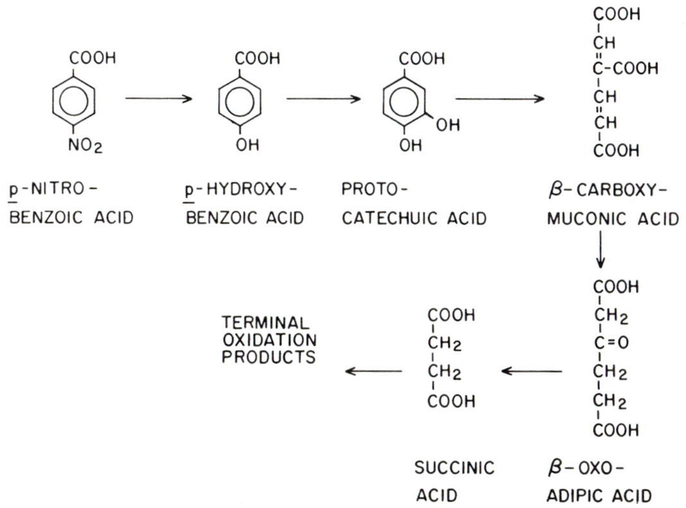

$$\underset{\text{p-NITRO-BENZOIC ACID}}{\text{COOH-C}_6\text{H}_4\text{-NO}_2} \rightarrow \underset{\text{p-HYDROXY-BENZOIC ACID}}{\text{COOH-C}_6\text{H}_4\text{-OH}} \rightarrow \underset{\text{PROTO-CATECHUIC ACID}}{\text{COOH-C}_6\text{H}_3(\text{OH})_2} \rightarrow \underset{\beta\text{-CARBOXY-MUCONIC ACID}}{\text{HOOC-CH=C(COOH)-CH=CH-COOH}}$$

$$\underset{\text{TERMINAL OXIDATION PRODUCTS}}{} \leftarrow \underset{\text{SUCCINIC ACID}}{\text{HOOC-CH}_2\text{-CH}_2\text{-COOH}} \leftarrow \underset{\beta\text{-OXO-ADIPIC ACID}}{\text{HOOC-CH}_2\text{-CO-CH}_2\text{-CH}_2\text{-COOH}}$$

(17)

and the enzyme system for its complete breakdown to CO_2 and NH_3. If *N. opaca* was grown on glucose, no oxidation of β-oxoadipate was seen. The author states that the first enzyme for the oxidation of β-oxoadipate is catechol-1,2-oxygenase and is not the same enzyme that is induced if *N. opaca* is grown on *o*-nitrobenzoic acid. If *N. opaca* is grown on *o*-nitrobenzoic acid, anthranilate appears in the medium at the beginning of the oxidation process and an anthranilate oxidase system is induced by its appearance so that at the end of the fermentation period no *o*-nitrobenzoate or anthranilic acid are seen. Evidently the anthranilate is produced as a side reaction in the pathway not as a true intermediate (18).

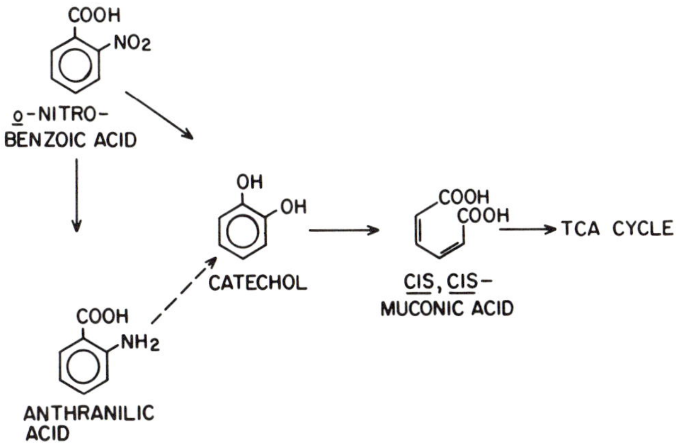

$$\underset{\underline{o}\text{-NITRO-BENZOIC ACID}}{} \searrow \underset{\text{CATECHOL}}{} \rightarrow \underset{\text{CIS,CIS-MUCONIC ACID}}{} \rightarrow \text{TCA CYCLE} \qquad (18)$$

$$\downarrow \nearrow$$

ANTHRANILIC ACID

Tranter and Cain (1967) used *N. erythropolis* cells to reduce 2-fluoronitrobenzoate (with restricted aeration) to 2-fluoroprotocatechuic acid. Further metabolism of the *o*-dihydroxy product by cell extract showed that the pathway was the same as that found when *p*-nitrobenzoate was the substrate. Later Cain *et al.* (1968) used washed cells of *N. erythropolis* grown on benzoate to oxidize 2-fluoro-*p*-nitrobenzoate to fluoroacetate (**19**).

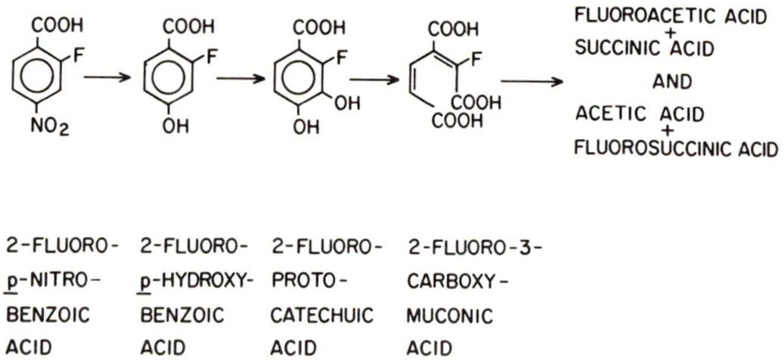

$$\underset{\substack{\text{2-FLUORO-}\\\text{\underline{p}-NITRO-}\\\text{BENZOIC}\\\text{ACID}}}{\text{COOH}/\text{F}/\text{NO}_2} \rightarrow \underset{\substack{\text{2-FLUORO-}\\\text{\underline{p}-HYDROXY-}\\\text{BENZOIC}\\\text{ACID}}}{\text{COOH}/\text{F}/\text{OH}} \rightarrow \underset{\substack{\text{2-FLUORO-}\\\text{PROTO-}\\\text{CATECHUIC}\\\text{ACID}}}{\text{COOH}/\text{F}/\text{OH}/\text{OH}} \rightarrow \underset{\substack{\text{2-FLUORO-3-}\\\text{CARBOXY-}\\\text{MUCONIC}\\\text{ACID}}}{\text{COOH}/\text{F}/\text{COOH}} \rightarrow \begin{array}{l}\text{FLUOROACETIC ACID}\\+\\\text{SUCCINIC ACID}\\\text{AND}\\\text{ACETIC ACID}\\+\\\text{FLUOROSUCCINIC ACID}\end{array}$$

(19)

The fluoro compounds proved more easily oxidized than did the chloro, bromo, or iodo derivatives.

Cain (1968) found that *Nocardia opaca* was a very unusual organism when anthranilate was used as the substrate. It had the ability to form the enzymes for the catechol pathway or the enzymes for the gentisate pathway in the breakdown of anthranilic acid (**20**).

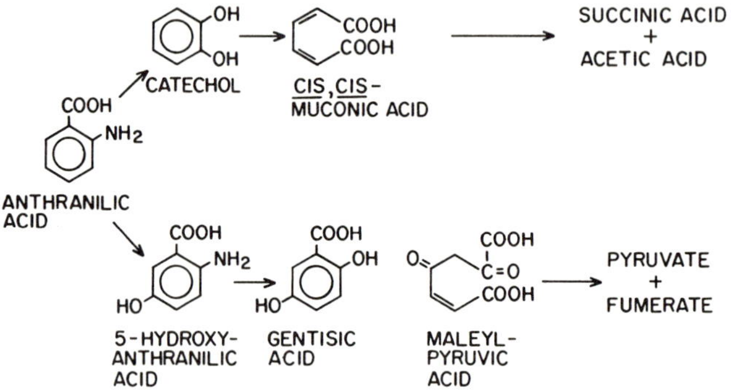

(20)

The major pathway appears to be by the catechol route with the gentisic route a secondary one. The 5-hydroxyanthranilate was oxidized to gentisate by cell-free extracts.

Villanueva (1960, 1961, 1964a) isolated a *Nocardia* that was capable of growing on various aromatic nitro compounds. The organism is

Nocardia V and can reduce some of the aromatic nitro compounds to their amino derivatives by a nitro-reductase. It would appear that the enzyme is not very specific, and also that there is an inorganic nitrato-reductase as well as the organic nitro-reductase. The nitro-reductase (Villanueva, 1964b) was found to be a constitutive enzyme and requires a sulfhydryl compound to be active and uses reduced pyridine nucleotides as the electron donor. Its optimal activity varies with the conditions of growth and life cycle of the organism.

The amidase activities of 75 strains of *Nocardia* were studied by Schneidau (1963) but because of the varied results the author concluded that this data could not be applied to classifying the genus. Watanabe *et al.* (1963) isolated a phenol-utilizing nocardia from activated sludge which could also metabolize several related compounds.

IV. Transformation of Steroids and Sterols

The microbial transformation of steroids and sterols has been the subject of many reports in the literature since Mamoli and Vercellone (1937) described the first microbiological conversion of steroids. The subject has been the topic of many reviews (Vischer and Wettstein, 1958; Peterson, 1959; Shull, 1956; Eppstein *et al.*, 1956; Fried *et al.*, 1955; Stoudt, 1960; Iizuka and Naito, 1967). The first conversions were reported by Turfitt (1948) and by Krámli and Horváth (1948, 1949). Turfitt using a *Nocardia* sp. reported the conversion of cholesterol to 4-dehydroetiocholanic acid. Krámli and Horváth using *N. roseus* were able to convert cholesterol to 7 ε-hydroxycholesterol. Since these first papers have appeared several workers have shown that the genus *Nocardia* is capable of performing many different types of transformations and these will be discussed separately.

The basic structure for the steroid hormones and sterols with their accepted scheme for numbering the carbon atoms and rings is given below (**21**) and these numbering systems will be used throughout the discussion.

BASIC CONFIGURATION
OF STEROIDS
- - - β CONFIGURATION

BASIC CONFIGURATION
OF STEROLS
— α CONFIGURATION

(**21**)

A. TRANSFORMATION OF STEROIDS

1. *Hydroxylation*

A report on the hydroxylation of the 1α and 2α carbons of 17α-ethynyl-17β-hydroxyandrost-4-en-3-one (17α-ethynyltestosterone) was made by Sax et al. (1965) using *Nocardia corallina* (ATCC 999). The mechanism for the formation of the 1α,2α-dihydroxy product was not detailed. However, the authors stated that the route probably involves the formation of 1,2-dehydro derivatives and then one or more mono-oxygenated intermediates. They speculated that the intermediates could include a monohydroxy-Δ^4-3-one, a monohydroxy-$\Delta^{1,4}$-3-one or a 1,2-epoxide. They did not isolate any of these postulated intermediates. The experiment was run using 30 liters of a 1% glucose, 0.4% peptone, 0.1% yeast extract, 0.4% beef extract, and 0.25% sodium chloride medium inoculated with 1 liter of a 24-hour culture of *N. corallina*. After 2 hours, 600 mg of androstenedione was added to the fermenting liquor and the fermentation was continued for 4½ hours. Six grams of substrate was added and the fermentation allowed to go to completion (no time was given). The 1α,2α-diol was obtained in 21% yield. The authors found that induction of the enzyme system to produce the 1α,2α-diol was necessary (**22**).

17α-ETHYNYL-17β-HYDROXY-ANDROST-4-EN-3-ONE → 17α-ETHYNYL-1α,2α-17β-TRIHYDROXYANDROST-4-EN-3-ONE   (22)

Hydroxylation at the C-6 has been reported by Lee and Sih (1964) when using *Nocardia restrictus*. After 24 hours growth of *N. restrictus* No. 545, 10.8 gm of 3β-hydroxy-5α,6α-oxidoandrostan 17-one was added. The fermentation was allowed to proceed for 19 hours after which time 150 mg of 6α-hydroxyandrost-4-ene-3,17-dione was recovered. If the experiment was allowed to go for 12 hours very little of the 6α-hydroxy product was recovered.

Many workers have reported on the hydroxylation of the C-9 of the steroid structure. Dodson and Muir (1958b) utilizing *Nocardia* sp. (A20-10) obtained 9α-hydroxy-4-androstene-3,17-dione from the fermentation of 4-androstene-3,17-dione. The first step in the degradation of Δ^4-androstene-3,17-dione using *N. restrictus* (Dr. Ruth Gordon) is the hydroxylation of the C-9 to produce 9α-hydroxy-4-androstene-

3,17-dione as reported by Sih (1962a). Sih et al. (1965a) using N. restrictus (ATCC 14887) showed the C-9 hydroxylation of 6,19-oxidoandrost-4-ene-3,17-dione to yield 9α-hydroxy-6,19-oxidoandrost-4-ene,3,17-dione (**23**).

6,19-OXIDOANDROST-4-ENE-3,17-DIONE → 9α-HYDROXY-6,19-OXIDOANDROST-4-ENE-3,17-DIONE   (23)

Chang and Sih (1964) showed that the 9α-hydroxylase system of Nocardia restrictus (Lee and Sih, 1964) was a very complicated one. When androst-4-ene-3,17-dione, pregn-4-ene-3,20-dione, and 21-hydroxypregn-4-ene-3,20-dione were exposed to the cell-free extract of N. restrictus the C-9 was hydroxylated. However, when 3-β-hydroxyandrost-4-en-17-one, cholest-5-en-3β-ol, cholest-4-en-3-one, and estra-1,3,5(10)-triene-3,17β-diol were used as substrates, no hydroxylation took place. The 9α-hydroxylase also catalyzed the epoxidation of androsta-4,9(11)-diene-3,17-dione. The authors feel there is little doubt that the hydroxylation and epoxidation take place on the same enzyme site but that the two systems are not identical. They hypothesize that all the enzymes and coenzymes in the epoxidation sequence plus one or more additional are needed for the 9α-hydroxylase to replace the 9α proton.

The C-16 of cortisol was hydroxylated as cited in Belgium Patent 620-272. No other references for the hydroxylation of any other carbons of steroid structure were found.

2. *Dehydrogenation*

Sax et al. (1965) isolated 9α-fluoro-11β, 17β-dihydroxy-17α-methylandrost-1,4-diene-3-one from the fermentation of 9α-fluoro-11β,17β-dihydroxy-17α-methylandrost-4-en-3-one (**24**) with *Nocardia corallina* (ATCC 999).

One hundred Erlenmeyer flasks with 100-ml portions of medium (1% glucose, 0.4% peptone, 0.4% beef extract, 0.25% sodium chloride, and 0.1% yeast extract) were inoculated and shaken for 16 hours at 28°C and then 1 ml of methanol containing 10 mg of substrate was added to each flask and the fermentation continued for

$$\text{9}\alpha\text{-FLUORO-11}\beta,\ \text{17}\beta\text{-DIHYDROXY-17}\alpha\text{-METHYLANDROST-4-EN-3-ONE} \longrightarrow \text{9}\alpha\text{-FLUORO-11}\beta,\ \text{17}\beta\text{-DIHYDROXY-17}\alpha\text{-METHYLANDROST-1,4-DIENE-3-ONE} \quad (24)$$

28 hours. The product was extracted and identified, a 12% yield was obtained (122 mg).

Further evidence of dehydrogenation has been shown by Stoudt et al. (1958) when they reported that N. blackwellii could carry out 1,4-dehydrogenation of 3-hydroxypregnane and 3-hydroxyallopregnane (25).

$$\text{PREGNANE-3}\alpha,\text{11}\beta,\text{17}\alpha,\text{21-TETROL-20-ONE} \longrightarrow \text{PREDNISOLONE} \quad (25)$$

The authors isolated 3-keto and Δ^4-3-keto as intermediates in the transformation process.

Sih and Bennett (1960) reported that cell-free extracts of *Nocardia* sp. could introduce double bonds in the A ring of steroids. Sih and Bennett (1962) in later experiments showed that the 1-dehydrogenase from *N. restrictus* was an adaptive enzyme and that its properties differed greatly from those of the 1-dehydrogenase of *Pseudomonas testosteroni*.

$$\text{2}\alpha\text{-HYDROXYTESTOSTERONE-2}\alpha,\text{17}\beta\text{-DIACETATE} \longrightarrow \text{2-HYDROXYANDROST-1-ENE-3,17-DIONE} \quad (26)$$

3. Reduction and Side-Chain Degradation

The only evidence that we could find of these types of reactions carried out by *Nocardia* was in patents. The reduction reaction is found in U.S. Patent (1963). *Nocardia corallina* (10 mg of lyophilized thalli) and 200 µg of 2α-hydroxytestosterone-2α,17β-diacetate was diluted to 1 ml with 0.05M tris buffer pH 7.6. After 21 hours, 100% of the steroid was converted to 2-hydroxy-androst-1-ene-3,17-dione (**26**).

British Patent (1959) claims the side-chain degradation of progesterone to 9α-hydroxytestosterone (**27**) by *N. corallina*.

PROGESTERONE → 9α-HYDROXYTESTOSTERONE (27)

4. Aromatization

Nocardia restrictus (Dr. Ruth Gordon) was used by Sih (1962a) to prove the pathway for the aromatization of the A ring of androst-4-en-3,17-dione. Steroid 1-dehydrogenase (10-fold purification) in the presence of an electron acceptor was exposed to the steroid and 9,10-seco-3-hydroxyandrosta-1,3,5 (10)-triene-9,17-dione was recovered (**28**).

ANDROST-4-EN-3,17-DIONE → 9,10-SECO-3-HYDROXYANDROSTA-1,3,5(10)-TRIENE-9,17-DIONE (28)

This work suggests that the A ring aromatization goes through the 1,2-dehydrogenation which is very similar to the formation of estrogens from androgenic steroids in mammals described by Ryan (1959). Dodson and Muir (1958a) used a *Nocardia* sp. to show that the aromatization of 1,4-androstadiene-9α-ol-3,17-dione follows a pathway that shows C_9–C_{10} cleavage to give 9,10-seco-3-hydroxy-1,3,5(10)-androstatriene-9,17-dione with the hydroxylation of the C-9 followed by the introduction of the Δ^1-double bond. This is just the opposite of the pathway found when employing a *Pseudomonas* culture.

Singh *et al.* (1970) used a *Nocardia* sp. to transform 360 mg of 3β, 19-dihydroxy-pregn-5-en-20-one 3-acetate to 165 mg of 3-hydroxy-

19-norpregna-1,3,5(10)-trien-20-one. Aromatization of the A ring was also obtained when the substrate was 19-hydroxyprogesterone or pregn-5-ene-3β,19,20β-triol 3 acetate. This reaction occurred without the cleavage of the side chain. The authors concluded that apparently the organism did not have the enzymatic system to degrade the C-17 side chain of 19-hydroxypregnanes. Since it did degrade the side chain of cholestane the enzymatic system for the side-chain degradation of cholestane and pregnane must be different.

Nocardia corallina (ATCC 13259) has been reported by Kluepfel and Vézina (1970) to aromatize both the A and B rings of androsta-1, 4,7-triene-3,17-dione to give equilin and equilenin (**29**). Yields of 1–2% of equilin and 2–3% of equilenin were obtained after 36 to 48 hours incubation time. No change in the ratio of products could be seen if the incubation time was extended.

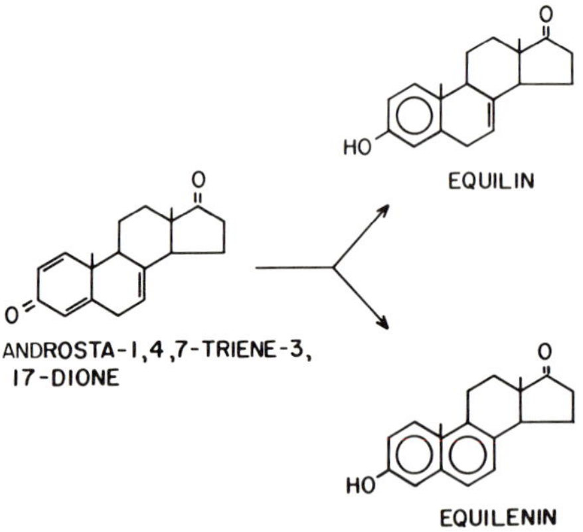

(29)

ANDROSTA-1,4,7-TRIENE-3, 17-DIONE

EQUILIN

EQUILENIN

5. *Expoxidation*

Sih (1962b) introduced 0.5 gm of androst-9(11)-ene-3,17-dione in 4 ml of dimethylformamide into shaken flasks of *Nocardia* sp. (ATCC 13934) grown on Difco nutrient broth for 24 hours at 25°C. The fermentation continued for 72 hours at which time 0.68 gm of a product was recovered and identified as androst-9α, 11-α-oxido-3,17-dione (**30**). Sih also used 9(11)-dehydrocortexolone which gave a 20% yield of 9-α,11-α-oxidocortexolone.

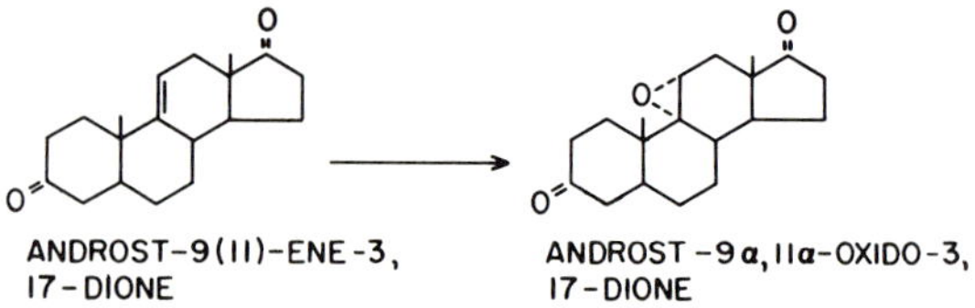

(30)

ANDROST-9(11)-ENE-3, 17-DIONE

ANDROST-9α,11α-OXIDO-3, 17-DIONE

The author states that this work supports the theory of Bloom and Shull (1955) that microbial enzymes transform isolated double bonds into oxides rather than alcohols.

6. Cleavage of the Steroid Skeleton

Sih and Wang (1963) reported that *Nocardia restrictus* No. 545 grown on Difco nutrient broth would convert 14.4 gm of androst-4-ene-3,17-dione to 210 mg of 3-hydroxy-9,10-seco-androsta-1,3,5(10)-triene-9,17-dione and 1.48 gm of 3aα-H-4-α-[3′-propionic acid]-7aβ-methylhexahydro-1,5-indanedione (31).

ANDROST-4-ENE-3,17-DIONE

3-HYDROXY-9,10-SECO-ANDROSTA-1,3,5(10)-TRIENE-9,17-DIONE

3Aα-H-4α-[3′-PROPIONIC ACID]-7Aβ-METHYLHEXAHYDRO-1,5-INDANEDIONE

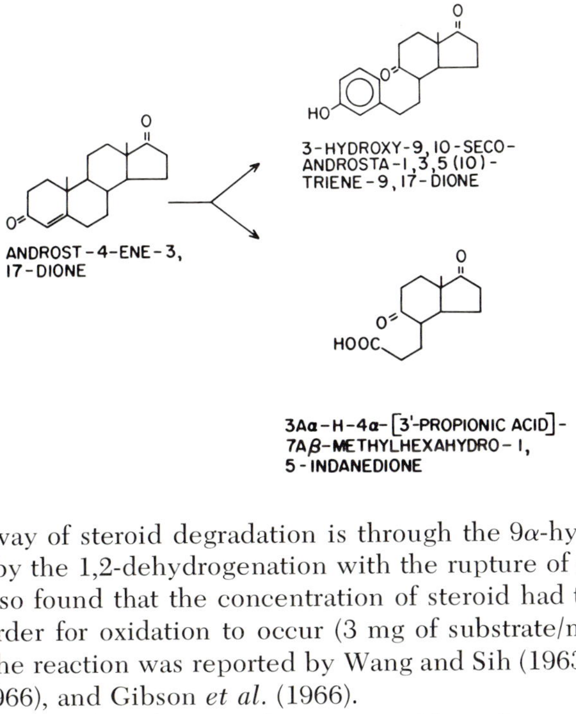

(31)

The pathway of steroid degradation is through the 9α-hydroxylation followed by the 1,2-dehydrogenation with the rupture of ring B. The authors also found that the concentration of steroid had to be rather high in order for oxidation to occur (3 mg of substrate/ml). Further work on the reaction was reported by Wang and Sih (1963), Sih *et al.* (1965b, 1966), and Gibson *et al.* (1966).

Coombe *et al.* (1966) exposed estrone to *Nocardia* sp. (E110) and demonstrated the cleavage of the A and B rings resulting in the products shown (32).

Schubert *et al.* (1968) used *N. opaca* to reduce the C and D rings of the sterane skeleton. The final acids that were recovered could enter the citric acid cycle. Kondo *et al.* (1969) using *N. opaca* were able to recover 7aβ-methyl-1,5-dioxo-7α-hydroxy-3aα-hexahydro-4-indanpropionic acid from 7aβ-methyl-1 β-acetoxy-5-oxo-3aα-hexahydro-4-indanpropionic acid.

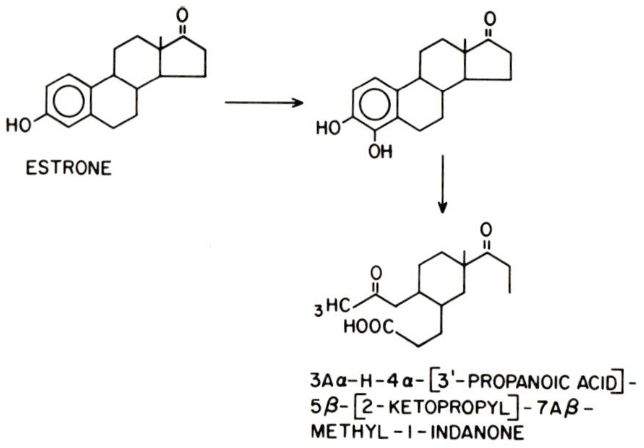

ESTRONE

3Aα-H-4α-[3'-PROPANOIC ACID]-
5β-[2-KETOPROPYL]-7Aβ-
METHYL-I-INDANONE

(32)

B. Transformation of Sterols

1. Dehydrogenation

Proactinomyces erythropolis (Nocardia erythropolis) has been shown to transform cholesterol to cholest-4-en-3-one by Turfitt (1948). Sih and Wang (1965) used *N. restrictus* (ATCC 14887) to show the same reaction (33).

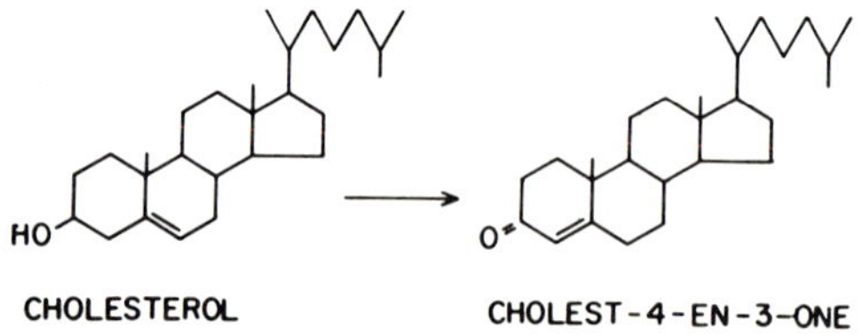

CHOLESTEROL   CHOLEST-4-EN-3-ONE

(33)

2. Side-Chain Degradation

Whitmarsh (1964) reported that when *Nocardia* sp. was grown in the presence of cholesterol several degradation products were obtained. Very small amounts of androst-4-ene-3,17-dione were recovered and larger amounts of 3-oxobisnorchol-4-enic acid and 3-oxobisnorchola-1,4-dienic acid were also recovered. Low yields of androsta-1,4-diene-3,17-dione could also be isolated. Turfitt (1948) employed *N. erythropolis* to degrade the side chain of cholesterol to produce 3-oxo-etiochol-4-enic acid (androst-4-en-3-one-17-carboxylic acid).

Sih and Wang (1965) used *Nocardia restrictus* (ATCC 14887) to reduce the side-chain of 19-hydroxycholest-4-en-3-one to estrone (34).

19-HYDROXYCHOLEST-4-EN-3-ONE → ESTRONE (34)

Afonso et al. (1966) working with the same organism showed the transformation of 19-norcholestra-1,3,5(10)-triene-3-ol to estrone. They recovered the product in an 8% yield after an incubation period of 240 hours.

Sih and Wang (1965) postulated that the major pathway for sterol degradation is the removal of the side chain first, which yields the C-19 steroid, and then ring fission proceeds. The organism that they did this work with was *N. restrictus* (ATCC 14887). Sih et al. (1968b) used cells of *N. restrictus* to cleave the side chain of cholesterol to give propionic acid and a C_{24} acid. Sih et al. (1968a) showed that C_{22} acids are important intermediates in the degradation of the cholesterol side chain.

V. Summary

Seventeen identified and many unidentified soil isolates of the genus *Nocardia* have been examined for their potential in transformations of hydrocarbons, nonhydrocarbon aromatics, herbicides, and steroids. Of the cultures surveyed, most appear to prefer paraffinic over aromatic hydrocarbons as a growth substrate. The *n*-alkanes (C_{6-28}) are readily oxidized and the primary end product which results is cells. With few exceptions the oxidation of alkanes parallels fatty acid metabolism, resulting in cell composition changes, particularly the lipid fraction, which are directly related to the initial structure of the growth substrate. In properly designed and operated systems high cell concentrations of *Nocardia* can be achieved in a reasonable time period. Cooxidation results in the transformation of a large number of nongrowth aromatic hydrocarbons to acids, alcohols, and catechols. The single most important factor in cell growth and transformation of hydrocarbons is maintenance of the proper concentration of the hydrocarbon in the fermentation system.

Degradation of nonhydrocarbon aromatics is very much influenced by the type, number, and position of substitution. Generally, the predominate ring cleavage mechanism is dihydroxylation followed by ortho-cleavage, resulting in a muconic acid.

Hydroxylation, dehydrogenation, aromatization, epoxidation, and cleavage of steroids have been observed in a relatively small number of *Nocardia* species. Yields and concentration of products are low in most of the studies reported.

REFERENCES

Abbott, B. J., and Casida, L. E., Jr. (1968). *J. Bacteriol.* **96**, 925-930.
Afonso, A., Herzog, H. L., Federbush, C., and Charney, W. (1966). *Steroids* **1**, 429-432.
Belgium Patent (1962). No. 620,272.
Bloom, B. M., and Shull, G. M. (1955). *J. Amer. Chem. Soc.* **77**, 5767-5768.
British Patent (1959). No. 862,701.
Cain, R. B. (1966a). *J. Gen. Microbiol.* **42**, 197-217.
Cain, R. B. (1966b). *J. Gen. Microbiol.* **42**, 219-235.
Cain, R. B. (1968). *Antonie van Leeuwenhoek J. Microbiol. Serol.* **34**, 417-432.
Cain, R. B., and Cartwright, N. J. (1960). *Biochim. Biophys. Acta* **37**, 197-213.
Cain, R. B., Tranter, E. K., and Darrah, J. A. (1968). *Biochem. J.* **106**, 211-227.
Cartwright, N. J , and Cain, R. B. (1959). *Biochem. J.* **71**, 248-261.
Chang, F. N., and Sih, C. J. (1964). *Biochemistry* **3**, 1551-1557.
Coombe, R. G., Tsong, Y. Y., Hamilton, P. B., and Sih, C. J. (1966). *J. Biol. Chem.* **241**, 1587-1595.
Davis, J. B. (1964a). *Appl. Microbiol.* **12**, 210-214.
Davis, J. B. (1964b). *Appl. Microbiol.* **12**, 301-304.
Davis, J. B. (1967). "Petroleum Microbiology," Elsevier, Amsterdam.
Davis, J. B., and Raymond, R. L. (1961). *Appl. Microbiol.* **9**, 383-388.
Dodson, R. M., and Muir, R. D. (1958a). *J. Amer. Chem. Soc.* **80**, 5004-5005.
Dodson, R. M., and Muir, R. D. (1958b). *J. Amer. Chem. Soc.* **80**, 6148.
Eppstein, S. H , Meister, P. D., Murray, H. C., and Peterson, D. H. (1956). *Vitam. Horm. (New York)* **14**, 359-432.
Erikson, D. (1949). *J. Gen. Microbiol.* **3**, 361-368.
Finnerty, W. R., Hawtrey, E., and Kallio, R. E. (1962). *Z. Allg. Mikrobiol.* **2**, 169-177.
Forro, J. R. (1965). Ph.D. Thesis. Pennsylvania State Univ., State College, Pennsylvania.
Foster, J. W. (1962). *In* "Oxygenases" (O. Hayaishi, ed.), p. 243. Academic Press, New York.
Fried, J., Thoma, R. W., Perlman, D., Herz, J. E., and Borman, A. (1955). *Recent Progr. Horm. Res.* **11**, 149-181.
Fukui, S., Shimizu, S., and Fujii, K. (1967). *Hakko Kogaku Zasshi* **46**, 530-540.
Gibson, D. T., Wang, K. C., Sih, C. J., and Whitlock, H., Jr. (1966). *J. Biol. Chem.* **241**, 551-559.
Gordon, R. E. (1966). *J. Gen. Microbiol.* **43**, 329-343.
Gordon, R. E., and Mihm, J. M. (1957). *J. Bacteriol.* **73**, 15-27.
Gordon, R. E., and Mihm, J. M. (1959). *J. Gen. Microbiol.* **21**, 736-748.
Hirsch, P., and Alexander, M. (1960). *Can. J. Microbiol.* **6**, 241-249.
Hosler, P., and Eltz, R. W. (1969). *In* "Fermentation Advances" (D. Perlman, ed.), p. 789-797. Academic Press, New York.
Iizuka, H., and Naito, A. (1967). "Microbial Transformation of Steroids and Alkaloids." Univ. of Tokyo Press, Tokyo and Penn State Univ. Press, University Park, Pennsylvania.
Jamison, V. W., Raymond, R. L., and Hudson, J. O. (1969). *Appl. Microbiol.* **17**, 853-856.

Jamison, V. W., Raymond, R. L., and Hudson, J. O. (1971). *Develop. Ind. Microbiol.* **12**, 99–105.
Kaufman, D. D. (1964). *Can. J. Microbiol.* **10**, 843–852.
Kluepfel, D., and Vézina, C. (1970). *Appl. Microbiol.* **20**, 515–516.
Kondo, E., Stein, B., and Sih, C. J. (1969). *Biochim. Biophys. Acta* **176**, 135–145.
Krámli, A., and Horváth, J. (1948). *Nature (London)* **162**, 619.
Krámli, A., and Horváth, J. (1949). *Nature (London)* **163**, 219.
Lee, S. S., and Sih, C. J. (1964). *Biochemistry* **3**, 1267–1271.
Linday, E. M., and Donald, M. B. (1961). *J. Biochem. Microbiol. Technol. Eng.* **3**, 219–233.
McClung, N. M., and Adams, J. N. (1962). *Rev. Latinoamer Microbiol.* **5**, Suppl. 9, 1–17.
McKenna, E. J. (1966). Ph.D. Thesis. Univ. of Iowa, Ames, Iowa.
Mamoli, L., and Vercellone, A. (1937). *Ber. Deut. Chem. Ges. B.* **70**, 470–471, 2079–2082.
Morikawa, H , and Kamikubo, T. (1969). *Hakko Kogaku Zasshi* **47**, 470–477.
Peterson, D. H (1959). *Proc. 4th Int. Congr. Biochem., Vienna, 1958* **4**, 83–119.
Ratledge, C. (1970). *Chem. Ind. (London)* pp. 843–854.
Raymond, R. L., and Davis, J. B. (1960). *Appl. Microbiol.* **8**, 329–334.
Raymond, R. L., Jamison, V. W., and Hudson, J. O. (1967). *Appl. Microbiol.* **15**, 357–365.
Raymond, R. L., Jamison, V. W., and Hudson, J. O. (1971). *Lipids* (in press).
Ryan, K. J. (1959). *J. Biol. Chem.* **234**, 268–272.
Sax, K. J., Holmlund, C. E., Feldman, L I., Evans, R. H., Jr., Blank, R. H , Shay, A. J., Shultz, J. S., and Dann, M. (1965). *Steroids* **5**, 345–359.
Schneidau, J. D., Jr. (1963). *Amer. Rev. Resp. Dis.* **88**, 563–564.
Schneidau, J. D., Jr., and Shaffer, M. F. (1957). *Amer. Rev. Tuberc. Pulm. Dis.* **76**, 770–788.
Schubert, K , Böhme, K. H., Ritter, R., and Hörhold, C. (1968). *Biochim. Biophys. Acta* **152**, 401–408.
Shull, G. M. (1956). *Trans. N.Y Acad. Sci.* **19**(2), 147–172.
Sih, C. J. (1962a). *Biochem. Biophys. Res. Commun.* **7**, 87–90.
Sih, C. J. (1962b). *J. Bacteriol.* **84**, 382.
Sih, C. J., and Bennett, R. E. (1960). *Biochim. Biophys. Acta* **38**, 378–379.
Sih, C. J., and Bennett, R. E. (1962). *Biochim. Biophys. Acta* **56**, 584–592.
Sih, C. J., and Wang, K C. (1963). *J. Amer. Chem. Soc.* **85**, 2135–2137.
Sih, C. J., and Wang, K. C. (1965). *J. Amer. Chem. Soc.* **87**, 1387–1388.
Sih, C. J., Lee, S. S., Tsong, Y. Y , and Wang, K. C. (1965a). *J. Amer. Chem. Soc.* **87**, 1385–1386.
Sih, C. J., Wang, K. C., Gibson, D. T., and Whitlock, H. W., Jr. (1965b). *J. Amer. Chem. Soc.* **87**, 1386–1387.
Sih, C. J., Lee, S. S., Tsong, Y. Y., and Wang, K. C. (1966). *J. Biol. Chem.* **241**, 540–550.
Sih, C J., Wang, K. C., and Tai, H. H. (1968a). *Biochemistry* **7**, 796–807.
Sih, C. J., Tai, H. H., Tsong, Y. Y., Lee, S. S., and Coombe, R. G. (1968b). *Biochemistry* **7**, 808–818.
Singh, K., Marshall, D. J., and Vézina, C. (1970). *Appl. Microbiol.* **20**, 23–25.
Stewart, J. E., and Kallio, R. E. (1959). *J. Bacteriol.* **78**, 726–730.
Stoudt, T. H. (1960). *Advan. Appl. Microbiol.* **2**, 183–222.
Stoudt, T. H., McAleer, W. J., Kozlowski, M. A., and Marlatt, V. (1958). *Arch. Biochem. Biophys.* **74**, 280–281.
Taylor, H. F., and Wain, R. L. (1962). *Proc. Roy. Soc., Ser. B* **268**, 172–186.
Tranter, E. K., and Cain, R. B. (1967). *Biochem. J.* **103**, 22p–23p.
Treccani, V. (1953). *Ann. Microbiol.* **5**, 232–237.

Treccani, V., and Bianchi, B. (1959). *Atti Congr. Naz. Soc. Ital. Microbiol., 10th, Bologna* p. 1-4.
Treccani, V., Walker, N., and Wiltshire, G. M. (1954). *J. Gen. Microbiol.* **11,** 341-348.
Treccani, V., Canonica, L., and deGirolamo, M. G. (1955). *Ann. Microbiol.* **6,** 183-199.
Turfitt, G. E. (1948). *Biochem. J.* **42,** 376-383.
U.S. Patent (1963). No. 3,087,864.
Villanueva, J. R. (1960). *Microbiol. Espan.* **13,** 387-391.
Villanueva, J. R. (1961). *Microbiol. Espan.* **14,** 157-162.
Villanueva, J. R. (1964a). *J. Biol. Chem.* **239,** 773-777.
Villanueva, J. R. (1964b). *Antonie van Leeuwenhoek J. Microbiol. Serol.* **30,** 17-32.
Vischer, E., and Wettstein, A. (1958). *Advan. Enzymol.* **20,** 237-282.
Wagner, F., Zahn, W., and Buhring, U. (1967). *Angew. Chem.* **6,** 359-360.
Wagner, F., Kleemann, T., and Zahn, W. (1969). *Biotechnol. Bioeng.* **11,** 393-408.
Waksman, S. A. (1967). "Actinomyces," Ronald Press, New York.
Wang, K. C., and Sih, C. J. (1963). *Biochemistry* **2,** 1238-1243.
Watanabe, S., Kumagaya, A., and Murooka, H. (1963). *J. Gen. Appl. Microbiol.* **9,** 363-376.
Webley, D. M., and DeKock, P. C. (1952). *Biochem. J.* **51,** 371-375.
Webley, D. M., Duff, R. B., and Farmer, V. C. (1955). *J. Gen. Microbiol.* **13,** 361-369.
Webley, D. M., Duff, R. B., and Farmer, V. C. (1956). *Nature (London)* **178,** 1467-1468.
Webley, D. M., Duff, R. B., and Farmer, V. C. (1957). *Nature (London)* **179,** 1130-1131.
Webley, D. M., Duff, R. B., and Farmer, V. C. (1958). *J. Gen. Microbiol.* **18,** 733-746.
Webley, D. M., Duff, R. B., and Farmer, V. C. (1959). *Nature (London)* **183,** 748-749.
Wegner, G. H. (1969). *Bacteriol. Proc.* p. 62.
Whitmarsh, J. M. (1964). *Biochem. J.* **90,** 23p-24p.
Wieland, T., Griss, G., and Haccius, B. (1958). *Arch. Mikrobiol.* **28,** 383-393.
Wodzinski, R. S., and Johnson, M. J. (1968). *Appl. Microbiol.* **16,** 1886-1891.

Microbial Transformations of Antibiotics[1]

OLDRICH K. SEBEK

Upjohn Research Laboratories, Kalamazoo, Michigan

AND

D. PERLMAN

The School of Pharmacy, University of Wisconsin, Madison, Wisconsin

I.	Introduction	123
II	Types of Microbial Transformations	126
	A. Acylation	126
	B. Phosphorylation	130
	C. Adenylylation and Ribonucleotide Formation	131
	D. Hydrolysis	132
	E. Reduction	136
	F. Oxidation	136
	G. Sulfoxidation	138
	H. Demethylation	139
	I. Deamination	140
	J. Nonspecific Degradation	142
III	Techniques Useful in Studying Microbial Transformations of Antibiotics	143
IV	Summary	145
	References	146

I. Introduction

Although more than 30 years have passed since the first reports of the therapeutic promise of penicillin, and the description of a microbial transformation of this antibiotic (Abraham and Chain, 1940), only a limited effort has been invested in using enzymes to produce new and potentially useful antibiotic derivatives. The signal success of the programs on chemical modification of a number of antibiotics including penicillin (Price, 1969), cephalosporin (Sassiver and Lewis, 1970), tetracycline (Blackwood and English, 1970), lincomycin (Magerlein, 1971), and rifamycin (Sensi *et al.*, 1967) has encouraged large-scale efforts in preparing modifications of other antibiotics by chemical means (Cron *et al.*, 1970a,b; Kinoshita *et al.*, 1970).

Many antibiotics have been shown to be susceptible to microbial attack. Their addition to soil has resulted in a complete loss of antimicrobial activity as shown by experiments with actinomycin, chloramphenicol, chlortetracycline, cycloheximide, griseofulvin, mycophenolic acid, oxytetracycline, patulin, penicillin, and streptomycin

---

[1] Manuscript completed in December, 1970.

(Brian, 1958; Pramer, 1958), which were found to be due to microbial degradation and chemical inactivation.

The successes in microbial transformations of the cyclopentanophenanthrene nucleus of steroids (Čapek et al., 1966; Charney and Herzog, 1967; Goll, 1966; Iizuka and Naito, 1968) have shown that microbial systems can carry out the following types of chemical reactions:

1. Oxidation (and dehydrogenation)
2. Reduction
3. Esterification
4. Hydrolysis

In addition to these four types of changes, the following are found in the literature on microbial modification of antibiotics:

a) Adenylylation
b) Phosphorylation
c) Demethylation
d) Deamination
e) Reduction

Antibiotics reported to undergo these transformations are listed in Table I.

TABLE I
Transformation of Some Antibiotics by Microorganisms

Antibiotic	Change noted	Reference (see list below table)
Actinomycin	Hydrolysis of lactone	1,2,3
Cephalosporin	Hydrolysis of lactam	4,5
Cephalosporin	Hydrolysis of peptide bond	6
Cephalosporin	Hydrolysis of ester	7
Chloramphenicol	Hydrolysis of amide	8,9
Chloramphenicol	Reduction of nitro group	8,10
Chloramphenicol	Acetylation of hydroxyl	11,12,13,61
Circulin	Hydrolysis of peptide ring	14
Clindamycin	Sulfoxide formation	15
Clindamycin	Phosphorylation	16
Clindamycin	N-Demethylation	15
Clindamycin	Ribonucleotidation	17
Cycloheximide	Acetylation	18
Colistin	Hydrolysis of peptide chain	19
Cordycepin	Deamination	20,21
Echinomycin	Hydrolysis of lactone	3

TABLE I (Continued)

Antibiotic	Change noted	Reference
Etamycin	Hydrolysis of lactone	3
Formycin B	Deamination	22
Formycin B	Oxidation	23
Fusidic acid	Oxidation of hydroxyl to ketone	24
Fusidic acid	Hydroxylation	25
Gentamicin A	Phosphorylation	26
Gramicidin S	Hydrolysis of peptide ring	27
Griseofulvin	Demethylation	28
Griseofulvin	Hydroxylation	29
Griseofulvin	Reduction (of dehydrogriseofulvin)	29
Kanamycin	Phosphorylation	31,32,33
Kanamycin	Acetylation	34
Lincomycin	Sulfoxide formation	35
Lincomycin	N-Demethylation and C-demethylation	35
Lincomycin	Phosphorylation	36
Mannosidostreptomycin	Hydrolysis	37,38
Mannosidostreptomycin	Adenylylation	39
Mycophenolic acid	Oxidation	40
Paromamine	Phosphorylation	41
Penicillin	Hydrolysis of lactam	42,43
Penicillin	Hydrolysis of peptide	44,45,46
Polymyxin	Hydrolysis of peptide	14
Rifamycin S	Acetylation	47
Rifamycin S	Esterification	47
Rifamycin B	Deacetylation	48
Staphylomycin S	Hydrolysis of lactone	1,3
Stendomycin	Hydrolysis of lactone	3
Spectinomycin	Adenylylation	49
Spiramycin	Acetylation	50
Streptomycin	Adenylylation	32,51,52,53
Streptomycin	Phosphorylation	53,54,55
5a,6-Anhydrotetracycline	Rehydration	56
12a-Deoxytetracycline	Hydroxylation	57,58
Toyocamycin	Hydrolysis	59
Tylosin	Reduction	60
T-2636 Antibiotics	Deacetylation	30
T-2636 Antibiotics	Dehydrogenation	30
T-2636 Antibiotics	Acylation	30

References Cited in Table I

1. Hou et al. (1970).
2. Hou and Perlman (1970).
3. Perlman and Hou (1969).
4. Demain et al. (1963).
5. Sabath et al. (1965).
6. Walton (1964a,b).
7. Claridge et al. (1963).
8. Smith and Worrel (1950).
9. Smith and Worrel (1953).
10. Egami et al. (1951).

11. Shaw (1967).
12. Shaw and Brodsky (1968b).
13. Suzuki and Okamoto (1967).
14. Warren and Neilands (1964).
15. Argoudelis et al. (1969).
16. Coats and Argoudelis (1971).
17. Argoudelis and Coats (1970a).
18. Howe and Moore (1968).
19. Ito et al. (1966).
20. Fukagawa et al. (1965).
21. Herr and Murray (1967).
22. Sawa et al. (1968).
23. Sawa et al. (1967).
24. Dvonch et al. (1966).
25. von Daehne et al. (1968).
26. Tanaka (1970).
27. Yukioka et al. (1966).
28. Boothroyd et al. (1961).
29. Andres et al. (1969).
30. Fugono et al. (1970).
31. Doi et al. (1969).
32. Takasawa et al. (1968).
33. Umezawa et al. (1968a).
34. Okamoto and Suzuki (1965).
35. Argoudelis and Mason (1969).
36. Argoudelis and Coats (1969).
37. Perlman and Langlykke (1948).
38. Demain and Inamine (1970).
39. Schwartz and Perlman (1970).
40. Jones et al. (1970).
41. Maeda et al. (1968).
42. Abraham and Chain (1940).
43. Abraham (1951).
44. Cole (1966, 1967).
45. Huang et al. (1963).
46. Hamilton-Miller (1966).
47. Lancini et al. (1969).
48. Lancini and Hengeller (1969).
49. Benveniste et al. (1970).
50. Ninet and Verrier (1961).
51. Harwood and Smith (1969a).
52. Umezawa et al. (1968b).
53. Yamada et al. (1968).
54. Ozanne et al. (1969).
55. Miller and Walker (1970).
56. McCormick et al. (1962).
57. Holmlund et al. (1959a,b).
58. Beck and Shull (1961).
59. Rao and Renn (1963) and Suhadolnik (1970b).
60. Feldman et al. (1964).
61. Argoudelis and Coats (1970b).

In most of the reports, major attention has been focused on the identification of the reaction (transformation) product and, in comparison, only a limited effort has been expended on the characterization of the enzymes involved. If the potential of microbial transformations of the antibiotics is to be realized more fully in the future, enzymatic mechanisms of these processes should be studied more extensively and be better understood. In this review we have attempted to collect pertinent information on microbial transformations of antibiotics; noted only a limited amount of success in the application of these data to the production of new useful antibiotics; and discussed ways whereby microorganisms suitable for this kind of work may be obtained.

II. Types of Microbial Transformations

A. ACYLATION

During the 1959 dysentery epidemic in Japan, most of the chloramphenicol-resistant shigellae isolated from clinical cases were found to inactivate the antibiotic. Species of other enteric bacteria including *Escherichia, Klebsiella,* and *Morganella* showed the same characteristics. Miyamura et al. (Okamoto and Suzuki, 1965) observed that the cultures of the R-factor carrying strains inactivated chloramphenicol more rapidly than the drug-sensitive strains or strains with an *in vitro* developed resistance to chloramphenicol. In other studies (Okamoto and Suzuki, 1965) it was shown that the multiple drug-resistant strains of R-factor carrying *Escherichia coli* inactivated chloramphenicol, kanamycin, dihydrostreptomycin, tetracycline, and sulfonamides enzymatically. The 100S fraction of the cell-free extracts from these organisms (which was further purified on DEAE-cellulose and Sephadex A-50 columns), inactivated chloramphenicol and kanamycin when supplemented with ATP, Mg^{++}, coenzyme A, and acetate. Since acetyl-coenzyme A and Mg^{++} could replace these supplements, it was concluded that acetylation was the inactivating mechanism. The enzyme system was found only in the R-factor carrying strains and not in those where antibiotic resistance was developed *in vitro*.

Further study (Suzuki and Okamoto, 1967) resulted in identification of the products of the reactions as chloramphenicol 3-acetate, and chloramphenicol 1,3-diacetate. Chloramphenicol 1-acetate which was also detected, was thought to be due to a nonenzymatic reaction, and the following sequence was proposed:

i) Chloramphenicol + acetyl-CoA → chloramphenicol 3-acetate + CoA-SH;

ii) Chloramphenicol 3-acetate → chloramphenicol 1-acetate;

iii) Chloramphenicol 1-acetate + acetyl-CoA → chloramphenicol 1,3-diacetate + CoA-SH

Structures of these and other chloramphenicol metabolites are shown in Fig. 1.

The enzyme catalyzing *i* was purified 120-fold and that catalyzing *iii*, 50-fold. The chloramphenicol-*O*-acetyltransferase (reaction *i*) showed substrate specificity (Suzuki and Okamoto, 1967), and the preparations obtained from *Escherichia* and *Proteus* resembled those of *E. coli* (Shaw, 1967).

An enzyme with the same acetylating ability and function was also found in non-R-factor carrying bacteria whose chloramphenicol resistance was most likely mediated by a different extrachromosomal element (plasmid): The inducible chloramphenicol acetyltransferase from *Staphylococcus aureus* had the same pH stability, electrophoretic mobility, substrate affinity, and immunologic reactivity, as the *Escherichia coli* enzyme, but the antiserum to the *E. coli* enzyme neither precipitated nor neutralized it (Shaw and Brodsky, 1968a,b; Suzuki et al., 1966). The enzyme activity was induced by the antibiotically inactive analog, 3-deoxychloramphenicol, and could be followed by

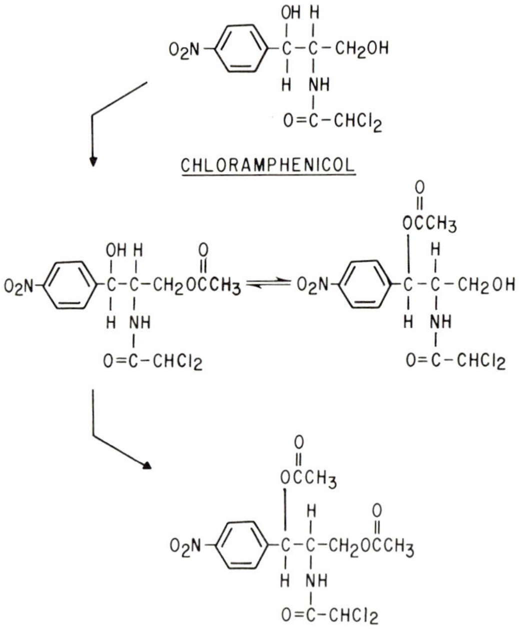

FIG. 1. Acetylation of chloramphenicol.

a simple spectrophotometric assay (Bonanchaud, 1967) or thin-layer chromatography (Piffaretti and Pitton, 1970). In subsequent studies with an enzyme from S. *epidermidis* (Shaw, 1970) it was shown that it has a molecular weight of 80,000 and is made up of four subunits of a molecular weight of 20,000 each. The S. *epidermis* enzyme also formed propionyl and butyryl (but not palmityl) esters of chloramphenicol (Shaw *et al.*, 1970). The same products, and chloramphenicol 3-isobutyrate, chloramphenicol 3-isovalerate, and chloramphenicol 1-acetate, were isolated from chloramphenicol-supplemented *Streptomyces coelicolor* fermentations (Argoudelis and Coats, 1970b).

The acetylation of kanamycin studied by Okamoto and Suzuki (1965) was further investigated by Umezawa *et al.* (1967b). The antibiotic product was shown to contain an acetyl group on the 6-amino-6-deoxy-D-glucose moiety of the kanamycin A. Kanamycin C (containing glucosamine instead of 6-amino-6-deoxyglucose), and neomycin and paromomycin (both containing 2,6-diamino-2,6-dideoxy-D-glucose) were not acetylated by this *E. coli*-derived enzyme. However, deoxykanamycin which was prepared by chemical synthesis and contains 6-amino-6-deoxy-D-glucose, was acetylated (Okanishi *et al.*, 1967) (see Fig. 2 for structures).

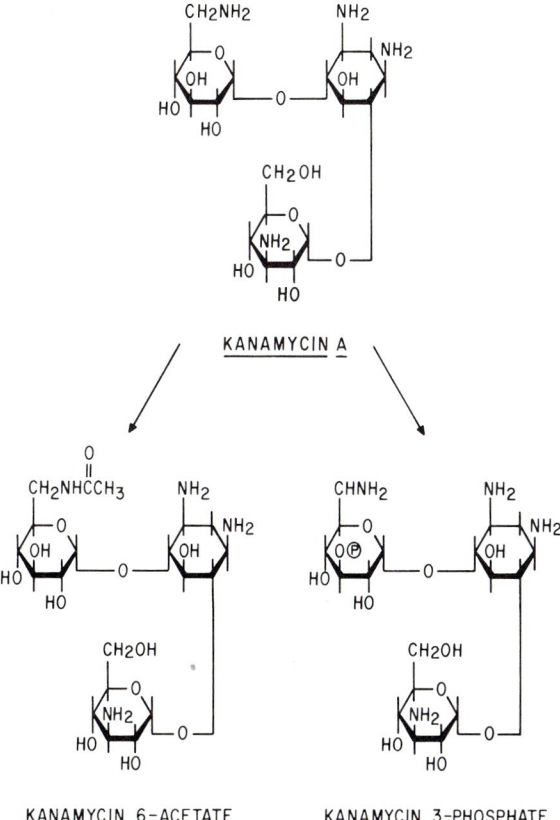

FIG. 2. Microbial inactivation of kanamycin.

An examination of the interconversions of the T-2636 antibiotics of which bundlin is a member (Kamiya et al., 1969; Harada et al., 1969) showed that acetylation of T-2636C and T-2636F by an enzyme from *Streptomyces rochei* var. *volubilis*, or crude enzymes from *Aspergillus sojae*, *A. niger*, and *Trametes sanguinea* occurred only when ethyl acetate was part of the incubation mixture (Fugono et al., 1970); ethyl formate and ethyl propionate were also suitable acyl donors.

Other acetylations reported to be carried out by growing organisms or resting cells include:
1) Conversion of spiramycin I to spiramycin II (also known as foromacidin B and acetylspiramycin) and to spiramycin III (propionylspiramycin) by cells of *Streptomyces ambofaciens* (Ninet and Verrier, 1961).
2) Acetylation of the hydroxyl group of cycloheximide by *Cunninghamella blakesleeana* (Howe and Moore, 1968).

3) Acetylation of rifamycin S (to rifamycin B) by *Streptomyces mediterranei*. This organism also deacetylates rifamycin B, as reported by Lancini *et al.* (1969) and Lancini and Hengeller (1969).

B. PHOSPHORYLATION

Phosphorylation was found to be another inactivation mechanism of aminoglycoside antibiotics by R-factor carrying bacteria (Umezawa *et al.*, 1967a; Kondo *et al.*, 1968; Okanishi *et al.*, 1968; Ozanne *et al.*, 1969; Davies *et al.*, 1969). Cell-free extracts of the R-factor carrying *E. coli* incubated with kanamycin or with paromamine in the presence of ATP and Mg^{++} resulted in phosphorylation at the C-3 hydroxyl of 6-amino-6-deoxy-D-glucose in kanamycin and of 2-amino-2-deoxy-D-glucose in paromamine (see Fig. 2).

Kanamycin was also phosphorylated by cell-free extracts from clinical isolates of drug-resistant *Staphylococcus aureus* (Doi *et al.*, 1968a) and from strains of *Pseudomonas aeruginosa* (Umezawa *et al.*, 1968a; Doi *et al.*, 1968a, 1968b, 1969).

Davies and associates (Ozanne *et al.*, 1969) obtained two enzymes located in the periplasmic region (the space between the cell wall and cell membrane) of the *E. coli* cells which phosphorylated aminoglycosides. One (streptomycin phosphotransferase) was specific for streptomycin (at the C-3 hydroxyl group of the 2-deoxy-2-methylamino-L-glucose) and the other phosphorylated neomycin B and C, paromomycin, neamine, paromamine, and kanamycins A and B. The evidence for the existence of two separate enzymes was based on single-step mutants of an *E. coli* strain carrying an R factor: the change from resistance to sensitivity was accompanied by loss of the ability to phosphorylate the antibiotics. In addition, the protein fraction which phosphorylated streptomycin, had no effect on the neomycins and kanamycins (Ozanne *et al.*, 1969). A second enzyme from this *E. coli* strain appeared to be identical with the kanamycin-phosphorylating enzyme reported by Umezawa *et al.* (Umezawa *et al.*, 1968; Kondo *et al.*, 1968; Okanishi *et al.*, 1968).

Although some reports have suggested phosphorylation of the gentamicin complex, this biological inactivation mechanism has not been conclusively demonstrated. Concomitant resistance and phosphorylation of components C_1, C_{1a}, and C_2 of the gentamicin complex by the enzymes of the R-factor carrying strain of *E. coli* have not been observed (Davies *et al.*, 1969; Ozanne *et al.*, 1969), but phosphorylation of gentamicin A was reported (Tanaka, 1970). Since gentamicin A contains the paromamine moiety (which is phosphorylated and forms paromamine 3-phosphate) and is absent in gentamicins C_1, C_{1a}, and C_2, these results are easily understood.

Streptomycin phosphate (with the phosphate group attached to the C-6 of streptidine) was isolated from *Streptomyces griseus* (Nomi *et al.*, 1967, 1969; Nimi *et al.*, 1970; Nomi and Nimi, 1969, 1970) and was also formed by cell-free extracts of *S. griseus* and *S. bikiniensis* (Miller and Walker, 1969, 1970). Neomycin phosphate has been similarly isolated from *S. fradiae* (Majumdar and Majumdar, 1970).

Lincomycin and its derivatives (4'-depropyl-4'-ethyllincomycin, S-demethyl-S-ethyllincomycin, 1'-demethyllincomycin) as well as the closely related celesticetin, were phosphorylated by *Streptomyces rochei*. The product of the lincomycin transformation was shown to be lincomycin 3-phosphate (Argoudelis and Coats, 1969). Clindamycin, the related analog, was found to be converted to clindamycin 3-phosphate by whole cells and cell-free extracts of *S. coelicolor* (Coats and Argoudelis, 1971), see Fig. 7.

C. Adenylylation and Ribonucleotide Formation

Another kind of inactivation of streptomycin by R-factor carrying *Escherichia coli* cells has been investigated in several laboratories (Okamoto and Suzuki, 1965; Umezawa *et al.*, 1968b; Okanishi *et al.*, 1968; Takazawa *et al.*, 1968; Yamada *et al.*, 1968; Davies *et al.*, 1969; Benveniste *et al.*, 1970; Harwood and Smith, 1969a,b; Smith *et al.*, 1970; Harwood *et al.*, 1969). As in the case of streptomycin phosphotransferase, the enzyme involved in this inactivation was located in the periplasma and was released from the cells suspended in sucrose-tris-EDTA buffer by osmotic shock. The biologically inactive product was identified as streptomycin adenylate in which the adenylic acid is linked to the 3-hydroxyl of the 2-deoxy-2-methylamino-L-glucose (Yamada *et al.*, 1968; Umezawa *et al.*, 1968b; Takasawa *et al.*, 1968).

The enzyme which catalyzes this reaction (streptomycin adenylate synthetase or adenylate transferase) also inactivates a few related antibiotics including dihydrostreptomycin, mannosidostreptomycin, didesaminido-dihydro-streptomycin, and bluensomycin. Although methylstreptobiosaminide was also adenylylated, streptidine, streptamine, neomycins B and C, neamine, paromomycin, paromamine, kanamycin, and gentamicin were not affected (Davies *et al.*, 1969). Mannosidostreptomycin was adenylylated at about 10 to 20% the rate of streptomycin by enzymes from *Escherichia coli, Klebsiella*, and *Salmonella typhimurium* (Schwartz and Perlman, 1970). However, the enzyme involved may differ from that which adenylylates streptomycin (Schwartz, unpublished observations).

The R-factor carrying *Escherichia coli* which adenylylates streptomycin (Davies et al., 1969) was also resistant to spectinomycin. The resistance was determined to be due to the adenylylation of the latter antibiotic, and the site of attachment of the adenylic acid to spectinomycin was located on the hydroxyl adjacent to the D-threo-methylamino hydroxyl of the actinamine moiety (Benveniste et al., 1970). Smith et al. (Harwood and Smith, 1969a,b; Harwood et al., 1969) confirmed the adenylylation of spectinomycin by a different strain of *E. coli*. These workers attempted to further define the functional group(s) responsible for the mode of action of aminoglycoside antibiotic by genetic means: On the basis of their observation that streptomycin adenylate does not cause phenotypic suppression *in vivo* and that, therefore, R-factor infected and streptomycin-adenylylating mutants of *E. coli* form clones only after genetic alteration, they developed a method of selecting streptomycin sensitive mutants of R factors which adenylylate streptomycin with 10 to 30,000 times higher frequency (Harwood et al., 1969). Of 100 natural isolates of drug-resistant enteric bacteria, 46 contained transferable R factors which mediated spectinomycin resistance, and cell-free extracts from the latter adenylylated not only spectinomycin as expected, but also dihydrospectinomycin, actinamine, streptomycin, and bluensomycin. If the adenylylation of streptomycin and spectinomycin is catalyzed by the same enzyme, the resistance in the R-factor-carrying strains to streptomycin should occur simultaneously with the resistance to spectinomycin. This was shown to be true in the 11 of the R-factor-carrying strains of *E. coli* tested. However, in those streptomycin-resistant strains which are spectinomycin-sensitive, the inactivation proceeded by phosphorylation (Ozanne et al., 1969; Benveniste et al., 1970).

Streptomyces coelicolor which was previously mentioned to phosphorylate clindamycin (Coats and Argoudelis, 1971) formed five additional biologically inactive products from this antibiotic. The purified material from clindamycin-supplemented *S. coelicolor* were identified as: clindamycin 5'-cytidylate; clindamycin 5'-adenylate; clindamycin 5'-uridylate; and clindamycin 5'-guanylate. In addition, clindamycin sulfoxide 5'-adenylate was isolated (Argoudelis and Coats, 1970a), see Fig. 7.

D. Hydrolysis

1. β-Lactam

As mentioned above, one of the first enzymatic transformations of antibiotics was the Abraham and Chain (1940) report on penicillinase which hydrolyzes the amide bond of the β-lactam ring of benzylpenicillin to penicilloic acid, an antimicrobially inactive substance. This reaction has a considerable importance in medicine, since it is the mechanism by means of which the penicillin-resistant staphylococci overcome the toxic effect of penicillin *in vivo*. All of the naturally occurring penicillins are susceptible to penicillinase, while many of the semisynthetic penicillins are resistant. Penicillinase is produced by a wide range of gram-positive and gram-negative bacterial genera. It has always been found extracellularly in the former, and cell-bound in the latter. It is inducible and its genetic determinants are located in the extrachromosomal elements (plasmids) which can be transferred by transduction (Novick and Richmond, 1965; Richmond, 1968) and by genetic transformation (Dubnau and Pollock, 1965). Excellent accounts of this topic have appeared (Pollock, 1960, 1964, 1965, 1967; Citri and Pollock, 1966) (see Fig. 3).

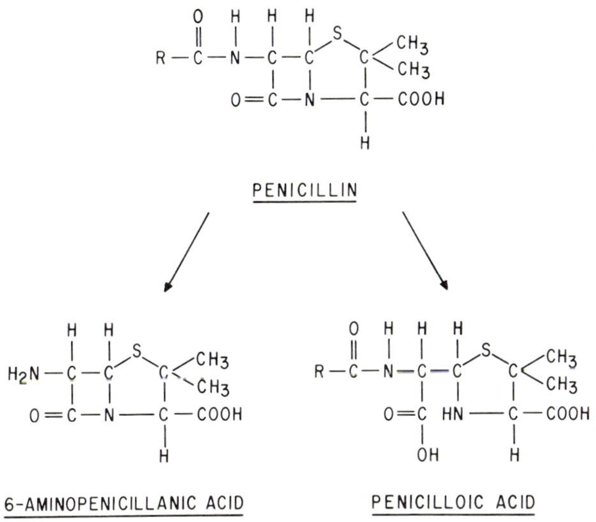

FIG. 3. Microbial inactivation of penicillin.

In an analogous way, cephalosporins are hydrolyzed to cephalosporoic acid by microbial cephalosporinase (Sabath *et al.*, 1965) which is clinically equally undesirable. As in the case of semisynthetic penicillins, novel semisynthetic cephalosporins have been prepared which are not susceptible to the action of β-lactamase (Sassiver and Lewis, 1970).

2. *Lactone*

A number of the heterodetic peptide antibiotics contain lactone bonds in addition to the peptide bonds. Enzymes from *Actinoplanes missouriensis* were found to hydrolyze these lactones in actinomycins (Fig. 4), echinomycin, etamycin, staphylomycin S, and stendomycin

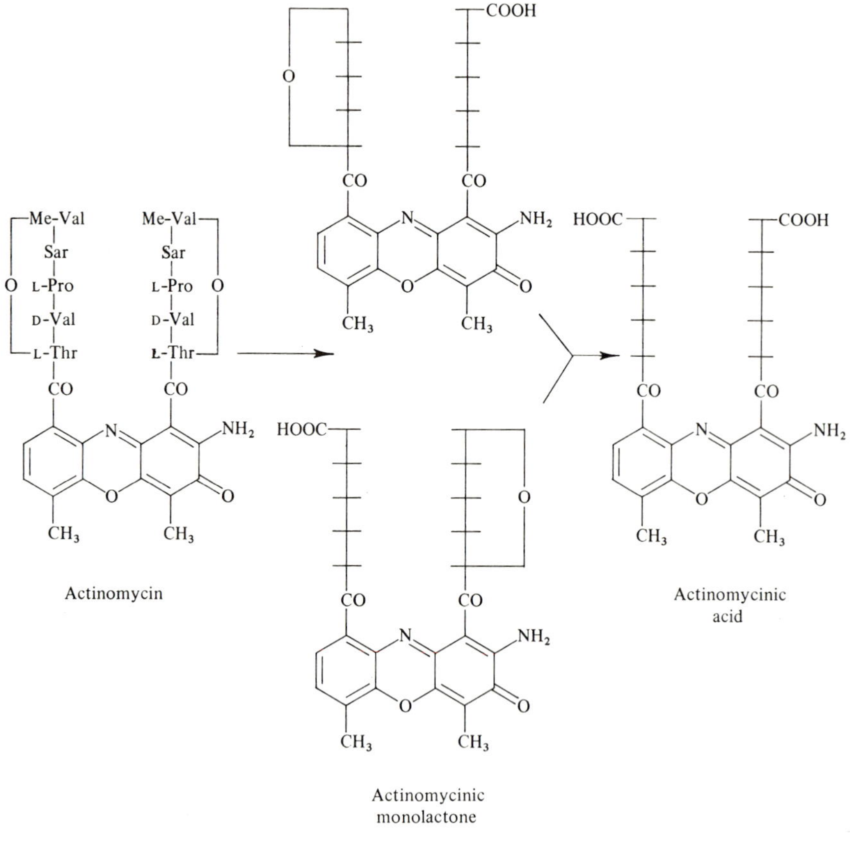

FIG. 4. Microbial oxidation of actinomycin.

(Hou and Perlman, 1970; Hou *et al.*, 1970; Perlman and Hou, 1969). The enzymes involved were purified 30–100 times and, in the case of the actinomycins, shown to be inducible.

3. Polysaccharide

Enzymatic conversion of mannosidostreptomycin to streptomycin by preparations from streptomycin-producing streptomycetes has been widely studied (Perlman and Langlykke, 1948; Shaw, 1963; Kollar, 1958; Inamine *et al.*, 1969; Hockenhull *et al.*, 1954; Demain and Inamine, 1970). Since mannosidostreptomycin may account for as much as 40% of the total antibiotic potency produced in fermentation beers, and its *in vitro* potency is only 20 to 25% that of streptomycin, the economic importance of a process converting the mannosidostreptomycin to the more useful streptomycin is obvious. Although mannosidostreptomycin was considered at one time an intermediate of streptomycin biosynthesis, more recent data obtained with mannosidase-negative mutants of *Streptomyces griseus* suggest that the antibiotic is a shunt product of streptomycin biosynthesis (Inamine *et al.*, 1969). The mannosidase itself is an inducible enzyme which is repressed by several metabolizable C sources and, under proper conditions, is synthesized toward the end of the fermentation cycle in the streptomycin-producing *S. griseus* (Hockenhull *et al.*, 1954; Kollár, 1958; Demain and Inamine, 1970).

4. Peptide

Enzymatic conversion of penicillins to the useful 6-aminopenicillanic acid (6-APA, see Fig. 3) was first reported by Sakaguchi and Murao (1950), and more extensively studied in other laboratories (Cole, 1966, 1967; Chiang and Bennett, 1967; Claridge *et al.*, 1963; Huang *et al.*, 1963; Murao and Kishida, 1961). The hydrolysis[1] has been of economic importance as the main source of 6-APA used in the preparation of the more than 10,000 semisynthetic penicillins made to date (Price, 1969). The occurrence of the enzyme (interchangeably called penicillin acylase, amidase, acyl transferase, or amidohydrolase) in bacteria and fungi is widespread, which suggests that its acyl transferring function may not be restricted to the penicillin-type structures but may be of a more general significance in microbial metabolism. In general, the enzyme exists in two distinct types: The acylases of bacterial origin hydrolyze benzylpenicillin more readily than phenoxymethylpenicillin while the reverse is true for acylases of fungal origin. Since the hydrolysis catalyzed by them, is reversible, the enzymatic synthesis of penicillin from 6-APA and suitable carboxylic acids can be carried out (cf. Cole, 1966, 1967; Demain, 1966; Hamilton-Miller, 1966; Jarvis, 1969; Jones, 1970).

In contrast, the distribution of the corresponding enzyme (cephalosporin C acylase) which hydrolyzes cephalosporin C to 7-aminocephalosporanic acid (7-ACA) has been limited only to a few species of *Brevibacterium, Achromobacter,* and *Flavobacterium* (Claridge

---

[1]Recently a chemical alternative for this hydrolysis has been described which may be of practical importance since it may displace the enzyme process (Weissburger and van der Hoeven, 1970).

et al., 1963; Demain *et al.*, 1963; Walton, 1964a,b) and none of the penicillin acylases tested carries out the cleavage. The 7-ACA required for the preparation of semisynthetic cephalosporins, is instead obtained by an efficient chemical hydrolysis of cephalosporin C (Morin *et al.*, 1962; Sassiver and Lewis, 1970).

Other hydrolyses of the antibiotic peptides have been reported which have some specificity: Nagarse was found to be effective for the cleavage of cyclic peptide gramicidin S (Yukioka *et al.*, 1966), but other peptidases were not able to attack this substrate. Hydrolysis of the peptide side chain of colistin by an enzyme from this antibiotic-producing organism (*Bacillus colistinus*) was reported (Ito *et al.*, 1966). Warren and Neilands (1964) found an enzyme in a pseudomonad which was claimed to hydrolyze circulin, bacitracin, and polymyxin into the constituent amino acids.

E. Reduction

Enzymatic reduction of the nitro group of chloramphenicol (Fig. 1) was the first microbial transformation of this molecule reported (Egami *et al.*, 1951; Smith and Worrel, 1949). The product has no antibacterial activity, and four other chloramphenicol metabolites are derived from it (Smith and Worrel, 1950), see Fig. 6.

Tylosin was converted to relomycin by washed cells of *Streptomyces hygroscopicus* and *S. griseospiralis*. This conversion (a reduction of an aldehyde) was carried out by the organisms which had the ability to hydroxylate steroids in the α position at C-16 (Feldman *et al.*, 1964).

F. Oxidation

Fusidic acid, a tetracyclic triterpene antibiotic produced by *Fusidium coccineum* was oxidized by *Corynebacterium simplex* to 3-ketofusidic acid (Dvonch *et al.*, 1966) which was also found in the fusidic acid-producing fermentation along with 3,11-diketofusidic acid (Godtfredsen *et al.*, 1966). The acid also undergoes oxidative modifications at C-6 and C-7 when incubated with *Acrocylindrium oryzae* (von Daehne *et al.*, 1968). These reactions are summarized in Fig. 5.

Hydroxylation is prominent in terminal reactions of tetracycline biosynthesis. Anhydrotetracycline is first oxidized (6-hydroxylated) by cell-free extracts of *Streptomyces aureofaciens and rimosus*, to 5a,11a-dehydrotetracycline which in turn is oxidized (5-hydroxylated) to 5a,11a-dehydro-5-hydroxytetracycline (and is then reduced to 5-hydroxytetracycline). The oxygen introduced at C-5 is most likely

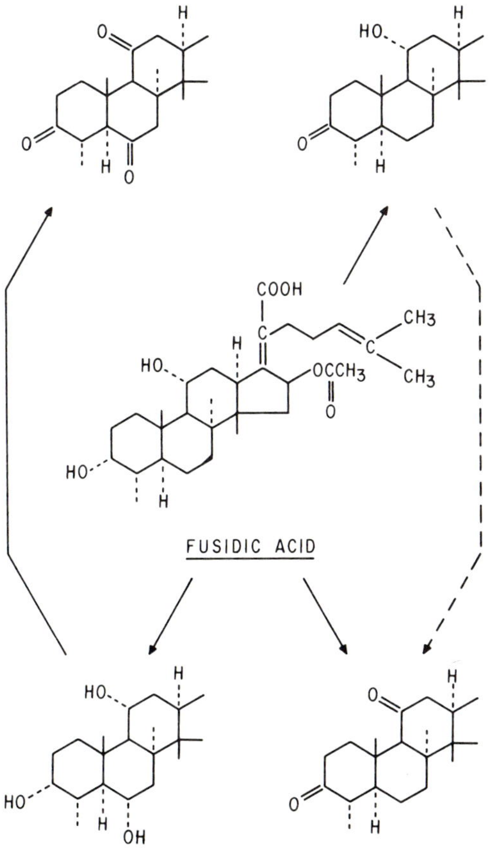

FIG. 5. Microbial oxidation of fusidic acid.

derived from molecular oxygen rather than from water since the hydroxylation required air (and $NADH_2$) to proceed (Miller et al., 1965). Essentially the same reaction sequence was found in the biosynthesis of 5-hydroxy-7-chlortetracycline from 5a,11a-dehydro-7-chlortetracycline (Mitscher et al., 1966; Martin et al., 1966).

12a-Hydroxylation of these molecules was achieved by several fungi which do not have the capability of biosynthesizing these antibiotics namely *Curvularia lunata, C. pallescens, Botrytis cinerea, Sporomia minima,* and *Thielavia terricola* (Holmlund et al., 1959b; Beck and Schull, 1961), and also by chemical means (Holmlund et al., 1959a). Several chlortetracycline-producing strains did not carry out this oxidation (Beck and Shull, 1961).

A number of bacteria oxidize the secondary hydroxyl groups of chloramphenicol (see Fig. 6) to the corresponding ketone (Smith and Worrel, 1950).

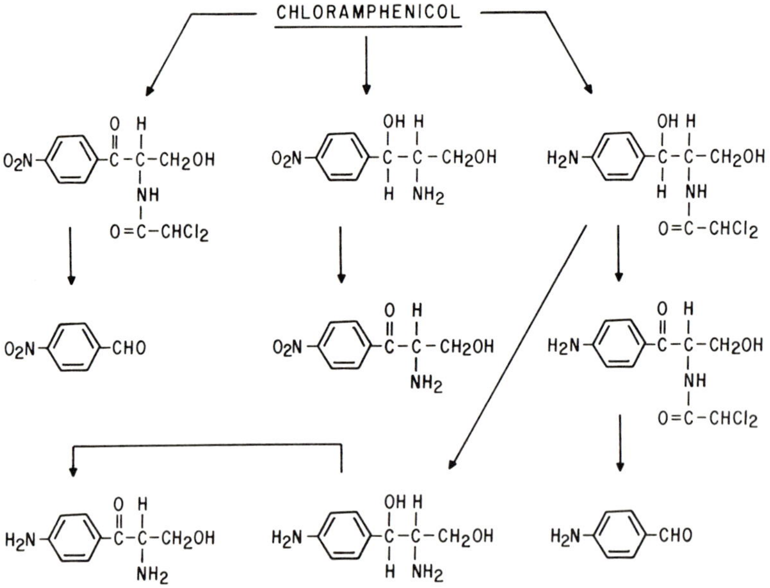

FIG. 6. Microbial degradation of chloramphenicol.

Jones et al. (1970) reported the following oxidative changes when the effect of 500 microorganisms on mycophenolic acid was investigated: (1) Oxygenation of the 4-methyl group giving an alcohol (fungi and bacteria) and an aldehyde; (2) oxygenation of the 3'-methyl group and lactonization giving a δ-lactone (fungi); (3) oxidation at C-3 giving a lactol (fungi and algae); (4) loss of the double bond and oxidation at C-4 giving a β-hydroxy acid (fungi); (5) oxygenation of the double bond giving a hydroxy-lactone (bacteria and fungi); (6) oxygenation and cyclization of the terpenoid substituent giving dihydrobenzofurans (fungi); (7) oxidative cyclization of the terpenoid substituent giving mycochromenic acid (fungi); and (8) combination of the carboxy-group with the amino groups of glycine and alanine giving amides (fungi).

Hydroxylation was also noted when griseofulvin was added to growing *Streptomyces cinereocrocatus* with the hydroxyl inserted in the 5'-position (see Fig. 8) (Andres et al., 1969).

G. SULFOXIDATION

An extended incubation of lincomycin-producing cultures of *Streptomyces lincolnensis* resulted in the formation of two products which had antibacterial spectra similar to that of lincomycin. One was identified as lincomycin sulfoxide and the other as 1-demethylthio-1-hydroxylincomycin (Argoudelis et al., 1969; Argoudelis and Mason, 1969). The sulfoxidation of clindamycin was similarly accomplished by incubating the antibiotic with S. *armentosus* and several other streptomycetes and fungi (Fig. 7; Argoudelis et al., 1969).

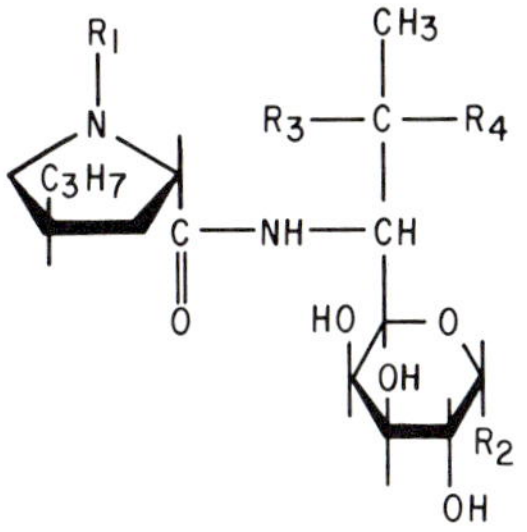

FIG. 7. Structures of lincomycin, clindamycin, and their microbially produced analogs. Lincomycin: $R_1 = CH_3$; $R_2 = SCH_3$; $R_3 = OH$; $R_4 = H$. Lincomycin sulfoxide: $R_1 = CH_3$; $R_2 = SCH_3$; $R_3 = OH$; $R_4 = H$. 1-Demethylthio-1-hydroxylincomycin: $R_1 =$
$$\downarrow$$
$$O$$
CH_3; $R_2 = R_3 = OH$; $R_4 = H$. Clindamycin: $R_1 = CH_3$; $R_2 = SCH_3$; $R_3 = H$; $R_4 = Cl$. N-Demethylclindamycin: $R_1 = R_3 = H$; $R_2 = SCH_3$; $R_4 = Cl$. Clindamycin sulfoxide: $R_1 = CH_3$; $R_2 = SCH_3$; $R_3 = H$; $R_4 = Cl$.
$$\downarrow$$
$$O$$

H. DEMETHYLATION

As mentioned in the preceding paragraph, an extended incubation of lincomycin with *Streptomyces lincolnensis* resulted in the demethylation and oxidation at C-1 of the antibiotic.

Incubation of clindamycin with *Streptomyces punipalus* brought about a demethylation at the N-1' site of the molecule and, as a result, N-demethylclindamycin was formed. Since several other streptomycetes and fungi carried out not only the N-demethylation but also sulfoxidation of clindamycin, it was postulated that the N-demethylation of this antibiotic involves oxidation of the methyl group rather than transmethylation (Fig. 7, Argoudelis et al., 1969).

Studies of the microbial transformation of griseofulvin (Boothroyd et al., 1961) showed demethylation reactions at positions C-4 (by *Microsporum canis*), C-2' (by *Botrytis allii*), and C-6 (by *Cercospora melonis*). Since the demethylated griseofulvins did not account for all of the metabolized substrate, it was concluded that they represented an initial stage of a more extensive degradation to less easily recognized products. A similar suggestion was made for the decomposition of the antibiotic by a pseudomonad and a few fungi including *Trichophyton mentagrophytes* and *T. persicolor* which are sensitive to griseofulvin (Abbott and Grove, 1959; El-Nakeeb and Lampen, 1965) (see Fig. 8 for structures).

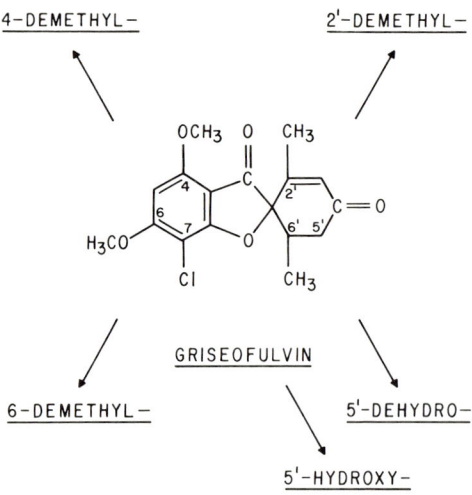

FIG. 8. Microbial modification of griseofulvin.

I. Deamination

Pyrrolopyrimidine nucleoside antibiotics which have unusual biological properties and antitumor activities, have been shown to be susceptible to microbial attack. Toyocamycin was converted to sangivamycin by toyocamycin nitrile hydratase, an enzyme isolated from the unidentified streptomycete producing both antibiotics (Rao and Renn, 1963). This enzyme was not found in *Streptomyces toyocanesis* or in *S. tubercidicus* (Suhadolnik, 1970a,b).

Puromycin aminonucleoside, a fragment of the puromycin molecule, has not antibacterial activity but is 3 to 4 times more active than puromycin itself in inhibiting *Trypanosoma equiperdum* and suppresses growth of transplantable mammary carcinoma in mice. This puromycin fragment is degraded by a ferrichrome-requiring *Arthrobacter flavescens*. The first step in the decomposition is the removal of the dimethylamine moiety from the 6-position of the purine, thus yielding 3-amino-3-deoxyriboside of hypoxanthine which in turn is believed to be deaminated and subsequently cleaved to hypoxanthine and ribose (Greenberg and Barker, 1962).

Formycins are C-nucleoside antibiotics which are deaminated[2] by enzyme mixtures prepared from certain streptomycetes and aspergilli, and by purified adenosine deaminase. Thus, formycin [7-amino-3-(β-D-ribo-furanosyl)-pyrazolo(4,3-d)pyrimidine] was converted to formycin B by *Nocardia interforma* (the formycin producer), by an *Aspergillus*, by streptomycetes, and by *Escherichia coli* as well as Takadiastase (commercial preparation of *A. oryzae*) (Umezawa et al., 1965). Formycin B may undergo further oxidation to 5-oxoformycin B when incubated with low formycin-producing strains of *N. interforma*, and by *Pseudomonas fluorescens, Xanthomonas oryzae*, or *Streptomyces kasugaensis*. It can also be reconverted to the original formycin A by *N. interforma* and, to a limited extent, by *X. oryzae*. Oxoformycin B appears to be an irreversible end product of these transformations (Fig. 9) since it was resistant to the action of several microorganisms (Sawa et al., 1968).

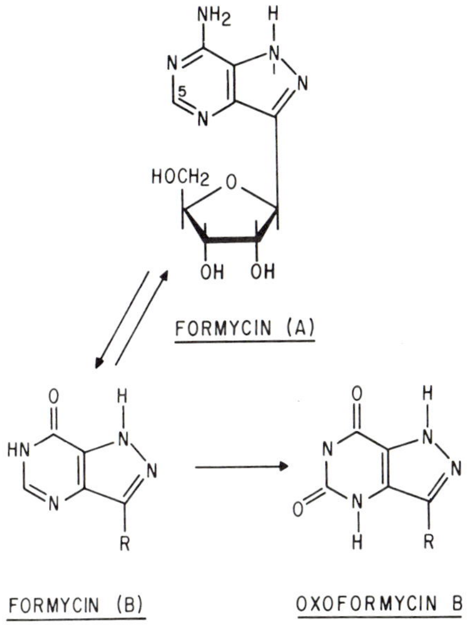

FIG. 9. Microbial transformation of formycin.

Purified adenosine deaminase (from *Aspergillus niger*) (Fukagawa et al., 1965) and whole cells of *Escherichia coli* (Herr and Murray, 1967) oxidized 3'-deoxyadenosine (cordycepin) to 3'-deoxyinosine.

---

[2] The term "deamination," namely the replacement of an aromatic amino group by a hydroxyl is not accurate since mechanistically the reaction (at least in the case of purines and pyrimidines) is hydrolytic but the net result may be considered an oxidation.

J. Nonspecific Degradation of Antibiotics

Several antibiotics have been reported to serve as the only source of C, N, and energy for bacterial growth, or to serve as growth factors for antibiotic-dependent bacteria.

Chloramphenicol served as the sole substrate for growth of an aerobic *Flavobacterium* which degraded the antibiotic through a sequence of several intermediates, of which eleven were identified (Lingens *et al.*, 1966), see Fig. 6.

Unidentified species of penicillia and fusaria utilized chloramphenicol, penicillin, and oxytetracycline for growth but no identification of the degradation products was made (Nissen, 1954).

Although the microbial inactivation of actinomycin was suggested in the early studies on this antibiotic, an enzymatic degradation of actinomycin D (and—to some extent—also of B and C) by growing and resting cells, and by acetone powder preparations of *Achromobacter* sp. was demonstrated in 1957. Crude solvent extractable materials were isolated but not identified (Katz and Pienta, 1957).

Streptomycin was shown to be a growth substrate for soil pseudomonads (Pramer and Starkey, 1951), and it was suggested that individual strains decomposed the antibiotic in different ways. Intermediates of the degradation were isolated and characterized as urea and streptamine (Klein and Pramer, 1961, 1962). It was evident that the bacterial dissimilation of streptomycin involved hydrolysis of the guanido groups of the streptidine.

In a study with a neamine-dependent *Staphylococcus aureus* it was noted that those antibiotics which contain aminohexoses and amino groups at C-2 of the hexose (e.g., streptomycin, paromomycin, zygomycin, and neomycin C) could substitute for the neamine, while those with an amino group at C-3 (e.g., kanamycin, erythromycin) or no aminohexoses (e.g., spectinomycin, vancomycin) did not support growth of the coccus. Similar results were obtained with streptomycin-dependent *S. aureus* and *Salmonella paratyphi* B but not with streptomycin-dependent strains of *Escherichia coli* and mycobacteria (Sokolski *et al.*, 1962). The components of the streptomycin molecule as well as streptomycin derivatives were also able to substitute for growth of streptomycin- and neamine-dependent mutants of *E. coli* (Szybalski and Cocito-Vandermeulen, 1958). These and other reports (Goldstein, 1954) are indicative of the modification of the respective antibiotics but they do not identify the changes in chemical or enzymatic terms.

Meyers and Smith (1962) reported that microbiological degradation of tetracyclines was accomplished by a *Xylaria*, but were unable to identify any of the products of the inactivation. Similarly, the mitomycin-producing streptomycete *(Streptomyces caespitosus)* elaborates enzymes capable of inactivating this antitumor antibiotic, but the degradation product has not been identified (Gourevitch, 1961).

III. Techniques Useful in Studying Microbial Transformations of Antibiotics

An inspection of reports on microbial transformations of antibiotics leads to the conclusion that a variety of techniques have been employed in selecting a suitable organism and monitoring the degradation. It is obvious that the selection of a particular technique depends on the objective of the proposed study, and on the background, training, preference as well as ingenuity, skill, and special interest of the investigator. It is then easily understood that an ecologist or a soil microbiologist will be likely to use techniques different from those employed by a microbial physiologist, enzymologist, biochemist, organic or physical chemist, or a molecular biologist. In each instance, however, the data collected will contribute some additional knowledge to the phenomenon of biological stability and susceptibility of the antibiotic.

The classical enrichment culture techniques (cf. Schlegel, 1965; Veldkamp, 1970) have been applied to antibiotic transformations. In most cases, the closed system has been used wherin antibiotics, the degradation of which was sought, were added to soil samples, incubated, and their potency was determined at definite time intervals by suitable methods (Witkamp and Starkey, 1957). If the bioactivity in these soils decreased more rapidly than in control (sterilized) soils, such change suggested microbial inactivation. Other factors, however, must be considered before more definitive conclusions are drawn. Some antibiotics (especially albidin, frequentin, gliotoxin, penicillin, viridin) have been reported to be inactivated due to their intrinsic instability at the natural pH of the particular soil, while others, e.g., streptomycin, were inactivated by absorption on clay minerals or organic matter of the soil (Pramer, 1958). Only griseofulvin, mycophenolic acid, and patulin appeared to be biologically degraded (Brian, 1958; Jefferys, 1952).

Since under the conditions of soil enrichment, antibiotics may serve a dual function, namely that of providing C, N, and energy for growth, and also interfering with the life processes of the cells themselves, enrichment of the open type is sometimes preferred. Under such conditions, nontoxic (subinhibitory) constant antibiotic concentrations are maintained while the metabolic products are continuously removed.

Investigators not primarily interested in ecologically important organisms, are usually satisfied to employ pure cultures for their degradative abilities. More or less randomly chosen microorganisms are grown in complex media which promote rapid and heavy cell growth. The antibiotic is then added, incubation is continued for various lengths of time, and the mixture is examined for any change in the characteristics (e.g., antimicrobial activity) that might have occurred. In some respects, this method is patterned after those used in the study of steroid transformations (Čapek et al., 1966; Charney and Herzog, 1967).

Since nearly all antibiotic-producing microorganisms have the ability to produce families of closely related antibiotics, it has been rewarding to search among these organisms for enzymatic ability to transform these antibiotics. Phosphorylation of streptomycin (Nomi and Nimi, 1970; Miller and Walker, 1970), phosphorylation of neomycin (Majumdar and Majumdar, 1970), acetylation of spiramycin (Ninet and Verrier, 1962), modification of lincomycin (Argoudelis and Mason, 1969; Argoudelis et al., 1969), and hydrolysis of mannosidostreptomycin (Perlman and Langlykke, 1948) are among the transformations carried out by the antibiotic-producing cultures. Thus instead of expenditure of time and effort to isolate organisms from soil and other natural habitats, it may be more profitable to search among the antibiotic-producing culture for mutants with the ability to degrade or transform the antibiotic of interest.

Biological activity of an antibiotic is due to its unique molecular structure with functional groups arranged in specific spatial arrangements. It is common experience that even a very minor chemical change of such structure results in a drastic decrease or complete loss of activity. Only rarely an increase in biopotency has been found. Hence, a convenient method to detect changes due to the microbial action is to monitor the bioactivity by a microorganism sensitive to the substrate antibiotic. Since the possibility exists that bioactive products may also be formed (cf. Argoudelis et al., 1969), the traditional agar diffusion disc-plate assay method should be supplemented or completely replaced by paper or thin-layer chromatography. Use of the latter results in separation and purification of the metabolic products for further investigation. If it is possible to obtain the substrate antibiotic in labeled form (^{14}C and ^{3}H), much time and effort can be saved by scanning the various spots on the chromatogram for radioactivity and discarding those which do not contain the label. Once isolated and purified, the compound may be identified and characterized by conventional chemical and physical methods.

Inactivation of the aminoglycoside antibiotics, e.g., streptomycin, kanamycin, etc., by means of cell-free preparations of the R-factor carrying bacteria has led to the expected conclusion that several closely related enzymes obtained from different organisms may carry out the same type of transformation. Each has some substrate specificity and it has been possible to obtain mutants which produce (or are devoid of) the desired enzyme.

There are other ways to study biological transformations of a chosen antibiotic: One can gather from the literature information on strains of microorganisms likely to carry out a certain type of transformation. If the basic structure of the antibiotic resembles a steroid, those organisms useful in transformations of steroids might be tested first. If the antibiotic is a peptide, organisms reported to modify peptides might be examined, etc. Alternatively, organisms known to be resistant to the antibiotic can be examined to determine whether their resistance is due to formation of a derivative or antibiotic degradation. And finally, organisms selected at random can be evaluated to determine whether they have unsuspected abilities. It is also possible to utilize the antibiotic-producing organisms themselves since some of them are known to elaborate enzymes which destroy the antibiotic they have synthesized.

IV. Summary

In spite of the marked success in utilizing microorganisms to prepare intermediates for the chemical synthesis of steroids, there has been less interest in using microbial transformations of known antibiotics to prepare new and therapeutically useful ones. A measure of success has been noted only in the hydrolysis of benzylpenicillin to 6-aminopenicillanic acid, and of mannosidostreptomycin to the clinically more interesting streptomycin. Many other antibiotics have been modified (and thereby inactivated) or degraded by microbial enzymes, but no practical application has been found for most of the products.

Once a suitable microorganism has been isolated, genetic techniques can be employed to select organisms with the desirable abilities. Thus, it should be possible to generate novel and chemotherapeutically valuable compounds, and to add a new chapter to the exciting story of antibiotic research.

References

Abbot, M. T. J., and Grove, J. F. (1959). *Exp. Cell. Res.* **17**, 105–113.
Abraham, E. P. (1951). *In* "The Enzymes" (K. Myrbäck and J. B. Sumner, eds.), 1st Ed., Ch. 37. Academic Press, New York.
Abraham, E. P., and Chain, E. (1940). *Nature (London)* **146**, 837.
Andres, W. W., MacGahren, W. J., and Kunstmann, M. P. (1969). *Tetrahedron Lett.* pp. 3777–3780.
Argoudelis, A. D., and Coats, J. H. (1969). *J. Antibiot.* **22**, 341–343.
Argoudelis, A. D., and Coats, J. H. (1970a). *Abstr. 10th Intersci. Conf. Antimicrob. Ag. Chemother.* pp. 23–24.
Argoudelis, A. D., and Coats, J. H. (1970b). *Abstr. 10th Intersci. Conf. Antimicrob. Ag. Chemother.* p. 24.
Argoudelis, A. D., and Mason, D. J. (1969). *J. Antibiot.* **22**, 289–291.
Argoudelis, A. D., Coats, J. H., Mason, D. J., and Sebek, O. K. (1969). *J. Antibiot.* **22**, 309–314.
Beck, D., and Shull, G. M. (1961). US. Patent No. 2,970,087.
Benveniste, R., Yamada, T., and Davies, J. (1970). *Infec. Immunity* **1**, 109–119.
Blackwood, R. K., and English, A. R. (1970). *Advan. Appl. Microbiol.* **13**, 237–266.
Bonanchaud, D. (1967). *Ann. Inst. Pasteur* **113**, 59–66.
Boothroyd, B., Napier, E. J., and Sommerfield, G. A. (1961). *Biochem. J.* **80**, 34–37.
Brian, P. W. (1958). *In* "Microbial Ecology" 17th Symposium of the Society for General Microbiology (R. E. O. Williams, and C. C. Spicer, eds.), pp. 168–188. Cambridge Univ. Press, London and New York.
Čapek, A., Hanč, O., and Tadra, M. (1966). "Microbial Transformations of Steroids." Academia, Prague.
Charney, W., and Herzog, H. L. (1967). "Microbial Transformations of Steroids. A Handbook." Academic Press, New York.
Chiang, C., and Bennett, R. E. (1967). *J. Bacteriol* **93**, 302–308.
Citri, N., and Pollock, M. R. (1966). *Advan. Enzymol. Relat. Subj. Biochem.* **28**, 237–323.
Claridge, C. A., Luttinger, J. R., and Lein, J. (1963). *Proc. Soc. Exp. Biol. Med.* **113**, 1008–1012.
Coats, J. H., and Argoudelis, A. D. (1971). *J. Bacteriol.* (in press).
Cole, M. (1966). *Process Biochem.* **1**(6), 334–338; **1**(7), 373–377.
Cole, M. (1967). *Process Biochem.* **2**(4), 35–41.
Cron, M. J., Godfrey, J. C., Hooper, I. R., Keil, J. G., Nettleton, D. E., Price, K. E., and Schmitz, H. (1970a). *Progr. Antimicrob. Anticancer Chemother.* **2**, 1069–1082.
Cron, M. J., Smith, R. E., Hooper, I. R., Keil, J. G., Ragan, E. A., Schreiber, R. H., Schwab, G., and Godfrey, J. C. (1970b). *Antimicrob. Ag. Chemother.* 1969, pp. 219–224.
Davies, J., Benveniste, R., Kvitek, K., Ozanne, B., and Yamada, T. (1969). *J. Infec. Dis.* **119**, 351–354.
Demain, A. L. (1966). *In* "Biosynthesis of Antibiotics" (J. F. Snell, ed.), Vol. 1, pp. 29–94. Academic Press, New York.
Demain, A. L., and Inamine, E. (1970). *Bacteriol. Rev.* **34**, 1–19.
Demain, A. L., Walton, R. B., Newkirk, J. F., and Miller, I. M (1963). *Nature (London)* **199**, 909–910.
Doi, O., Kiyamoto, M., Tanaka, N., and Umezawa, H. (1968a). *Appl. Microbiol.* **16**, 1282–1284.

Doi, O., Ogura, M , Tanaka, N., and Umezawa, H. (1968b). *Appl. Microbiol.* **16**, 1276-1281.
Doi, O., Kondo, S., Tanaka, N., and Umezawa, H. (1969). *J. Antibiot.* **22**, 273-282.
Dubnau, D. A., and Pollock, M. R. (1965). *J. Gen. Microbiol.* **41**, 7-21.
Dvonch, W., Greenspan, G., and Alburn, H. E. (1966). *Experientia* **22**, 517.
Egami, F., Ebata, M., and Sato, R. (1951). *Nature (London)* **167**, 118-119.
El-Nakeeb, M. A., and Lampen, J. O. (1965). *J. Gen. Microbiol.* **39**, 285-293.
Feldman, L. I., Dill, I. K., Holmlund, C. E., Whaley, H.A., Patterson, E. L., and Bohonos, N. (1964). *Antimicrob. Ag. Chemother.* 1963, pp. 54-57.
Fugono, T., Higashide, E., Suzuki, T., Yamamoto, H., Harada, S., and Kishi, T. (1970). *Experientia* **26**, 26.
Fukagawa, Y , Sawa, T., Takeuchi, T., and Umezawa, H. (1965). *J. Antibiot.* **18**, 191.
Godtfredsen, W. O., von Daehne, W., Tybring, L., and Vangedal, S. (1966). *J. Med. Chem.* **9**, 15-22.
Goldstein, A. (1954). *J. Pharmacol. Exp. Ther.* **112**, 326-340.
Goll, P. H. (1966). *Proc. Biochem.* **1**(4), 201-205.
Gourevitch, A. (1961). *Arch. Biochem. Biophys.* **93**, 283-285.
Greenberg, J., and Barker, H. A. (1962). *Biochim. Biophys. Acta* **61**, 71-74.
Hamilton-Miller, J. M. T. (1966). *Bacteriol. Rev.* **30**, 761-771.
Harada, S., Migashide, E., Fuguno, T., and Kishi, T. (1969). *Tetrahedron Lett.* pp. 2239-2244.
Harwood, J. H., and Smith, D. H. (1969a). *J. Bacteriol.* **97**, 1262-1282.
Harwood, J. H., and Smith, D. H. (1969b). *Genet. Res.* **14**, 259-273.
Harwood, J. H , Janjigian, J., and Smith, D. H. (1969). *Genet. Res.* **14**, 323-327.
Herr, M. E., and Murray, H. C. (1967). Unpublished observations.
Hockenhull, D. J. D , Ashton, G. C., Fantes, K H , and Whitehead, B. K (1954). *Biochem. J.* **57**, 93-98.
Holmlund, C. E., Andres, W. W., and Shay, A. J. (1959a). *J. Amer. Chem. Soc.* **81**, 4748-4749.
Holmlund, C. E., Andres, W. W., and Shay, A. J. (1959b). *J. Amer. Chem. Soc.* **81**, 4750-4751.
Hou, C. T., and Perlman, D. (1970). *J. Biol. Chem.* **245**, 1289-1295.
Hou, C. T., Perlman, D., and Schallock, M. R. (1970). *J. Antibiot.* **23**, 35-42.
Howe, R., and Moore, R. H. (1968). *Experientia* **24**, 904.
Huang, H T., Seto, T. A., and Shull, G. M. (1963). *Appl. Microbiol.* **11**, 1-6.
Iizuka, H., and Naito, A. (1968). "Microbial Transformation of Steroids and Alkaloids." Univ. of Tokyo Press, Tokyo.
Inamine, E., Lago, B. D., and Demain, A. L. (1969). *In* "Fermentation Advances" (D. Perlman, ed.), pp. 199-221. Academic Press, New York.
Ito, M., Aida, T., and Koyama, Y. (1966). *Agr. Biol. Chem.* **30**, 1112-1118.
Jarvis, B. (1969). *Chem. Ind. (London)* pp. 1721-1724.
Jefferys, E. G. (1952). *J. Gen. Microbiol.* **7**, 295-312.
Jones, D. F., Moore, R. H., and Crawley, G. C. (1970). *J. Chem. Soc. C* pp. 1725-1737.
Jones, R. D. (1970). *Amer. Sci.* **58**, 404-411.
Kamiya, K., Harada, S., Wada, Y., Nishikawa, M., and Kishi, T. (1969). *Tetrahedron Lett.* pp. 2245-2248.
Katz, E., and Pienta, P. (1957). *Science* **126**, 402-403.
Kinoshita, S., Uzu, K., Nakano, K., Shimizu, M., Takahashi, T., Wakaki, S., and Matsui, M. (1970). *Progr. Antimicrob. Anticancer Chemother.* **2**, 1058-1068.
Klein, D., and Pramer, D. (1961). *J. Bacteriol.* **82**, 505-510.
Klein, D., and Pramer, D. (1962). *J. Bacteriol.* **83**, 309-313.

Kollár, G. (1958). *Acta Microbiol.* **5,** 11-17.
Kondo, S., Okanishi, M., Utahara, R., Maeda, K , and Umezawa, H (1968). *J. Antibiot.* **21,** 22-29.
Lancini, G. C., and Hengellar, C. (1969). *J. Antibiot.* **22,** 637-638.
Lancini, G. C., Gallo, G. G., Sartori, G., and Sensi, P. (1969). *J. Antibiot.* **22,** 369-372.
Lingens, F., Eberhardt, H., and Oltmanns, O. (1966). *Biochim. Biophys. Acta* **130,** 345-354.
McCormick, J. R. D., Miller, P. A., Johnson, S., Arnold, N., and Sjolander, N. O. (1962). *J. Amer. Chem. Soc.* **84,** 3023-3025.
Maeda, K., Kondo, S., Okanishi, M., Utahara, R., and Umezawa, H. (1968). *J. Antibiot.* **21,** 458-459.
Magerlein, B. J. (1971). *Advan. Appl. Microbiol.* **14.**
Majumdar, M. K., and Majumdar, S. K. (1970). *Biochem. J.* **120,** 271-278.
Martin, J. H., Mitscher, L. A., Miller, P. A., Shu, P., and Bohonos, N. (1966). *Antimicrob. Ag. Chemother.* 1966, pp. 563-567.
Meyers, E., and Smith, D. A. (1962). *J. Bacteriol.* **84,** 797-802.
Miller, A. L., and Walker, J. B. (1969). *J. Bacteriol.* **99,** 401-405.
Miller, A. L., and Walker, J. B. (1970) *J. Bacteriol.* **104,** 8-12.
Miller, P. A., Hash, J. A., Lincks, M., and Bohonos, N. (1965). *Biochem. Biophys. Res. Commun.* **18,** 325-331.
Mitscher, L. A., Martin, J. H., Miller, P. A., Shu, P., and Bohonos, N. (1966). *J. Amer. Chem. Soc.* **88,** 3647-3648.
Morin, R. B., Jackson, B. G., Flynn, E. H., and Roeske, R. W. (1962). *J. Amer. Chem. Soc.* **84,** 3400-3401.
Murao, S., and Kishida, Y. (1961). *Nippon Nogei Kagaku Kaishi* **35,** 607-610.
Nimi, O., Kiyohara, H., Mizoguchi, T., Ohata, Y , and Nomi, R. (1970). *Agr. Biol. Chem.* **34,** 1150-1156.
Ninet, L., and Verrier, J. (1961). Fr. Patent No. 1,262,571.
Nissen, T. V. (1954). *Nature (London)* **174,** 226-227.
Nomi, R., and Nimi, O. (1969). *Agr. Biol. Chem.* **33,** 1459-1463.
Nomi, R., and Nimi, O. (1970). *Progr. Antimicrob. Anticancer Chemother.* **2,** 1088-1093.
Nomi, R., Nimi, O., Miyazaki, T., Matsuo, A., and Kiyohara, H (1967). *Agr. Biol. Chem.* **31,** 973-978.
Nomi, R., Nimi, O., and Katho, T. (1969). *Agr. Biol. Chem.* **33,** 1454-1458.
Novick, R. P., and Richmond, M. H. (1965). *J. Bacteriol.* **90,** 467-480.
Okamoto, S., and Suzuki, Y. (1965). *Nature (London)* **208,** 1301-1303.
Okanishi, M., Kondo, S., Suzuki, Y., Okamoto, S., and Umezawa, H. (1967). *J. Antibiot., Ser. A* **20,** 132-135.
Okanishi, M , Kondo, S., Utahara, R., and Umezawa, H. (1968). *J. Antibiot.* **21,** 13-21.
Ozanne, B., Benveniste, R., Tipper, D., and Davies, J. (1969). *J. Bacteriol.* **100,** 1144-1146.
Perlman, D., and Hou, C. T. (1969). *Appl. Microbiol.* **18,** 272-273.
Perlman, D., and Langlykke, A. F. (1948). *J. Amer. Chem. Soc.* **70,** 3968.
Piffaretti, J. C., and Pitton, J. S. (1970). *Chemotherapy* **15,** 84-98.
Pollock, M. R. (1960). *In* "The Enzymes" (P. D. Boyer, H. A. Lardy, and K. Myrbäck, eds.), 2nd Ed., Vol. 4, pp. 269-278. Academic Press, New York.
Pollock, M. R. (1964). *Antimicrob. Ag. Chemother.* 1964, pp. 292-301.
Pollock, M R. (1965). *Biochem. J.* **94,** 666-675.
Pollock, M. R. (1967). *Brit. Med. J.* **4,** 71-77.
Pramer, D. (1958). *Appl. Microbiol.* **6,** 221-224.
Pramer, D., and Starkey, R. L (1951). *Science* **113,** 127.

Price, K. E. (1969). *Advan. Appl. Microbiol.* **11**, 17–75.
Rao, K. V., and Renn, R. W. (1963). *Antimicrob. Ag. Chemother.* 1963, pp. 77–79.
Richmond, M. H. (1968). *Adv. Microbial Physiol.* **2**, 43–88.
Sabath, L. D., Jago, M, and Abraham, E. P. (1965). *Biochem. J.* **96**, 739–752.
Sakaguchi, K., and Murao, S. (1950). *Nippon Nogei Kagako Kaishi* **23**, 411–413.
Sassiver, M. L., and Lewis, A. (1970). *Advan. Appl. Microbiol.* **13**, 163–236.
Sawa, T., Fukugawa, Y., Homma, I., Takeuchi, T., and Umezawa, H (1967). *J. Antibiot., Ser. A* **20**, 317–321.
Sawa, T., Fukagawa, Y, Homma, I., Wakashiro, T., Takeuchi, T., and Hori, M. (1968). *J. Antibiot.* **21**, 334–339.
Schlegel, H. G., ed. (1962). "Anreicherungskultur und Mutantenauslese." Fischer, Stuttgart.
Schwartz, J. L, and Perlman, D. (1970). *J. Antibiot.* **23**, 254.
Sensi, P., Maggi, N., Füresz, S., and Maffii, G. (1967). *Antimicrob. Ag. Chemother.* 1966, pp. 697–714.
Shaw, R. K. (1963). *J. Appl. Bacteriol.* **26**, 249–252.
Shaw, W. V. (1967). *J. Biol. Chem.* **242**, 687–693.
Shaw, W. V. (1970). *Progr. Antimicrob. Anticancer Chemother.* **2**, 552–555.
Shaw, W. V, and Brodsky, R. F. (1968a). *Antimicrob. Ag. Chemother.* 1967, pp. 257–263.
Shaw, W. V., and Brodsky, R. F. (1968b). *J. Bacteriol.* **95**, 28–36.
Shaw, W. V., Bentley, D. W., and Sands, L. (1970). *J. Bacteriol.* **104**, 1095–1105.
Smith, D. H., Janjigian, J. A., Prescott, N., and Anderson, P. W. (1970). *Infec. Immunity* **1**, 120–127.
Smith, G. N., and Worrel, C. S. (1949). *Arch. Biochem.* **24**, 216–223.
Smith, G. N., and Worrel, C. S. (1950). *Arch. Biochem.* **28**, 232–241.
Smith, G. N., and Worrel, C. S. (1953). *J. Bacteriol.* **65**, 313–317.
Sokolski, W. T., Yeager, R. L., and Chidester, C. G. (1962). *Nature (London)* **196**, 776–777.
Suhadolnik, R. J. (1970a). *Abstr. 1st Symp. Genet. Ind. Microorganisms* p. 162. Czechoslovak Academy of Sciences, Prague.
Suhadolnik, R. J. (1970b). "Nucleoside Antibiotics," p. 327. Wiley (Interscience), New York.
Suzuki, Y., and Okamoto, S. (1967). *J. Biol. Chem.* **242**, 4722–4730.
Suzuki, Y., Okamoto, S., and Kono, M. (1966). *J. Bacteriol.* **92**, 798–799.
Szybalski, W., and Cocito-Vandermeulen, J. (1958). *Bacteriol. Proc.* p. 37.
Takasawa, S., Utahara, R., Okanishi, M, Maeda, K, and Umezawa, H. (1968). *J. Antibiot.* **21**, 477–484.
Tanaka, N. (1970). *J. Antibiot.* **23**, 469–471.
Umezawa, H, Sawa, T., Fukagawa, Y., Koyama, G., Murase, M., Hamada, M., and Takeuchi, T. (1965). *J. Antibiot., Ser. A* **18**, 178–181.
Umezawa, H, Okanishi, M, Kondo, S., Hamana, K., Utahara, R., Maeda, K., and Mitsuhashi, S. (1967a). *Science* **157**, 1559–1561.
Umezawa, H., Okanishi, M., Utahara, R., Maeda, K., and Kondo, S. (1967b). *J. Antibiot., Ser. A.* **20**, 136–141.
Umezawa, H., Doi, O., Ogura, M, Kondo, S., and Tanaka, N. (1968a). *J. Antibiot.* **21**, 154-155.
Umezawa, H., Takazawa, S., Okanishi, M., and Utahara, R. (1968b) *J. Antibiot.* **21**, 81–82.
Veldkamp, H. (1970). *In* "Methods in Microbiology" (J. R. Norris and D. W. Ribbons, eds.), Vol. 3A, pp. 305–361.

von Daehne, W., Lorch, H., and Godtfredsen, W. O. (1968). *Tetrahedron Lett.* pp. 4843–4846.
Walton, R. B. (1964a). Fr. Patent No. 1,357,977.
Walton, R. B. (1964b). *Develop. Ind. Microbiol.* **5,** 349–353.
Warren, R. A. J., and Neilands, J. B. (1964). *J. Gen. Microbiol.* **35,** 459–470.
Weissburger, H. W. O., and van der Hoeven, M. G. (1970). *Rec. Trav. Chim. Pay-Bas Belg.* **89,** 1081–1084.
Witkamp, M., and Starkey, R. L. (1957). *Soil Sci. Soc. Amer., Proc.* **20,** 500–504.
Yamada, T., Tipper, D., and Davies, J. (1968). *Nature (London)* **209,** 288–291.
Yukioka, M., Saito, Y., and Otani, S. (1966). *J. Biochem. (Tokyo)* **60,** 295–302.

In Vivo Evaluation of Antibacterial Chemotherapeutic Substances

A. KATHRINE MILLER

*Merck Institute for Therapeutic Research,
Rahway, New Jersey*

I.	Historical Background	151
II.	*In Vivo* Test Procedures	152
	A. General Principles	152
	B. Types of Tests	165
III.	Special Tests	171
	A. Synergy	171
	B. Development of Resistance *in Vivo*	175
	C. Others (Dental Caries, L-Forms, Aerosol Challenges)	178
IV.	Epilog	179
	References	179

I. Historical Background

Chemotherapy as an art has been practiced for centuries. Herbal remedies for human illnesses made use of plants, some of which contained drugs prescribed today. The development of medical practice in the Middle Ages was accompanied not only by the use of many strange and probably useless remedies, but also by the introduction of metallic agents as medicines. These became almost too popular and their overuse undoubtedly resulted in additional illness! Eventually the germ theory of disease was established and some scientific methods were practiced to assess the value of drugs in the treatment of disease. By the late 1800s, chemotherapy gradually changed from an art to a science. Although many workers contributed significantly to this change, Paul Ehrlich has been credited as the founder of the science of chemotherapy (Work and Work, 1948; Sophian *et al.*, 1952; Hawking, 1963). Certainly he developed criteria for determining the effectiveness of chemical compounds being tested *in vivo*. He sought for agents that had a stronger affinity for the microorganism than for the host tissue, and those that had the greatest spread between the curative and the toxic dose (Ehrlich, 1913).

Ehrlich worked mostly with protozoa. Morganroth and Levy (1911) were probably the first to report the cure of an experimental bacterial infection. They found that the quinine derivative ethylhydrocuprein protected mice against pneumococci. Later, other bacterial infections were used experimentally, particularly as the development of sulfonamides and antibiotic agents stimulated studies of the chemotherapy of systemic infections (Hoerlein, 1937; Herrell, 1970).

The work that has been done with specific infections or with certain therapeutic problems or substances has been the subject of many reviews. However, there seem to be no reviews of the many test procedures that have been developed during the course of these studies. The present paper attempts to provide such a survey. Admittedly, not every test method, or even every type of test procedure that has been described has been included, and many fine references undoubtedly have been overlooked. We regret and apologize for any omissions. Hopefully, these references that have been mentioned, and the discussions they illustrate, will be of some value to those involved in this exciting field of research.

II. *In Vivo* Test Procedures

A. GENERAL PRINCIPLES

The search for new antimicrobial chemotherapeutic agents occupies the attention of a large number of workers. The approaches to this problem are as many as there are laboratories concerned. Many such survey programs start with an *in vitro* test in an effort to determine the activity of a compound. Should an agar plate method measuring zones of inhibition be used, then questions of diffusion through the media must be kept in mind. The use of a tube test with fluid media and incubation periods of more than an hour or so require the testing of sterile materials. In every test procedure the influence of the test medium itself must be considered. Every worker in this field knows that *in vitro* activity disappointingly often is associated with *in vivo* inactivity. The reverse also may happen, particularly in the case of materials that must be metabolized to an active form in the body. Regardless of *in vitro* activity, the final test must be made *in vivo*, preferably by more than one route for therapy and against more than one test organism.

In present-day survey programs the purpose of this initial *in vivo* test is to uncover any possible evidence of activity. If it is hoped to develop an agent for use in human medicine, the test animal for such preliminary evaluation usually is the mouse. This is a small, relatively inexpensive mammal about which a great deal is known and for which there is a large background of information on infections and therapy. The infecting agent for preliminary tests often is one known to be inhibited *in vitro* by the test compound. To make the *in vivo* test as sensitive as possible, therapy starts at the time of, or very shortly after, infection (measuring prevention rather than cure), and the therapy and infection are given by the same route, usually intraperitoneally. If this procedure is described as "using the mouse as a test tube," then the tube must be qualified as being lined with living tissue through which both the organism and the drug may be removed from the cavity and into which white blood cells and host defense mechanisms may migrate. A very special tube indeed!

The number and timing of the therapeutic treatments to be used must be a balance between the course and speed of the infection and considerations of the possible toxicity and pharmacology of the test compound. The initial treatment of a rapidly fatal infection obviously cannot be delayed too long. On the other hand, a rapidly excreted drug given as a single treatment at the time of infection may have been eliminated from the animal before a slow infection has begun to develop. Several injections of small amounts of a compound may avoid the possibility of drug toxicity, but may not provide the infected animal with the protective concentration reached by fewer injections of larger quantities of drug. Tests to evaluate compounds whose toxicity and pharmacology may not be known usually take a middle course by beginning therapy at the time of infection and repeating it at least once during the period before the expected death of virulence control mice. In addition, several uninfected mice are given the drug by the same route and schedule in order to determine whether it is toxic under these test conditions.

If the compound protects mice in this first intraperitoneal trial, this activity is interpreted to mean that the compound can act against this organism in the presence of living metabolizing tissues in an intact animal. For the next trial, infection again is given by the intraperitoneal route but therapy by the subcutaneous or intramuscular route. Acitivity here demonstrates the ability of the compound to migrate from the point of injection to the infecting organisms, controlling their spread through the blood and tissues of the animal before it itself is metabolized and eliminated. The next route of therapy to be tried is the oral one to determine whether the compound can be transported in an active form from the gastrointestinal tract to the blood and tissues containing the invading organisms.

Using any successful route of therapy, treatment then may be delayed to determine curative rather than prophylactic ability. Withholding infection can determine how long a protective level of compound will remain in the animal after treatment (Miller *et al.*, 1949).

Since all these tests would have included adequate controls to check the toxicity of the material at least at levels used in the therapeutic trials, it can be seen that considerable preliminary pharmacological information can be learned about a test compound during these initial tests. This is a particularly valuable bonus in the case of materials of unknown structure and for which no assay methods have as yet been developed.

1. *The Infection*

For programs searching for broad-spectrum antibacterial chemotherapeutic agents, the initial *in vivo* evaluation often uses infections that kill the test animal rapidly. With such acute infections, usually given intraperitoneally, untreated virulence control mice may die within 24 to 36 hours. Some organisms such as highly virulent strains of *Streptococcus pyogenes, Diplococcus pneumoniae,* or *Pasteurella multocida* are injected suspended in broth or saline. The intraperitoneal injection of only a few bacterial cells may result in the death of the animal several days later. Other organisms, such as certain strains of *Proteus*, will kill mice only if injected in relatively large numbers suspended in mucin. Some organisms may be used either in large numbers in broth or in small numbers in mucin (Miller, 1964).

The effect of mucin in enhancing the virulence of intraperitoneally, but not intravenously injected bacteria was described by Nungester *et al.* (1932) and C. P. Miller (1933). It should be noted that while mucin may enhance the virulence of potentially pathogenic organisms it will not endow virulence on a nonpathogenic strain or organism (Nungester *et al.*, 1936).

There have been many studies to determine the mechanism of such virulence enhancement by mucin. Olitzki (1948) reviews the literature up to that time. He concluded that most of the observations made with mucin in intraperitoneal infections (the inhibition of phagocytosis and of intraphagocytic digestion; the inhibition of antibody uptake and bactericidal action) could be ascribed to the coating effect of mucin on bacteria. Between 1950 and 1953, in a series of seven papers, the senior author Smith described the studies from which he concluded there is no single mucin component that could be called the virulence-enhancing factor (Smith, 1950, 1951; Smith *et al.*, 1951, 1952, 1953). Rather, this activity is due to a synergistic combination of a viscous medium, a particulate residue, and a third factor composed of a highly active heparin (polysaccharide A), a chondroitin sulfuric acid (polysaccharide D), and the neutral blood-group mucoid (polysaccharide C).

The intraperitoneal injection of 0.5 ml sterile mucin into mice certainly stresses the animal in some way. Animals so treated will appear distressed about 6–8 hours later. Their fur is ruffled and they tend to huddle in groups. Indeed, the injection of any material, or even the insertion and withdrawal of a sterile needle may act as a stressing factor. The greater the number of injections, the greater the stress. Infected animals given multiple, subtherapeutic doses of drug often die before untreated challenge controls. For this reason, virulence control mice should be handled and injected with sterile saline or broth each time the test animals are handled or injected.

That broth used as a suspending medium for the challenging organism also may affect host resistance is suggested by the work of Cameron (1957) who isolated a virulence-enhancing peptone from broth medium.

Each mouse test should include a virulence control of the infecting organism. This usually consists of two 10-fold dilutions of the challenge preparation. The number of organisms actually injected in the challenge and these dilutions is determined by plate count. From the survival records of these three groups of infected, nonmedicated animals, the number of organisms lethal to 50% of the untreated mice in that day's test (the LD_{50}, or the lethal dose 50%) can be calculated. From this figure the number of LD_{50} doses in the challenge preparation used to infect the test animals is determined. For initial tests the challenge often is kept low deliberately, so that the compounds will have the best possible chance to show activity. Interpretation of the test results is simplified if the challenge used was 100% fatal for untreated mice, and for this reason, many workers plan to use at least 5 to 10 LD_{50} doses in the challenge. This however is subject to the objectives of the test, and procedures have been described using 100, 1000, or even 10,000 times the LD_{50} dose, while other workers routinely use small multiples of an LD_{50} dose for their tests. The severity of the infection will influence the test results. Table I shows that increasing the number of LD_{50} doses in the challenge may well increase the amount of chemotherapeutic agent required to protect 50% of the infected animals (ED_{50}).

It can be seen from the data listed in Table I that when the number of organisms in the challenging dose of *Pasteurella multocida* was increased from 35 to 350,000, this 10,000-fold increase in the number of LD_{50} doses injected caused a 27-fold increase in the amount of fosfomycin[1] and a 5-fold increase in the amount of penicillin required to protect 50% of the animals against infection. A similar increase in the size of the infecting dose of *Streptococcus pyogenes* raised the ED_{50} of fosfomycin 14 times and of penicillin 15 times. In the case of the *Staphylococcus aureus* and *Proteus mirabilis* tests, the large numbers of organisms required to establish infection did not permit

---

[1]Formerly called phosphonomycin.

TABLE I
INFLUENCE OF THE NUMBER OF INFECTING ORGANISMS
ON THE AMOUNT OF ANTIBIOTIC REQUIRED FOR PROTECTION[a]

Test organism[b]			ED_{50} in μg/dose × 2	
	Challenge			
Name	Number	Approx. LD_{50}	Fosfomycin (po)	Penicillin G (po)
Pasteurella multocida	35	20	372	149
	3,500	2,000	1,250	283
	350,000	200,000	10,100	744
Streptococcus pyogenes	32	20	592	5
	3,200	2,000	1,680	12
	320,000	200,000	8,280	74
Staphylococcus aureus	10,000,000	10		19
	100,000,000	100		58
			Streptomycin (sc)	Cephalothin (sc)
Proteus mirabilis	100,000	14	6	237
	10,000,000	1,400	47	2,810

[a] Data either unpublished or mentioned in Miller *et al.* (1969).

[b] Challenge given intraperitoneally as 0.5 ml of a broth dilution of the appropriate culture except that *Proteus mirabilis* was given suspended in mucin. Therapy was given by the indicated route at the time of infection and again 6 hours later. Five mice were used at each of the 4-fold drug dilutions.

studying so wide a range of challenge doses. The increase in ED_{50} values was correspondingly reduced. Although the increase in the ED_{50} is not in direct proportion to the increase in the LD_{50}, it is obvious that the lighter infections should permit the detection of smaller amounts of activity.

When the challenge is kept constant the amount of antibiotic required for protection also is quite constant. Table II lists ED_{50} values for three different antibiotics given by three routes of therapy against two different infections. It is noted that only five mice were used at each test level (and therefore each mouse represented 20% of the group's response) and 4-fold drug dilutions were used. In no instance did an individual value vary as much as 4-fold from the mean, and in only two cases did the replicates vary more than 4-fold.

a. Virulence and Pathogenicity

Watson and Brandly (1949) in their review of this subject point out that although "pathogenicity" usually is defined as the ability of a

TABLE II

REPRODUCIBILITY OF THE MEDIAN EFFECTIVE DOSE (ED_{50}) OF SEVERAL ANTIBIOTICS
USED TO PROTECT MICE AGAINST INFECTION WITH *Salmonella schottmuelleri* OR *Staphylococcus aureus*[a]

Antibiotic name	Route	ED_{50} in μg/dose × 2[b] S. schottmuelleri	Mean ED_{50}	ED_{50} in μg/dose × 2 S. aureus	Mean ED_{50}
Streptomycin	ip	13, 5, 12, 25, 11,	12	3, 2, 2, 7,	3
	sc	22, 50, 30	32	25, 13, 13	16
	po	2070, 2360, 2650	2350	250, 414, 471	340
Tetracycline	ip	10, 30, 15, 29, 15,	18	31, 14, 6, 14, 8	12
	sc	58, 57, 62, 40	54	31, 15, 24	22
	po	234, 805, 691, 653, 504	545	250, 311, 250, 110	215
Chloramphenicol	ip	40, 62, 100	63	1600, 910	1210
	sc	101, 125, 390	170	1100, 1030	1060
	po	141, 276	197	1610, 569	957

[a] A. K. Miller (unpublished observations).
[b] Mice were infected intraperitoneally with about 10-30 LD_{50} doses of test organisms suspended in broth. Therapy was given by the indicated route at the time of infection and again 6 hours later. Five mice were used at each of the 4-fold drug concentrations tested.

parasite to cause disease, and "virulence" expresses the degree of pathogenicity, the present tendency is to use the terms interchangeably. MacLeod and Bernheimer (1965) in their discussion of the properties of bacteria which enable them to cause disease, comment that although in the mouse spontaneous pneumococcal disease never has been observed, the pneumococcus is pathogenic for this animal. They point out that the mouse virulence of most hemolytic streptococci of group A and some pneumococci isolated from man is relatively low on primary isolation. By repeated animal passage, however, virulence of these organisms for mice often may be increased from a minimum lethal dose of 100,000 organisms or more to a point where only one or two bacteria suffice to bring about a fatal infection. It seems probable that enhancement of virulence by animal passage is the result of selection of virulent mutants present in the original culture. Animal passage of fully virulent cultures does not seem to affect their killing power. During a 2-year period, a strain of *Escherichia coli* was passaged 45 times in guinea pigs by intraperitoneal injection of diluted homogenates of infected spleens. The lethal dose remained at 10^8 organisms (Bullen *et al.*, 1968).

Virulence also is said to be increased by certain artificial methods such as injecting bacteria suspended in mucin, as was mentioned previously. This increase might be more apparent than real since one theory holds that large numbers of such organisms are required to kill the animal. Almost as many living as dead cells are lethal, which implies that the bacteria themselves are toxic. When a few such cells are suspended in mucin for injection, this adjuvant probably protects the organisms so that they can multiply sufficiently to constitute a lethal toxic dose (MacLeod and Bernheimer, 1965).

The ability of iron compounds to increase the virulence of certain bacteria for mice, rats, or guinea pigs has been described by several workers. It is suggested that iron-binding serum proteins such as transferrin may play an important role in host-defense mechanisms. Saturating such proteins with iron permits the rapid *in vivo* growth of certain bacteria. The lethal dose of a strain of *Escherichia coli* for guinea pigs was increased approximately 100,000-fold by the injection of iron compounds (Bullen and Rogers, 1969).

Cameron (1957) found that streptococci in broth were more virulent than they were when injected in saline. An enhancing factor was isolated from the peptone. When injected into control mice, the peptone caused shock. Neither it nor any of the virulence-enhancing preparations by themselves killed mice.

Virulence can be decreased by passage of the culture *in vitro*. Such a decrease often is associated with morphological changes or the loss of toxin production. Braun (1965) states that such a loss of virulence often is associated with a change in colonial form.

If the test culture and animals are handled consistently, the infecting dose is remarkably constant. Table III lists LD_{50} values calculated during a 2-year period for five gram-negative and two gram-positive organisms. The virulence controls of such tests consisted of five mice at each of three 10-fold dilutions of the challenge organism. Before each test, the culture was restored from the lyophilized state and passaged once through broth. It can be seen that the number of organisms required to kill 50% of the animals did not vary greatly.

TABLE III
REPRODUCIBILITY OF THE MEDIAN LETHAL DOSE (LD_{50}) OF THE TEST ORGANISM[a]

Test organism	LD_{50} in number organisms per mouse[b]				
	Test 1	Test 2	Test 3	Test 4	Test 5
Escherichia coli	4.8×10^6	2.1×10^6	3.9×10^6	2.7×10^6	2.5×10^6
Klebsiella pneumoniae	2.6×10^6	4.5×10^6	4.7×10^6	2.0×10^6	2.0×10^6
Proteus vulgaris	3.0×10^3	2.3×10^3	1.2×10^3	3.7×10^3	9.3×10^2
Pseudomonas aeruginosa	8.4×10^3	7.3×10^4	8.2×10^4	4.0×10^4	7.5×10^3
Salmonella schottmuelleri	3.0×10^5	7.3×10^5	1.3×10^6	1.1×10^6	2.3×10^5
Staphylococcus aureus	1.7×10^6	3.1×10^6	7.3×10^5	3.0×10^5	7.8×10^5
Streptococcus pyogenes	1.0	0.5	1.8	0.9	0.5

[a] A. K. Miller (unpublished observations).
[b] Calculated from tests performed over a period of 1–2 years. The cultures were restored from the lyophilized state before each test. All were passaged once through broth and given as broth dilutions of such a culture except that *Proteus vulgaris* is used suspended in mucin. The test animals were Charles River CD 1 female mice, weighing between 20 and 23 gm at the time of challenge. Five mice were used at each of three 10-fold dilutions of the culture.

b. Preservation of Cultures

As stated above, passage of a potentially pathogenic bacterial culture through mice may be used to maintain the organism in a virulent state. To avoid the possible loss of such a culture and to minimize the changes that may accompany such *in vitro* or *in vivo* passaging, many laboratories use some means of preserving cultures. The most universally used method is that of lyophilization. Although many of the individual cells may not survive the lyophil procedure (Tanguay, 1959), such dehydrated cultures if properly prepared and stored remain viable 3 to 5 years and some have been restored 10, 18, or even 35 years after lyophilization (Engley, 1956). There are suggestions that the lyophilization process may cause some genetic alterations in the organisms. Servin-Massieu reports on changes in pigment production, oxygen uptake, and production of an enzyme in a strain of *Serratia marcescens* (Servin-Massieu and Cruz-Camarillo, 1969) and a *Staphylococcus aureus* (Servin-Massieu, 1961). Virulence was not investigated.

Heckley (1961, p. 57–59) in his comprehensive review on the preservation of bacteria by lyophilization, includes many references reporting the maintenance of virulence for years, as well as some showing a loss of virulence after a year or more of storage of such freeze-dried cultures. It appears to be important to permit a reconstituted culture to metabolize by subculturing in ordinary medium before testing it for virulence.

Cultures also may be stored under liquid nitrogen. This procedure has been used mostly for tissue cells and protozoa that do not readily survive lyophilization. Some workers, however, preserve bacterial cultures this way. Sokolski and his co-workers describe the preservation of suspensions of *Lactobacillus leichmannii* (Sokolsky et al., 1964) and of *Sarcina lutea* and *Staphylococcus aureus* (Stapert et al., 1964) that on thawing were used as direct inocula in vitamin or antibiotic assays. They claim excellent reproducibility of assay results and the advantage that once the inoculum is prepared and stored, the test organism may be regarded as another "reagent." Stewart and Wright (1970) studied the effect of storage conditions on the survival of meningococcal and streptococcal L forms. Maximum survival for both strains was found in specimens stored in liquid nitrogen ($-196°C$). Lyophilized cultures of these same organisms showed a loss in viability that apparently occurred during the initial drying period, not during storage. The authors suggest that for long-term storage, if this initial loss can be accepted, lyophilization may be a less expensive and more reliable method of preservation than liquid nitrogen.

Refrigerators for liquid nitrogen can be obtained with a vapor phase canister so that the vials of frozen cells may be stored over, rather than in, the liquid. Such storage in the vapor phase lessens the danger of the explosion of an improperly sealed vial that occurs when it is removed from liquid nitrogen.

Methods of preservation are available that can be used in a laboratory not equipped for lyophilization or nitrogen storage. Hunt et al. (1958) described a method for preserving bacterial cultures on porcelain beads. Cultures restored after 10 months were said to have the original characteristic colony form, mouse virulence, hemolysin, coagulase, pigment production, fermentation reactions, and antibiotic resistance. Rather than glass beads, granules of silica gel also have been used (Grivell and Jackson, 1969).

Other workers prepare concentrated stock pools of their virulent organisms and distribute these in ampules for quick freezing and storage at $-70°C$. Yurchenko and co-workers (1954, 1967) claim there is no loss of viability in such frozen pools and that these materials gave reproducible mortality rates for as long as 2 years.

Grover et al. (1967) reported no loss of viability or infectivity by the respiratory route of *Mycobacterium tuberculosis* H 37_{Rv} suspended in Dubos broth and stored 1 year at $-70°C$. Such suspensions are said to assure reproducibility of the amount of infection and permit comparisons between experiments.

2. The Host

For most antibacterial chemotherapy tests the mouse is used as the test animal. Many supply houses provide animals of known breeding. Reputable firms control the strains and housing conditions rigidly in order to provide animals as closely matched as possible, and of equal susceptibility and resistance. Strains are available for many purposes and should be chosen with care. Ielasi and Kotlarski (1969) comment on examples of strain variation in susceptibility to infection, and they themselves are studying the species variation (mice versus rats) in susceptibility to *Salmonella typhimurium*.

The proper animal housing in the laboratory is extremely important and the best possible conditions should be provided. Many studies have shown the importance of caging, lighting, feeding, and handling of both stock and test animals. Animal care books and bulletins are available for the general laboratory worker but the actual running of animal quarters is of sufficient importance to be placed in the hands of a veterinarian trained for this purpose.

When planning for tests, consideration should be given to publications showing the influence on test results of the sex of the animal used, and on, for example, circadian variation. Triyanond et al. (1969), working with male mice housed under standardized conditions, showed they were more susceptible to infection with a highly virulent strain of pneumococcus and died more rapidly when they were challenged at 4 p.m. than when they were challenged at 4 a.m. The *in vivo* generation time of organisms injected at 4 p.m. was less than that for the 4 a.m. injected organisms. Since the animals died within 4 to 5 hours after the bacteremia reached 10^8 organisms per milliliter, the shortened generation time could account for the faster death of animals challenged in the afternoon.

Glenn and Becker (1969), using female mice, concluded that the ability of the animals to respond immunologically did not vary during a 24-hour period. Such a response, however, was statistically greater in mice housed 6 to a cage than individually-housed mice. They suggest this represents a psychophysiological influence on immunological response.

Under certain conditions, such as the use of small challenge doses and suboptimal therapy, with some but not all chemotherapeutic drugs, the female mouse is more resistant to bacterial infections than is the male (Wheater and Hurst, 1961). Working with rats Ashby *et al.* (1969) showed no difference in the rate of healing of peritoneal defects in male and female animals.

3. Interpreting the Results

Statistical procedures are used to determine the end points of antibacterial chemotherapy tests. As a rule the PD_{50} or ED_{50} value (the amount of agent required to protect, or the effective dose for 50% of the infected animals) is calculated. Such a figure is considered more valid and reproducible than the ED_{100} if only because the ED_{50} usually is found in the linear part of the normally S-shaped dose response curve. One of the most common methods of calculation is that of Reed and Muench (1938). Many other methods are available, and books concerning statistical procedures for biological testing discuss in detail methods for determining the significance of prolonged survival times when an ED_{50} can not be calculated, and for determining 95% confidence limits and other measurements of reliability of the data. The number of animals to use, the increment between drug concentrations to be tested, randomizing animals by weight before putting them on test, adjusting the quantity of injected drug to the actual weight of each animal are all factors to be considered in terms of the information desired. The services and advice of a statistician are most helpful.

No matter how statistically significant the animal test results may be, however, they are, after all, only animal tests. The application of such observations to treatment of human infection is directive at best. Physiological differences between species may influence test results. For example, streptomycin may be used orally in mouse tests because the rodent, unlike man, absorbs a fair amount of this antibiotic from its alimentary tract (Rake and Donovick, 1949). Even here, however, mouse protection tests may be used as a guide. Table IV shows some ratios of systemic and oral ED_{50} values obtained for presentation in 1954 (Miller *et al.*, 1954). It can be seen that penicillin and tetracycline, that are used clinically by the oral route, have relatively low po/ip ratios. Bacitracin and streptomycin have considerably higher po/ip ratios in these tests; neither is used orally to treat sys-

TABLE IV
EFFECT OF THE ROUTE OF THERAPY ON THE THERAPEUTIC DOSE

Antibiotic	ED_{50} in µg/dose × 7[a]			ED_{50} Ratios		
	ip	im	po	im/ip	po/im	po/ip
Penicillin	1	14	14	14	1	14
Tetracycline	15	23	113	1.5	5	8
Bacitracin	116	119	12,160	1	102	105
Streptomycin	4	13	244	3	19	60
Streptogramin	2	3525	1509	1762	0.4	755

[a]Mice were infected intraperitoneally with *Staphylococcus aureus* suspended in mucin. Therapy was given at the time of infection and twice a day for the following 3 days. See Miller et al. (1954).

temic infections. Indeed, the poor absorption of these agents made them candidates for gastrointestinal therapeutic drugs. Streptogramin (Verwey et al., 1958) is included as an example of a drug highly active in the presence of living tissue (compare the intraperitoneal ED_{50} with that of penicillin), but one whose im/ip ratio indicates it probably does not leave the site of injection, and whose poor oral absorption is reflected by the high po/ip ratio. The low po/im ratio emphasizes the necessity of considering all the data before drawing final conclusions.

The mouse species also is unique in that its serum has been reported to lack an *in vitro* bactericidal effect on gram-negative organisms (Marcus et al., 1954). Muschel and Muto (1956) showed this lack of bactericidal action to be attributable to deficiencies of the $C'2$ and $C'3$ components of complement, but later Carey et al. (1960) presented evidence that an active mouse complement system may be demonstrated in freshly drawn blood.

Another example of species difference is the fact that guinea pigs are much more sensitive to repeated injections of penicillin than are mice or rabbits (Hamre et al., 1943). DeSomer et al. (1955) and Farrar and Kent (1965) showed that penicillin therapy caused the replacement of the normal gram-positive intestinal flora of the guinea pig with gram-negative organisms. This was followed by gram-negative bacteremia leading to the death of the animal. The intestinal microflora of man and most other animals differs from that of the guinea pig. Had toxicity tests of this antibiotic been performed only in this animal, penicillin might have been considered too toxic for use in man!

In addition, it must be emphasized that the usual protection test employs an acute, rapidly fatal infection that in no way resembles the illnesses normally needing treatment in the human patient. In the preliminary tests in mice, the intraperitoneal infection results in a rapid overwhelming bacteremia. The organisms disseminate quickly into the various organs of the body and, for many infections, death of the untreated mouse occurs in 24 to 48 hours. Not all organisms, however, cause such a rapid death. In the case of some of the *Salmonella*, for example, deaths do not occur until 4 to 5 or more days after infection. It has been shown that following infection with *Salmonella typhimurium* death generally occurs when the bacterial population of the whole animal reaches about 10^8 (Berry, 1955) or 10^9–10^{10} cells (Schewe, 1958) regardless of the time of death. In the case of the organisms that grow only slowly *in vivo*, deaths may be spread over a wide period of time. Hence the time after infection used for the termination of the experiment always should be specified, since calculated ED_{50} values may vary, depending on the day of the calculation. The time of death also, of course, will be influenced by the number and scheduling of the therapeutic treatments. "Suboptimal" therapy, particularly of bacteriostatic agents, may well delay death even beyond the normal cutoff time of the test. Holding the test for an extra week may materially influence the calculated ED_{50} values. For acute infections, as a rule, a period of several days after therapeutic treatments have stopped and after all virulence control animals should have died is considered a valid test length.

Daily records should be made of deaths and survival. At the completion of the test these should be examined carefully. If too few mice have survived to permit the calculation of an ED_{50} value, borderline activity may be evidenced by the significant prolongation of survival time of treated mice. Or, an amount of drug not toxic to uninfected mice might kill animals stressed by the infection. This can sometimes be detected by comparing the time of death of animals receiving the highest quantity of drug with the death time of those receiving lesser quantities.

Animals surviving the infection because of therapy are not necessarily cleared of all organisms, nor would such persisting organisms cause the eventual death of the animal. We have isolated *Salmonella schottmuelleri* from the spleens of mice as long as 8 weeks after infection and treatment with sulfonamides. That these organisms were fully virulent was readily demonstrated by injecting them into normal mice. The animals from which they were recovered, however, appeared healthy and apparently controlled the organisms by virtue of immune mechanisms. Such animals were shown to possess agglutinins against this *Salmonella* strain and were immune to rechallenge with this organisms for as long as 1 year after the original challenge and treatment (Miller and FitzPatrick, 1965).

B. Types of Tests

1. General

The acute infections discussed above in Section I,A,1, are the most commonly used means of evaluating agents to be used in the treatment of systemic bacterial infections. Such procedures may not be suitable for programs with special objectives. For example, should a search for topical agents be the focus of the experiments, other procedures will be considered. These may include a rabbit ear test, using one ear as a control and the second as a test site; infections of abraded guinea pig or rabbit skin; the use of contaminated sutures in mice, and intradermal infections. Eye infections usually require rather drastic damage to the cornea, although there are a few real conjunctivitis tests described for some, but not all organisms. Special techniques also are used in models for pyelonephritis, kidney, bladder, or respiratory tract infections. Such tests will be discussed later.

Infection by the intranasal, intravenous, or the intracranial route are used in special cases. The mouse is not always the test animal of choice. For example, the only tests closely mimicking *Shigella* infection in man are those using the guinea pig or the monkey. Rabbits, rats, and hamsters also are used in some procedures.

2. Chronic Systemic Infections

a. Tuberculosis and Other Mycobacterial Infections

A comprehensive review of the experimental chemotherapy of tuberculosis and other mycobacterial infections is that of Youmans and Youmans (1964). In earlier work, the guinea pig was the test animal of choice, and it still plays a significant role in these studies. In 1945, however, Youmans and McCarter (1945) showed that mice infected with the human strain of *Mycobacterium tuberculosis* would indeed respond to streptomycin, and since that time the mouse largely has replaced the guinea pig as the experimental animal. Any of a number of routes of infection may be used, but all result in involvement of the lung. The most rapid and uniform response followed infection by the intravenous route or by inhalation (see Section III,C). At the time of Youman's review, the intravenous route was most popular.

Studies using *Mycobacterium lepraemurium* in rodents are described by Eisman (1964). The experimental transmission of rat leprosy has been successfully accomplished by several routes of infection in both rats and mice. The golden hamster is susceptible but the Chinese hamster and guinea pigs seem to be resistant to infection.

Only recently has the causative organism of leprosy in man, *Mycobacterium leprae*, been used successfully in an experimental animal infection. Rees *et al.* (1969) showed that inoculation of this organism into footpads of normal mice produced a long-term infection consistent with the pattern of development of human leprosy as far as this is known.

b. Urinary Tract Infections

Both rats and mice commonly are used for investigating urinary tract infections. Konopka *et al.* (1968) describe details of a procedure adapted from Miraglia *et al.* (1968) to produce pyelonephritis in rats. For this purpose the bladder of an anesthetized male rat was exposed surgically and irritated by the insertion of a surgical clip. A strain of *Proteus mirabilis* then was injected directly into the bladder, and the wound closed. At the termination of the experimental period, 21 days after infection, untreated control animals showed massive calculi in the bladder and occasionally in the kidneys. These latter organs showed gross signs of acute pyelonephritis and, in most cases, were enlarged and had necrotic lesions.

The production of a renal parenchymal infection in rats is described by Rocha *et al.* (1969). Bacteria were injected into the renal medulla of an exposed kidney. At the conclusion of the experiment, infection was determined by counting the number of organisms present in homogenates of the kidneys. Similarly, bladder infections were established by direct inoculation after completion of a partial cystectomy to reduce bladder capacity. Urine collected from animals sacrificed at the end of the experiment was examined for bacterial content. The number of organisms counted per gram of kidney or per milliliter of urine was used as the criterion of urinary tract infection.

The use of the intravenous route of infection, followed by renal massage, is described by Lipman *et al.* (1966) whose paper includes references to earlier work on the development of this method to produce acute pyelonephritis in the rat. Urine from the infected animals was cultured and, at the termination of the experiment, the kidneys were examined grossly and microscopically for lesions, and a part of the kidney was cultured for bacterial growth. Lipman's work showed the danger of assuming that the elimination of bacteria from the urine following therapy indicated a cure of pyelonephritis. Many of his treated animals had negative urine cultures, but positive kidney cultures.

When the mouse is used, infection with either gram-positive or gram-negative organisms usually is initiated by the intravenous injection of the bacterial culture (FitzPatrick, 1966). Indeed, Gorrill and DeNavasquez (1964) say that the lesions seen in such experimental mouse infections are similar to those seen in human pyelonephritis, an opinion shared by Harris who studied infections using *Pseudomonas aeruginosa* (Harris, 1968a), *Streptococcus faecalis* (Harris, 1968b), or both of these organisms (Harris, 1968c).

The effect of the age of the animal on susceptibility to urinary tract infection was studied by Freedman (1969) in rats. The organism was injected intravenously in the tail vein to determine susceptibility of the animal to infection and, in another set of tests into the lumen of the bladder to follow the clearance of bacteria from that organ. His studies showed that rats are most susceptible to urinary infections in old age and that at least two mechanisms were involved: increased susceptibility to intravenous inoculation of bacteria and decreased effectiveness of bacterial clearance from the bladder.

One of the problems in this field of work is the establishment of a reproducible infection. Pitsch *et al.* (1961) studied in mice a renal infection that followed the intravenous inoculation of rabbit erythrocytes and *Escherichia coli* cells. If the *Escherichia coli* were injected without erythrocytes, the infection was inconsistent and transient. Freedman *et al.* (1961) comment on the fact that in mice and rabbits mortality is appreciable following production of extensive pyelonephritis with virulent staphylococci. Moreover, there is a natural tendency of the infection to heal itself. These workers used the rat as the animal of choice for such an infection since here the mortality is negligible. They showed, however, that the size of the infecting dose was critical: reduction by one log resulted in a sharp decrease in the percentage of animals infected. Moreover, they suggest that many of the histological changes thought to be typical of renal infection may be produced by obstruction alone. This must be kept in mind in procedures that include ligation of the ureter, for example, as used by Pitsch *et al.* (1961), to study complicated infections involving anatomic derangement.

Interpretation of test results must keep in mind that in human therapy, uncomplicated acute urinary tract infections often are more readily amenable to treatment, whereas chronic infections, often associated with such anatomical defects, are relatively refractory to therapy.

c. Intestinal Tract Infections

Experimental enteric infections in mice are difficult to establish and rarely result in a dysentery. Small inocula of orally-introduced microorganisms tend to pass rapidly through the gut without multiplying. Such organisms can be induced to remain in the gastrointestinal tract by the prior preparation of the animal in some way. C. P. Miller and Bohnhoff (1962) working mostly with *Salmonella enteritidis* reduced the number of organisms required to produce infection orally by administering morphine immediately after infection in order to arrest peristalsis, thus detaining the small inoculum within the intestine long enough to permit its multiplication. The withholding of food before infection was believed to reduce host resistance and therefore increase susceptibility to infection. An even greater enhancement of susceptibility followed the administration by mouth of a single large dose (50 mg) of streptomycin (Bohnhoff et al., 1954). Fewer than 10 organisms sufficed to infect 50% of the streptomycin-treated mice, while approximately 10^6 cells were required to infect half the untreated controls. The infecting organism was made highly resistant to streptomycin so that it was not affected by residual antibiotic in the intestine, and it could be identified readily. The criteria for infection were the recovery of the test organisms from the feces on the sixth day after inoculation and from the spleen during the second or third week (Miller and Bohnhoff, 1962). Although bacteremia was commonly observed, few mice died of this oral infection. The ability of streptomycin or penicillin to enhance the susceptibility of mouse intestinal tract to infection with *Salmonella* or to the *Staphylococcus* was said to be due to the change in the ecology of the enteric microflora, particularly the *Bacteroides* (Bohnhoff and Miller, 1962; Miller and Bohnhoff, 1963). Such oral treatment with streptomycin did not enhance *Salmonella enteritidis* infection initiated intraperitoneally or subcutaneously.

Meynell and Subbaith (1963), working with *Salmonella typhimurium*, studied the antibacterial mechanisms of the mouse gut. They concluded that in the normal mouse the fall in the viable count of orally-inoculated organisms is due to a bacteriostatic and weakly bactericidal mechanism and to the elimination of the organism in the feces. The bactericidal activity may be attributed to the presence of volatile fatty acids in the cecum. Streptomycin reduces the normal flora, the concentration of volatile fatty acids falls, and conditions are favorable for the growth of the *Salmonella* (Meynell, 1963). Bohnhoff et al. (1964) also studied the interactions of pH, Eh, and fatty acids in the mouse colon. Following streptomycin treatment, the pH rose and the fatty acid concentration fell. These conditions had returned to normal 3 days after streptomycin treatment. Mice still were susceptible to oral infection on the third day, however, because lactic acid had accumulated in the colon in concentrations sufficient to counteract the fatty acid inhibition of *Salmonella* growth. Other carboxylic acids also antagonized the inhibitory activity of fatty acids, but glucose did not.

There have been a number of test systems describing the use of *Shigella* as the oral infecting agent. Freter found that fasting and the oral administration of antibiotics to guinea pigs made them susceptible to enteric infections. His papers in 1956 describe the method used to inhibit the normal enteric flora of mice and guinea pigs to permit the establishment of long-term asymptomatic enteric infections with *Shigella flexneri* or *Vibrio cholerae*. The organisms had to be made drug resistant, for in addition to pretreatment with antibiotics, the animals were maintained on antibiotics given in the drinking water throughout the experiment. The introduction of an antibiotic-resistant strain of *Escherichia coli* into the intestinal tract of the experimental animals resulted in a rapid elimination of the pathogen and increased the resistance of guinea pigs to the fatal enteric cholera.

This antagonism of normal intestinal flora to the growth of *Shigella* was said (Hentges and Freter, 1962) to be associated with *Escherichia coli*, *Aerobacter aerogenes*, and *Proteus*, but not *Pseudomonas*. In none of the models so far described have intestinal lesions been reported.

Formal and his group have developed an oral infection with *Shigella flexneri* in the guinea pig (Formal *et al.*, 1958) and the monkey (LaBrec *et al.*, 1964). Organisms are given by stomach tube to starved guinea pigs that then are injected with opium to inhibit intestinal peristalsis. Virulent organisms penetrate the epithelial cells of the intestinal wall and reach the lamina propria where they multiply, and ulcerative lesions are found in the large intestine (LaBrec *et al.*, 1964). In the rhesus monkey the oral administration of a virulent culture produces diarrheal symptoms and intestinal lesions. These models probably are most like *Shigella* infections in man. It should be mentioned that the ability of a strain of *Shigella flexneri* to cause a fatal infection when injected intraperitoneally into mice does not necessarily correlate with its ability to produce an oral infection in the guinea pig or monkey.

3. *Topical Infections*
   a. *Skin and Muscle*

Browning (1964) in his review makes reference to many of the earlier methods of testing for "local (surface) chemotherapy." These include infections of surgically-prepared wounds with organisms that, if untreated, produced sufficient toxin, or multiplied and invaded the tissues to cause death of test animals. Untreated infections of the peritoneal cavity also usually resulted in the death of animals that could be protected by prompt intraperitoneal local therapy.

In many of the topical test systems described, the infected wounds are self-healing, and the criterion of cure compares the rate of healing of the treated and untreated infections. West et al. (1953) describe a rabbit ear test in which the wound on one ear acts as a control for the other wound. Their report includes the statistical evidence that the wound on the right ear always healed before the one on the left, so half the animals of a group had the left and half the right ear used for the treated wound! Frost and Valiant (1964) injected penicillin-resistant staphylococci intradermally into depilated mouse skin to produce localized lesions. Although these were eventually self-healing, they could be used to measure the effect of therapeutic treatment. This was done by comparing the extent of the lesions produced in untreated control mice with those produced in treated animals. The concentration of compound calculated to produce a 50% reduction in the extent of the lesion was determined statistically.

Goldschmidt (1970) describes reproducible topical infections in rats that were laparotomized, sutured, and inoculated with *Staphylococcus aureus*. Topical antimicrobials can be evaluated for their ability to prevent stitch abscesses. A stitch and wound infection also is described by Howe (1969) who injected sublethal doses of *Klebsiella pneumoniae* intravenously into guinea pigs to cause contamination of clean surgical wounds.

Sutures contaminated with *Pseudomonas aeruginosa* or *Staphylococcus aureus* were used to infect wounds made on depilated mouse skin (Russell et al., 1967). These same authors describe a rat burn model using *Pseudomonas aeruginosa* as the infecting organism.

Selbri and Simon (1952) described a technique of injecting *Staphylococcus aureus* into the leg muscle of a mouse and measuring the increase in muscle size as evidence of edema and infection. The influence of penicillin therapy on such an infection was described in a separate paper (Selbri, 1954). Some of the modifications of this method are listed by Harrison and Fuquay (1964) who themselves refined the edema measurements by the use of a plethysmograph.

b. Eye Infections

Experimental corneal infections have been described for several animal species, but rabbits are perhaps the most commonly used host. The cornea may be abraded before infection (Cassady, 1959; Furgiuele, 1968; Wiggins, 1952), cut through the epithelium and into the stroma to form an incision for inoculation (McMeel and Wood, 1960), or injected by needle and syringe (Kohn et al., 1963). Hessburg et al. (1962) discuss their reasons for feeling that neither the epithelial injury nor the intracorneal inoculation technique closely mimic the clinical situation. They consistently induced a pseudomonal keratitis by using a corneal needle to draw a contaminated thread through the corneal lamella. Details and some modifications of their treatment methods are given in papers by Truant et al. (1961) and Hessburg et al. (1966).

A keratoconjunctivitis was induced in guinea pigs by depositing Shigella organisms on the underside of the eyelids that then were gently massaged to distribute the organisms over the eye. The technique did not result in infection with other enterobacteriaceae (Mackel et al., 1961). Pugh et al. (1968) produced in mice a Moraxella bovis keratoconjunctivitis they felt was similar to that seen in pink-eye of cattle. The infection was produced by irradiating the eye with a sunlamp for 20 minutes before flooding it with a suspension of the organism. The irradiation treatment was repeated daily. This technique also allowed the organism to become established in the eyes of sheep although the disease was not produced. Guinea pigs, rats, and rabbits were not susceptible to this infection.

The role of endotoxin in eye damage from keratoconjunctivitis caused by Shigella or Escherichia coli was studied by Cross and Nakamura (1970). They feel that the bacteria invade the eye tissue to produce the endotoxin that then produces the damage. It is of interest that Fisher and Allen (1958) reported that cell-free extracts of Pseudomonas aeruginosa inoculated intracorneally into rabbits produced corneal ulceration. They found correlation between the proteolytic capacity of the organisms and eye damage observed.

III. Special Tests

A. SYNERGY

There are many arguments as to the value of using more than one drug in the treatment of infection. Although the use of fixed combinations of chemotherapeutic drugs generally is disapproved by the Government Food and Drug Administration and by research clinicians, there is not disapproval of the properly controlled use of more than one drug. A recent editorial by Herrell (1969) discusses this problem and lists both desirable and undesirable combinations. Jawetz (1968) discusses this subject in greater detail in his recent review. An even earlier report by Dowling (1957) includes several discussions of the correlation between laboratory testing and clinical observations of the use of antibiotic combinations.

In general, the arguments for the use of combinations are (a) to cover the patient against several infectious organisms until the actual causative agent can be determined; (b) in known mixed infections to protect against both organisms; (c) to reduce the incidence or intensity of adverse reactions; (d) to delay the emergence of resistant strains; and (e) to obtain a result greater than the effect of either drug used alone (synergism).

Arguments against the use of combinations include (a) there may be a false sense of security which would discourage efforts toward a specific diagnosis. This may lead to (b) inadequate dosing; (c) the use of combinations may *enhance* the likelihood of adverse effects such as toxicities; (d) using a drug unnecessarily may expose the patient and organism to an antibiotic resulting in later hypersensitivity or emergence of resistance; and (e) possible antagonism.

The term *synergy* is interpreted in several ways; here we mean that the use of two materials, each effective by itself, produces a result greater than that expected by the use of an equal quantity of either one alone. Conversely, antagonism results when the action of one drug is less in the presence of a second drug than would have been seen had one of the drugs been used alone. If the effect of the combination is neither greater nor less than equivalent amounts of the individual drugs, then the effect is said to be additive.

Investigators over the years have tried to evaluate the clinical significance of laboratory demonstrations of synergy or antagonism between two agents. Dowling (1957) states "it must be emphasized, however, that synergism in animals even if established unequivocally, does not necessarily mean that synergism will be observed in human infections. Furthermore, dosage proportions that show synergism in animals may not be the same as those that show synergism in human infection. In fact, they may show indifference or even antagonism when used in humans." In this regard it is of interest to quote Jawetz and Gunnison (1952) who say of *in vitro* tests, "If a given combination of antibiotics was synergistic against a certain bacterium, the phenomenon was demonstrable over a wide range of concentrations. In no instance could synergism be converted to antagonism or vice versa, by a change in doses of drugs."

There is general agreement that only clinical evidence can be used to determine the interaction of combination therapy for human diseases. Here the best evidence of synergy is for the use of penicillin and streptomycin in the treatment of bacterial endocarditis caused by *Streptococcus faecalis* (Koenig and Kaye, 1961). Jawetz (1968) points out, however, that such synergistic drug effects were specific for given strains and that occasionally other drugs had to be used, such as penicillin with neomycin or kanamycin to clear the infection. Other reports of occasional claims of clinical synergism are included in the two review papers quoted above. C. B. Smith *et al.* (1969) list the clinical reports of the use of gentamicin with other agents in the treatment of gram-negative infecting organisms. Their own work reports on three patients who showed clinical improvement of pseudomonal infections following combined therapy of gentamicin with carbenicillin.

Clinically-demonstrated antagonism in antibiotic therapy also is relatively rare. Jawetz (1968) lists the references, including the oft-quoted one of Lepper and Dowling (1951) showing antagonism between chloramphenicol and penicillin in the treatment of pneumococcal meningitis. He (Jawetz, 1967) presents arguments to show that antagonism cannot be expected as a frequent outcome of clinical antimicrobial therapy because it occurs only under sharply defined time-dose relationships. Crofton (1969) is another to state that antagonism between chemotherapeutic agents is of little clinical importance.

Most of the literature-described studies of combined antimicrobial action have been performed *in vitro,* but some *in vivo* work is reported. Ahern *et al.* (1952) confirmed Jawetz' earlier studies on the time-dose relationship required to show antagonism between penicillin and chloramphenicol. In these experiments, mice infected intraperitoneally were given a single therapeutic treatment at the time of, or shortly after infection. Antagonism of penicillin protection by chloramphenicol was shown. It was not shown, however, when several doses a day were given for several days. The authors felt that in patients treated with antibiotics, effective levels are present in body fluids for a prolonged period. When similar conditions were established in the mouse by multiple dosing, the interfering action of chloramphenicol was suppressed. These experiments also were said to imply that the phenomenon of antibiotic interference is seldom of clinical significance.

Bliss *et al.* (1952) used three therapeutic treatments within 24 hours. Their protocol included a set of animals receiving the combination to be tested and four control sets, each antibiotic used alone at the concentration of the combination and also double this amount. Test results were too variable to allow them to draw more than inferences, but these were that synergism was seen more often than antagonism, and that the most commonly seen response was merely additive. Reactions differed, depending on the organisms tested and the concentration of the components used.

The effect of the test method on results is illustrated by the fact that various responses of penicillin and sulfonamide-combined therapy can be shown experimentally. Soo-Hoo and Schnitzer (1964) confirmed a reported *in vivo* synergy (Ungar, 1943) between a sulfonamide and penicillin used in mice infected with streptococci. Ungar gave both therapeutic agents twice a day intraperitoneally for 4 days. Soo-Hoo and Schnitzer used a single subtherapeutic dose of penicillin, given subcutaneously, at the time of infection and multiple subtherapeutic oral doses of sulfonamide, starting at the time of infection and continuing over the following 3 days. Gunnison *et al.* (1951) used sulfadiazine in subtherapeutic doses and penicillin in 95–100% curative doses against a streptococcal infection in mice. When the two drugs were given subcutaneously 1 hour after infection, the sulfonamide had no effect on the curative action of penicillin. Marked interference of this protection occurred when sulfadiazine was given 1 hour after infection and penicillin 4 or 5 hours later. The authors point out that there is no evidence that such interference occurs in the treatment of natural infections in man. Miller and Verwey (1954) gave penicillin and also a triple sulfonamide mixture orally at the time of infection and again 24 hours later. They determined the 50% protective dose of penicillin alone and in the presence of a constant quantity of triple sulfonamides. Combinations of penicillin and sulfonamides were synergistic in these tests and the addition of probenecid to such a mixture further enhanced protection.

Apparently, the timing and the dosages are important in determining the interaction of the bacteriostatic sulfonamides with penicillin that is most effective against rapidly multiplying bacteria. Similar effects of timing on response have been reported for other combinations of agents. For example, β-diethylaminoethyldiphenylpropylacetate hydrochloride in itself has no antibacterial activity but may prolong the activity of a variety of drugs in the body by impeding their metabolic degradation. This compound has been shown to increase or decrease the antibacterial therapeutic effect of penicillin in the opposite direction to that of chlortetracycline, depending on the timing of the administration of the drug. It had no effect on the sulfonamide therapy of a streptococcal infection in mice. It reduced the antiviral activity of mepacrine and enhanced the antimalarial action of this drug and also that of proguanil (Thorp *et al.*, 1960).

In the interest of completeness, another test system should be mentioned, one as used, for example by Böhni (1969). She tests the individual agents alone and also determines the 50% curative dose of a fixed combination of the two agents; for example, a 1:2 or a 1:5 mixture. Using such a test, she showed greater activity for mixtures of trimethoprim and sulfamethoxazole than for the individual agents alone used against infections in the mouse with *Haemophilus influenzae, Diplococcus pneumoniae,* or *Escherichia coli.*

It would seem from these many references that there is no general consensus as to the value of tests for synergy or antagonism. Much of the confusion may result from the interpretation of the words themselves. It is necessary always to define the meaning intended when writing or speaking of this subject. Perhaps the greatest hesitancy in making use of laboratory-obtained data is the fact that results of tests using currently-described protocols are so obviously dependent upon the exact poising of a number of conditions. Such conditions used to demonstrate the synergistic or antagonistic response in an animal test are not realistic in terms of clinical therapy since patients are not given marginal therapy.

B. DEVELOPMENT OF RESISTANCE *in Vivo*

Early work set up to demonstrate the *in vivo* development of resistance of bacteria to drugs has been reviewed by Schnitzer and Grunberg (1957). They point out that two methods have been used: (1) the serial passage of bacteria causing acute infections in mice or other animals treated with the drug and (2) the continuous exposure of bacteria causing chronic infections in animals by multiple treatments.

For bacteria causing acute infections, the general procedure consisted of treating infected mice, recovering the organism by bleeding moribund animals, and using this blood or a culture of it to infect the next set of animals for treatment and bleeding. After a variable number of such passages the development of resistance was demonstrated by a number of workers for pneumococci (i.e., to sulfonamides by Schmidt *et al.,* 1942), for the meningococcus (to penicillin by Miller and Bohnhoff, 1948), and in one case by a streptococcus (to a gold compound by Schnitzer, 1938). Schnitzer points out hemolytic streptococci do not respond as well as the other cocci in such a test, and he lists several descriptions of unsuccessful attempts with the β-hemolytic streptococcus. Rake *et al.* (1944) were not successful in developing staphylococcal resistance to penicillin *in vivo.* (Such resistance can be developed *in vitro* though the resulting culture does not produce penicillinase.) Streptomycin-fast *Escherichia coli* and

Aerobacter aerogenes emerged after passage through treated eggs (Silver and Kempe, 1947).

Streptomycin has been given over prolonged periods of time to animals infected with *Mycobacterium tuberculosis*. Increased streptomycin-resistance was shown in organisms recovered from treated guinea pigs by Feldman *et al.* (1948) and from treated mice by Youmans *et al.* (1949). Using this test system Williston and Youmans (1950) and Lenert and Hobby (1949) were unable to confirm reports of *in vitro* tests and clinical observations that the use of para-aminosalicylic acid along with streptomycin prevents or retards the development of resistance. Wolinsky and Steenken (1953) comment on the difficulties of studying the *in vivo* emergence of resistance experimentally. Working with mice, rabbits, and guinea pigs treated for prolonged periods, only rarely did they find moderately or highly streptomycin-resistant strains of tubercle bacilli. They concluded that the microorganisms behave differently in experimental animals than in human beings with pulmonary tuberculosis, and suggest that perhaps host factors played a part in the emergence of resistance *in vivo*. To explain the discrepancy between their studies and those of Youmans *et al.* (1949) and Williston and Youmans (1950) they point out the possibility that larger infecting inocula will be associated with a higher incidence of recoverable drug-resistant population of bacteria. In the light of more recent knowledge of the genetics of drug resistance, this indeed would be true.

Lenert and Hobby (1949) isolated streptomycin-dependent tubercle bacilli from 13 of 196 mice infected with *Mycobacterium tuberculosis* and given prolonged streptomycin treatment. These strains were highly resistant *in vitro* and could not be propagated repeatedly in the absence of streptomycin. On passage *in vivo* such strains produced only streptomycin-resistant or streptomycin-dependent strains. One or two of six tested were as virulent as the original parent strain. Virulence was not enhanced by streptomycin.

Bliss and Alter (1959) attempted to demonstrate the *in vivo* development of resistance to streptomycin by staphylococci. The organisms were given to mice intravenously to establish a kidney infection and the animals were treated over a 2-week period. Organisms were recovered from the treated animals and tested for their sensitivity to the antibiotic. Streptomycin resistance occurred only rarely. Of 456 mice infected with *Staphylococcus*, 6 yielded substrains with high resistance and 7 yielded substrains with low resistance. About half of the mice tested had received no treatment with streptomycin and from these, two of the very resistant strains were derived. The authors concluded that high resistance is a result of spontaneous mutation. Since all the strains showing low resistance came from treated mice, the authors suggest that this may be attributable to the presence of low concentrations of streptomycin in the tissues of the host.

Griffith and co-workers (1961) treated intraperitoneally *Proteus*-infected mice with actinospectacin. The organisms isolated from treated animals did not show a change in sensitivity after animal passage and 5 days of contact with the antibiotic. The authors suggest that the development of *in vivo* resistance may be slower than *in vitro*.

Konopka *et al.* (1968) used a urinary tract infection in rats as described by Miraglia *et al.* (1968). A strain of *Proteus mirabilis* was used that, on injection directly into the bladder, caused a developing acute pyelonephritis. Therapy with rifampin cleared the infecting organism from the kidneys of many of the animals, but in some that did not receive a full course of treatment, organisms were recovered and these were resistant to the antibiotic. Although nalidixic acid did not clear the kidneys of organisms in this test, the cultures recovered failed to show the development of resistance to this agent.

There seems to be no one test that can be used to show the experimental *in vivo* development of resistance to drugs by microorganisms. One point of discussion concerning such tests has been whether resistant organisms retain the virulence and pathogenicity of the parent strain. For example, Schmidt *et al.* (1942), compared the "superficial properties" of the sensitive parent and the *in vitro* and *in vivo* developed sulfonamide-resistant pneumococci substrains. They reported them to be identical in, among other things, virulence and invasiveness for mice. C. P. Miller and Bohnhoff (1948) stated that maximal virulence was retained by the meningococcal strains made resistant *in vivo* to penicillin, although morphologically these organisms were larger and had less affinity for the counterstain in the gram method than did the parent.

Cooper and Keller (1942) say that a strain of Flexner *Shigella*, made resistant to sulfathiazole *in vitro*, retained its virulence for mice, but was sensitive to the sulfonamide *in vivo*. A Sonne strain, also made resistant *in vitro*, became nonvirulent. No details are given in this note — it is not stated whether the Flexner strain was recovered from mice and found still to be resistant *in vitro*.

Rake *et al.* (1944), working with pneumococci, streptococci, and the Smith strain of *Staphylococcus* developed *in vitro* strains resistant to penicillin. These organisms grew more slowly and showed altered cultural characteristics. They also showed marked decrease in virulence for mice.

Schnitzer and Grunberg (1957, pp. 100–101) describe an experiment in which an isoniazid-resistant culture of *Mycobacterium tuberculosis* was isolated after therapy. The resistance of this culture then was confirmed *in vivo*. While not specifically stating that virulence had been lost, it is noted that infection with the resistant strain was made with an undiluted culture, while the original, parent strain, had been used at a 10^{-1} dilution. In her review on drug resistance in tuberculosis, Hobby (1962, 1963) states that the decreased virulence for guinea pigs of isoniazid-resistant tubercle bacilli often has been implied to mean decreased pathogenicity for man. Nevertheless, she says, there are reports that drug-resistant strains frequently are isolated from persons with newly diagnosed, previously untreated tuberculosis.

Kiser *et al.* (1969) quote a number of studies from which they conclude that a mutation to resistance may lead to no change or to a decrease but not to an increase in the virulence of a pathogen.

C. OTHERS (DENTAL CARIES, L-FORMS, AEROSOL CHALLENGES)

There have been recent excellent reviews of these subjects listing many references. Gold (1969) traces the work that unequivocally established the necessity of bacteria for tooth decay, and that provided a method by which cultures could be tested for cariogenic potential.

Feingold (1969) reviews the biology and pathogenicity of microbial spheroplasts and L-forms, including experimental observations of the pathogenicity of these microorganisms. He suggests that the reason it is so difficult to produce disease on challenge with L-forms is that very special conditions may be required to elicit their pathogenicity. A number of papers on the role of these organisms on experimental and clinical disease are included in the text edited by Guze (1968).

The airborne route of infection has long been recognized. In 1961 a conference on "Airborne Infection" reviewed its importance, epidemiological significance, and the techniques that had been developed to study these questions (McDermott, 1961). Bacterial, viral, and fungal diseases were discussed. Detailed descriptions of equipment and procedures used in the aerosol challenge of animals with bacteria or viruses are given by Jemski and Phillips (1965). They feel that infection by intratracheal or intranasal instillation does not simulate the naturally acquired infection nearly so much as does infection by inhalation. They quote, among others, Fukui *et al.* (1957) who showed that *Pasteurella pestis* cells were cleared much more rapidly from lungs of guinea pigs when inhaled as an aerosol than when introduced intratracheally. Speck and Wolochow (1957) needed 20,000 inhaled

Pasteurella pestis cells as an LD_{50} dose in monkeys in contrast to a reported 100 cells given intratracheally.

D. W. Smith *et al.* (1966) published their reasons for discontinuing the subcutaneous challenge in favor of an airborne challenge experiment using *Mycobacterium tuberculosis* infections in guinea pigs. The airborne challenge permitted infection by a natural route with small numbers of bacteria and the number of primary lesions, the rate of spread from the lesions and the rate of multiplication of the bacteria all could be used to evaluate the progress of the disease.

Undoubtedly the improvement of equipment designs and of apparatus insuring the safety of the experiments has led to a greater use of inhalation studies. It is of interest that in 1952 Middlebrook described an apparatus to study airborne infections in mice but said it was not suitable for large-scale testing. In 1961 he described work using a larger model and pointed out that despite "intensive, almost daily use over a period of 8 years, none of the tuberculin-skin-test negative personnel working in the area where the apparatus has been located have become tuberculin-skin-test positive."

IV. Epilog

It is obvious there are many procedures being used for the *in vivo* testing of antibacterial chemotherapeutic substances. In addition to those reviewed here many more will be found by checking the references quoted. This brief chapter does not pretend to exhaust the subject matter. No mention has been made, for example, of the use of hen's eggs or the place of germ-free animals in *in vivo* testing. The dynamics of the experimental infections used and the influence of therapy on the course of these infections, the biochemical aspects of microbial pathogenicity, and many other phases of the infection and host interrelationships have not been discussed. It is hoped that the papers to which reference has been made have been interpreted with reasonable accuracy. We shall be satisfied if we have called attention to interesting observations and literature, both old and new. We shall be happy if we have encouraged work to improve and enlarge the scope of *in vivo* testing. We shall be most pleased if we have stimulated the study of an infection in order to understand its progress in the test animal and to interpret the significance of its therapy in terms of medical practice.

REFERENCES

Ahern, J. J., Burnell, J. M., and Kirby, W. M. M. (1952). *Proc. Soc. Exp. Biol. Med.* **79**, 568–571.

Ashby, E. C., Mott, T. J., and Ellis H. (1969). *Brit. J. Exp. Pathol.* **50**, 76–77.

Berry, L. J. (1955). *Ann. N.Y. Acad. Sci.* **62**, 327–348.
Bliss, E. A., and Alter, B. McD. (1959). *J. Bacteriol.* **78**, 671–674.
Bliss, E. A., Warth, P. T., and Long, P. H. (1952). *Bull. Johns Hopkins Hosp.* **90**, 149–169.
Böhni, E. (1969). *Postgrad. Med. J.* **45**, 18–20.
Bohnhoff, M., and Miller, C. P. (1962). *J. Infec. Dis.* **111**, 117–127.
Bohnhoff, M., Drake, B. L., and Miller, C. P. (1954). *Proc. Soc. Exp. Biol. Med.* **86**, 132–137.
Bohnhoff, M., Miller, C. P., and Martin, W. R. (1964). *J. Exp. Med.* **120**, 805–828.
Braun, W. (1965). "Bacterial Genetics," 2nd Ed., p. 158. Saunders, Philadelphia, Pennsylvania.
Browning, C. H. (1964). *In* "Experimental Chemotherapy" (R. J. Schnitzer and F. Hawking, eds.), Vol. 2, pp. 1–36. Academic Press, New York.
Bullen, J. J., and Rogers, H. J. (1969). *Nature (London)* **224**, 380–382.
Bullen, J. J., Leigh, L. C., and Rogers, H. J. (1968). *Immunology* **15**, 581–588.
Cameron, J. (1957). *Nature (London)* **180**, 1136–1137.
Carey, W. F., Muschel, L. H., and Baron, L. S. (1960). *J. Immunol.* **84**, 183–188.
Cassady, J. V. (1959). *Amer. J. Ophthalmol.* **48**, 741–747.
Cooper, M. L., and Keller, H. M. (1942). *Proc. Soc. Exp. Biol. Med.* **51**, 238.
Crofton, J. (1969). *Brit. Med. J.* **ii**, 137–141, 209–212.
Cross, W. R., and Nakamura, M. (1970). *Bacteriol. Proc.* p. 94.
DeSomer, P., Van de Voorde, H., Eyssen, H., and Van Dijck, P. (1955). *Antibiot. Chemother. (Washington, D.C.)* **5**, 463–469.
Dowling, H. F. (1957). *J. Amer. Med. Ass.* **164**, 44–48.
Ehrlich, P. (1913). *Lancet* **ii**, 445–451.
Eisman, P. C. (1964). *In* "Experimental Chemotherapy" (R. J. Schnitzer and F. Hawking, eds.), Vol. 2, pp. 501–558. Academic Press, New York.
Engley, F. B. (1956). *Tex. Rep. Biol. Med.* **14**, 114–203.
Farrar, W. E., Jr., and Kent, T. H. (1965). *Amer. J. Pathol.* **47**, 629–642.
Feingold, D. (1969). *New Engl. J. Med.* **281**, 1159–1170.
Feldman, W. H., Karlson, A. G., and Hinshaw, H. C. (1948). *Amer. Rev. Tuberc.* **57**, 162–174.
Fisher, E., Jr., and Allen, J. H. (1958). *Amer. J. Ophthalmol.* **46**, 21–27.
FitzPatrick, F. K. (1966). *Proc. Soc. Exp. Biol. Med.* **123**, 336–339.
Formal, S. B., Dammin, G. J., LaBrec, E. H., and Schneider, H. (1958). *J. Bacteriol.* **75**, 604–610.
Freedman, L. R. (1969). *Yale. J. Biol. Med.* **42**, 30–38.
Freedman, L. R., Werner, A. S., Beck, D., and Paplanus, S. (1961). *Yale J. Biol. Med.* **34**, 40–51.
Freter, R. (1956a). *J. Exp. Med.* **104**, 411–418.
Freter, R. (1956b). *J. Exp. Med.* **104**, 419–426.
Frost, B. M., and Valiant, M. E. (1964). *J. Pathol. Bacteriol.* **88**, 125–136.
Fukui, G. M., Lawton, W. D., Janssen, W. A., and Surgalla, M. J. (1957). *J. Infec. Dis.* **100**, 103–107.
Furgiuele, F. P. (1968). *Amer. J. Ophthalmol.* **66**, 276–279.
Glenn, W. G., and Becker, R. E. (1969). *Bacteriol. Proc.* p. 68.
Gold, W. (1969). *Advan. Appl. Microbiol.* **11**, 135–157.
Goldschmidt, F. K. (1970). *Bacteriol. Proc.* p. 96.
Gorrill, R. H., and DeNavasquez, S. J. (1964). *J. Pathol. Bacteriol.* **87**, 79–87.
Griffith, L. J., Ostrander, W. E., Beswick, D. E., and Havens, D. W. (1961). *Antimicrob. Ag. Chemother.* pp. 516–519.

Grivell, A. R., and Jackson, J. F. (1969). *J. Gen. Microbiol.* **58**, 423–425.
Grover, A. A., Kim, H. K., Wiegeshaus, E. H., and Smith, D. W. (1967). *J. Bacteriol.* **94**, 832–835.
Gunnison, J. B., Speck, R. S., Jawetz, E., and Bruff, J. A. (1951). *Antibiot. Chemother. (Washington, D.C.)* **1**, 259–266.
Guze, L. B., ed. (1968). "Microbial Protoplasts, Spheroplasts and L-Forms." Williams & Wilkins, Baltimore, Maryland.
Hamre, D. M., Rake, G., McKee, C. M., and MacPhillamy, H. B. (1943). *Amer. J. Med. Sci.* **206**, 642–652.
Harris, D. M. (1968a). *J. Pathol. Bacteriol.* **96**, 77–87.
Harris, D. M. (1968b). *Brit. J. Exp. Pathol.* **49**, 128–135.
Harris, D. M. (1968c). *J. Pathol. Bacteriol.* **96**, 421–430.
Harrison, E. F., and Fuquay, M. E. (1964). *Antimicrob. Ag. Chemother.* pp. 764–766.
Hawking, F. (1963). *In* "Experimental Chemotherapy" (R. J Schnitzer and F. Hawking, eds.), Vol. 1, pp. 1–2. Academic Press, New York.
Heckley, R. J. (1961). *Advan. Appl. Microbiol.* **3**, 1–76.
Hentges, D., and Freter, R. (1962). *J. Infec. Dis.* **110**, 30–37.
Herrell, W. E. (1969). *Clin. Med.* **76** (Oct.), 11–14.
Herrell, W. E. (1970). *Clin. Med.* **77** (July), 10–17.
Hessburg, P. C., Truant, J. P., and Penn, W. P. (1962). *Amer. J. Ophthalmol.* **53**, 359–364.
Hessburg, P. C., Truant, J. P., and Penn, W. P. (1966). *Amer. J. Ophthalmol.* **61**, 49–54.
Hobby, G. L. (1962). *Amer. Rev. Resp. Dis.* **86**, 839–846.
Hobby, G. L. (1963). *Amer. Rev. Resp. Dis.* **87**, 29–36.
Hoerlein, H. (1937). *Practitioner* **139**, 635–649.
Howe, C. W. (1969). *Brit. J. Exp. Pathol.* **50**, 456–460.
Hunt, G. A., Gourevitch, A., and Lein, J. (1958). *J. Bacteriol.* **76**, 453–454.
Ielasi, A., and Kotlarski, I. (1969). *Aust. J. Exp. Biol. Med. Sci.* **47**, 689–699.
Jawetz, E. (1967). *Pharmacol. Physicians* **1**, 1–6.
Jawetz, E. (1968). *Annu. Rev. Pharmacol.* **8**, 151–170.
Jawetz, E., and Gunnison, J. B. (1952). *Antibiot. Chemother. (Washington, D.C.)* **2**, 243–248.
Jemski, J. V., and Phillips, G. B. (1965). *In* "Methods of Animal Experimentation" (W. I. Gay, ed.), Vol. 1, pp. 273–341. Academic Press, New York.
Kiser, J. S., Gale, G. O., and Kemp, G. A. (1969). *Advan. Appl. Microbiol.* **11**, 77–100.
Koenig, M. G., and Kaye, D. (1961). *New Engl. J. Med.* **264**, 257–264.
Kohn, S. R., Gershenfeld, L , and Barr, M. (1963). *J. Pharm. Sci.* **52**, 967–974.
Konopka, E. A., Zoganas, H. C., Lewis, L., and Gelzer, J. (1968). *Antimicrob. Ag. Chemother.* pp. 519–523.
LaBrec, E. H., Schneider, H., Magnani, T. J., and Formal, S. B. (1964). *J. Bacteriol.* **88**, 1503–1518.
Lenert, T. F., and Hobby, G. L. (1949). *Amer. Rev. Tuberc.* **59**, 219–220.
Lepper, M. H., and Dowling, H. F. (1951). *Arch. Intern. Med.* **88**, 489–494.
Lipman, R. L., Tyrell, E., Small, J., and Shapiro, A. P. (1966). *J. Lab. Clin. Med.* **67**, 546–558.
McDermott, W. (1961). *Bacteriol. Rev.* **25**, 173–377.
Mackel, D. C., Langley, L. F., and Venice, L. A. (1961). *Amer. J. Hyg.* **73**, 219–223.
MacLeod, C. M., and Bernheimer, A. W. (1965). *In* "Bacterial and Mycotic Infections in Man" (R. J. Dubos and J. G. Hirsch, eds.), 4th Ed., p. 147, 158. Lippincott, Philadelphia, Pennsylvania.

McMeel, J. W., and Wood, R. W. (1960). *Trans. Amer. Acad. Ophthalmol. Otolaryngol.* **64,** 486-490.
Marcus, S., Esplin, D. W., and Donaldson, D. M. (1954). *Science* **119,** 877.
Meynell, G. G. (1963). *Brit. J. Exp. Pathol.* **44,** 209-219.
Meynell, G. G., and Subbaith, T. V (1963). *Brit. J. Exp. Pathol.* **44,** 197-208.
Middlebrook, G. (1952). *Proc. Soc. Exp. Biol. Med.* **80,** 105-110.
Middlebrook, G. (1961). *Bacteriol. Rev.* **25,** 331-346.
Miller, A. K. (1964). *Chemotherapia* **8,** 154-162.
Miller, A. K., and FitzPatrick, F. K. (1965). *Bacteriol. Proc.* p. 71.
Miller, A. K., and Verwey, W. F. (1954). *Antibiot. Chemother. (Washington, D.C.)* **4,** 169-172.
Miller, A. K., Verwey, W. F., and Wilmer, D. L. (1949). *Proc. Soc. Exp. Biol. Med.* **70,** 313-315.
Miller, A. K., West, M. K., and Verwey, W. F. (1954). *Bacteriol. Proc.* p. 82.
Miller, A. K., Frost, B. M., Valiant, M. E., Kropp, H., and Hendlin, D. (1969). *Antimicrob. Ag. Chemother.* pp. 310-315.
Miller, C. P. (1933). *Science* **78,** 340-342.
Miller, C. P., and Bohnhoff, M. (1948). *J. Infec. Dis.* **83,** 256-261.
Miller, C. P., and Bohnhoff, M. (1962). *J. Infec. Dis.* **111,** 107-116.
Miller, C. P., and Bohnhoff, M. (1963). *J. Infec. Dis.* **113,** 59-66.
Miraglia, G. J., Renz, K. J., and Gadebusch, H. H. (1968). *Bacteriol. Proc.* p. 99.
Morganroth, J., and Levy, R. (1911). *Berlin Klin. Wochenschr.* **48,** 1979-1983.
Muschel, L. H., and Muto, T. (1956). *Science* **123,** 62.
Nungester, W. J., Wolf, A. A., and Jourdonais, L. F. (1932). *Proc. Soc. Exp. Biol. Med.* **30,** 120-121.
Nungester, W. J., Jourdonais, L. F., and Wolf, A. A. (1936). *J. Infec. Dis.* **59,** 11-21.
Olitzki, L. (1948). *Bacteriol. Rev.* **12,** 149-172.
Pitsch, B., Herbert, T., and Carey, W. F. (1961). *Antimicrob. Ag. Chemother.* pp. 54-60.
Pugh, G. W., Jr., Hughes, D. F., and McDonald, T. J. (1968). *Amer. J. Vet. Res.* **29,** 2057-2061.
Rake, G., and Donovick, R. (1949). *In* "Streptomycin, Nature and Practical Applications" (S. A. Waksman, ed.), pp. 237-238. Williams & Wilkins, Baltimore, Maryland.
Rake, G., McKee, C. M., Hamre, D. M., and Houck, C. L. (1944). *J. Immunol.* **48,** 271-289.
Reed, L. J., and Muench, H. (1938). *Amer. J. Hyg.* **27,** 493-497.
Rees, R. J. W., Weddell, A. G. M., and Pearson, J. M. H. (1969). *Brit. Med. J.* **iii,** 216-217.
Rocha, H., Teles, E. da S., and Barros, M. (1969). *Appl. Microbiol.* **18,** 547-549.
Russell, H. E., Gutekunst, D. P., and Chamberlain, R. E. (1967). *Antimicrob. Ag. Chemother.* pp. 497-501.
Schmidt, L. H., Sesler, C., and Dettwiler, H. A. (1942). *J. Pharmacol. Exp. Ther.* **74,** 175-189.
Schnitzer, R. J. (1938). Quoted in Schnitzer and Grunberg (1957). p. 94.
Schnitzer, R. J., and Grunberg, E. (1957). "Drug Resistance of Microorganisms," pp. 89-101. Academic Press, New York.
Schewe, E. (1958). *J. Infec. Dis.* **102,** 275-293.
Selbie, F. R. (1954). *Brit. Med. J.* **i,** 1350-1353.
Selbie, F. R., and Simon, R. D. (1952). *Brit. J. Exp. Pathol.* **33,** 315-326.
Servin-Massieu, M. (1961). *J. Bacteriol.* **82,** 316-317.

Servin-Massieu, M., and Cruz-Camarillo, R. (1969). *Appl. Microbiol.* **18**, 689–691.
Silver, H. K., and Kempe, C. H. (1947). *J. Immunol.* **57**, 263–272.
Smith, C. B., Dans, P. E., Wilfert, J. N., and Finland, M. (1969). *J. Infec. Dis.* **119**, 370–377.
Smith, D. W., Wiegeshaus, E., Navalkar, R., and Grover, A. A. (1966). *J. Bacteriol.* **91**, 718–724.
Smith, H. (1950). *Biochem. J.* **46**, 352–356, 356–363.
Smith, H. (1951). *Biochem. J.* **48**, 441–447.
Smith, H., Harris-Smith, P. W., amd Stanley, J. L. (1951). *Biochem. J.* **50**, 211–216.
Smith, H., Gallop, R. C., and Stanley, J. L. (1952). *Biochem. J.* **52**, 15–23, 24–32.
Smith, H., Zwartouw, H. T., Gallop, R. C., and Harris-Smith, P. W. (1953). *Biochem. J.* **53**, 673–678.
Sokolski, W. T., Stapert, E. M., and Ferrer, E. B. (1964). *Appl. Microbiol.* **12**, 327–329.
Soo-Hoo, G., and Schnitzer, R. J. (1944). *Arch. Biochem.* **5**, 99–106.
Sophian, L. H., Piper, D. L., and Schneller, G. H. (1952). "The Sulfapyrimidines," pp. 10–11. A. Colish Press, New York.
Speck, R. S., and Wolochow, H. (1957). *J. Infec. Dis.* **100**, 58–69.
Stapert, E. M., Sokolski, W. T., Kaneshiro, W. M , and Cole, R. J. (1964). *J. Bacteriol.* **88**, 532–533.
Stewart, R. H., and Wright, D. N. (1970). *Cryobiology* **6**, 529–532.
Tanguay, A. E. (1959). *Appl. Microbiol.* **7**, 84–88.
Thorp. J. M., Hurst, E. W., and Martin, A. R. (1960). *J. Med. Pharm. Chem.* **2**, 15–30.
Triyanond, C., Sukapanit, S., and Sawyer, W. D. (1969). *Bacteriol. Proc.* p. 68.
Truant, J. P., Hessburg, P. C., and Penn, W. P. (1961). *Antimicrob. Ag. Chemother.* pp. 100–108.
Ungar, J. (1943). *Nature (London)* **152**, 245–246.
Verwey, W. F., West, M. K., and Miller, A. K. (1958). *Antibiot. Chemother. (Washington, D.C.)* **8**, 500–505.
Watson, D. W., and Brandly, C. A. (1949). *Annu. Rev. Microbiol.* **3**, 195–220.
West, M. K., Verwey, W. F., and McConville, C. (1953). *Exp. Med. Surg.* **11**, 131–140.
Wheater, D. W. F., and Hurst, E. W. (1961). *J. Pathol. Bacteriol.* **82**, 117–130.
Wiggins, R. L. (1952). *Amer. J. Ophthalmol.* **35**, 83–100.
Williston, E. H., and Youmans, G. P. (1950). *Amer. Rev. Tuberc.* **62**, 156–159.
Wolinsky, E., and Steenken, W., Jr. (1953). *J. Bacteriol.* **66**, 229–237.
Work, T. S., and Work, S. (1948). "The Basis of Chemotherapy," pp. 1–18. Wiley (Interscience), New York.
Youmans, G. P., and McCarter, J. C. (1945). *Amer. Rev. Tuberc.* **52**, 432–439.
Youmans, G. P., and Youmans, A. S. (1964). *In* "Experimental Chemotherapy" (R. J. Schnitzer and F. Hawking, eds.), Vol. 2, pp. 393–499.
Youmans, G. P., Williston, E. H , and Osborne, R. R. (1949). *Proc. Soc. Exp. Biol. Med.* **70**, 36–37.
Yurchenko, J. A., Piepoli, C. R., and Yurchenko, M. C. (1954). *Appl. Microbiol.* **2**, 53–55.
Yurchenko, J. A., Hopper, M. W., and Warren, G. H. (1967). *Antimicrob. Ag. Chemother.* pp. 602–608.

Modification of Lincomycin

BARNEY J. MAGERLEIN

Research Laboratories, The Upjohn Company
Kalamazoo, Michigan

I.	Introduction	185
II.	Lincomycin-Related Antibiotics	186
	A. Celesticetin	186
	B. Lincomycins	188
III.	Chemical Modification of Lincomycin	189
	A. Lincomycin Esters	189
	B. Synthetic Lincomycins	195
IV.	Microbial Modification of Lincomycin and Clindamycin	224
	A. Sulfoxide Formation	225
	B. 1'-Demethylation	227
	C. Phosphorylation	227
V.	Summary	227
	References	227

I. Introduction

The fermentation broth of *Streptomyces lincolnensis* var. *lincolnensis*, a new species of streptomycete isolated from a soil sample collected in Gering, Nebraska, was found to contain a new antibiotic, named lincomycin.[1] The announcement by Mason and co-workers (1963) of the discovery of lincomycin coincided with chemical and physical characterization of the crystalline hydrochloride (Herr and Bergy, 1963). Initial reports indicated that lincomycin possessed good *in vitro* and *in vivo* potency against a variety of gram-positive microorganisms (Lewis *et al.*, 1963). It was not cross-resistant with known antibiotics and possessed a low order of toxicity (Mason *et al.*, 1963).

Lincomycin has won clinical acceptance as a major antibiotic for the treatment of diseases caused by gram-positive microbes. The reports of extensive experimental and clinical studies with lincomycin were the subject of a recent comprehensive monograph by Herrell (1969).

A study of the structure of lincomycin by Hoeksema and co-workers (1964), employing both classical chemical degradation and nuclear magnetic resonance data, led to the announcement of the structure of lincomycin (**1**). Details of the chemical characterization and nuclear magnetic resonance studies of lincomycin and its degradation products

---

[1]Lincocin® is the registered trademark of the Upjohn Company for lincomycin.

Lincomycin
(1)

subsequently appeared (Herr and Slomp, 1967; Schroeder et al., 1967; Slomp and MacKellar, 1967; Magerlein et al., 1967b).

Lincomycin proved to be a member of a new class of antibiotics characterized by an alkyl 6-amino-6,8-dideoxy-1-thio-D-*erythro*-α-D-*galacto*-octopyranoside joined with a proline moiety by an amide linkage. The availability of this interesting antibiotic presented an excellent opportunity for chemical and microbiological modification.

Before reviewing the chemical and microbiological modification of lincomycin, a summary of the other naturally occurring members of the lincomycin family of antibiotics is in order.

II. Lincomycin-Related Antibiotics

A. CELESTICETIN

Although lincomycin was the first member of its family to have complete structure assignment, a related antibiotic, celesticetin, was reported some years earlier to be produced from the culture broth of a new actinomycete species, *Streptomyces caelestis* (DeBoer et al., 1955). This antibiotic was isolated by Hoeksema and co-workers (1955). Employing some of the techniques used in the structural studies on lincomycin, Hoeksema (1968b) assigned structure (2) to celesticetin. Desalicetin, the alkaline hydrolysis product from celesticetin, was shown to possess structure (3). Both of these antibiotics proved to be less effective than lincomycin when assayed *in vitro* and *in vivo* against a spectrum of microorganisms. These assay data are summarized in Tables I and II (Mason and Lewis, 1965).

TABLE I

In Vitro Antibacterial Activities of Lincomycin and Lincomycin-Related Antibiotics

Compound	Standard curve assay with *Sarcina lutea*[a] ATCC 9341 UC 130 μg/ml	Serial dilution minimal inhibitory concentration[b]					
		Staphylococcus aureus UC 80	*Staphylococcus aureus* ATCC 151 UC 70	*Streptococcus faecalis* UC 3235	*Escherichia coli* ATCC 26 UC 51	*Proteus vulgaris* ATCC 8427 UC 93	*Klebsiella pneumoniae* ATCC 10031 UC 57
1	1000	0.8	0.2	0.4	600	>1000	60
2	80–320	1.6	0.8	0.8	>1000	>1000	50
3	5–12	64	64	64	>1000	>1000	>1000
4	250	1.4	0.8	1.6	>100	>100	>100
5	1300	0.4	0.2	0.4	>100	>100	>100
6	—	—	0.2	25	50	>200	125
7	60	6.0	3.0	1.6	>100	>100	>100

[a] Hanka et al. (1963).
[b] Determinations were made in Brain Heart Infusion medium (Difco); inocula consisted of about 10⁵ organisms; twofold dilutions of the antibiotic were used in each sensitivity determination. End points were read at 20 hours and expressed in μg/ml as the minimal inhibitory concentration of compound.

(2) R = [structure: HO-C6H4-C(=O)-]

(3) R = H

B. LINCOMYCINS

In addition to lincomycin a very closely related antibiotic was isolated from fermentations of *Streptomyces lincolnensis* var. *lincolnensis*. Structural studies indicated this compound to be 4'-depropyl-4'-ethyllincomycin (**4**) (Argoudelis *et al.*, 1965a).

The introduction of various chemicals to the fermentation of *Streptomyces lincolnensis* var. *lincolnensis* was found to induce the formation of lincomycin-related antibiotics. Thus, the addition of DL-ethionine led to the isolation of 1-demethylthio-1-ethylthiolincomycin (**5**) (Argoudelis and Mason, 1965) and 1'-demethyl-1-demethylthio-1'-ethyl-1-ethylthiolincomycin (**6**) (Argoudelis *et al.*, 1970a); methyl thiolincosaminide induced formation of 1'-demethyllincomycin (**7**) (Argoudelis *et al.*, 1965b); and ethyl thiolincosaminide yielded 1'-demethyl-1-demethylthio-1-ethylthiolincomycin (**8**) (Ar-

(4) $R_1 = CH_3$; $R_2 = CH_3$; $R_3 = C_2H_5$
(5) $R_1 = C_2H_5$; $R_2 = CH_3$; $R_3 = C_3H_7$
(6) $R_1 = C_2H_5$; $R_2 = C_2H_5$; $R_3 = C_3H_7$
(7) $R_1 = CH_3$; $R_2 = H$; $R_3 = C_3H_7$
(8) $R_1 = C_2H_5$; $R_2 = H$; $R_3 = C_3H_7$

goudelis *et al.*, 1964). In each case lincomycin was also noted in the fermentation extracts.

The lincomycin-related antibiotics show good activity against gram-positive bacteria, but indifferent activity to gram-negative bacteria (Mason and Lewis, 1965). *In vitro* and *in vivo* testing data are included in Tables I and II, respectively.

III. Chemical Modification of Lincomycin

Chemical modification of lincomycin has evolved into two distinct avenues of research. Each has different, but occasionally overlapping, objectives. The first avenue seeks to enhance the pharmaceutical acceptance of lincomycin by improving its taste, absorption, and depot action. This method envisions the preparation of a derivative of lincomycin whose antibacterial potency is due to the *in vivo* regeneration of lincomycin. Compounds falling into this category are the various esters of lincomycin. The second avenue is directed toward the synthesis of an intrinsically active analog of lincomycin. The desired characteristics of such a synthetic lincomycin are an extended antibacterial spectrum, increased potency, and variation in the rate and degree of absorption.

TABLE II
MOUSE-PROTECTION ASSAY[a] OF LINCOMYCIN
AND LINCOMYCIN-RELATED ANTIBIOTICS

Compound	*Staphylococcus aureus* OSU-284, UC 76 CD_{50} (mg/kg)	
	Subcutaneous	Oral
1	7.1	16
2	200	489
3	246	400
4	7.5	8.1
5	16.8	27
7	53	160

[a] Ten-mouse groups experimentally infected with approximately 100 lethal doses of *S. aureus* and treated with four daily doses of compound. Results established 7 days after infecting.

A. LINCOMYCIN ESTERS

Multiple esters of lincomycin, such as the tetra- and triacylates, possessed a low order of *in vitro* and *in vivo* antibacterial activity (Herr and Slomp, 1967; Hoeksema *et al.*, 1967).

Lincomycin 2-, 3-, 4-, and 7-monoesters, in which the ester group is enzymatically cleaved *in vivo*, possess interesting and varied antibacterial potency as well as different taste and absorption characteristics. These esters are discussed in the succeeding sections.

1. Lincomycin 2-Phosphate

The synthesis of lincomycin 2-phosphate (**9**), described by Morozowich and co-workers (1969a) is outlined in Chart I. 3,4-0-Anisylidene lincomycin (**10**) (Taraszka and Morozowich, 1968) was tritylated selectively at the 7-hydroxyl to afford (**11**). Treatment of (**11**) with phosphorus oxychloride formed the dichlorophosphate intermediate (**12**). Hydrolysis to the ionic phosphate and removal of the blocking groups gave lincomycin 2-phosphate (**9**).

Oesterling and Rowe (1970) showed that lincomycin 2-phosphate (**9**) possessed maximum stability at pH 6–10 in an aqueous solution. The predominate degradative routes above and below this pH range were phosphate ester and thioglycosidic hydrolysis, respectively. Hanka *et al.* (1963) indicated by a microbiological assay with *Sarcina lutea* that lincomycin 2-phosphate (**9**) was less than 2% as active as lincomycin (**1**). When administered either orally or subcutaneously to *Staphylococcus aureus* infected mice (Lewis *et al.*, 1963), 2-phosphate (**9**) was therapeutically equivalent to lincomycin. Since antibacterial activity of 2-phosphate (**9**) probably depends on lincomycin levels, the observed *in vivo* potency is most likely due to the very rapid *in vivo* hydrolysis of the phosphate group.

Taste evaluation of lincomycin 2-phosphate (**9**) in pharmaceutical formulation indicated that statistically it was less bitter than lincomycin (Morozowich *et al.*, 1969a).

2. Lincomycin 2-Acylates and 2-Carbonates

Acylation of 7-0-trityl-3,4-0-anisylidene lincomycin (**11**) with a given acid chloride or chlorocarbonate followed by removal of the blocking groups gave a series of 2-monoesters (**13**) and 2-monocarbonates (**14**) shown in Table III,A.

Standard curve plate assay with *Sarcina lutea* reached a maximum with C_4 to C_8 esters. When administered orally in the mouse protection assay against *Staphylococcus aureus* good activity was obtained with esters of chain length C_4 to C_{16}. When administered subcutaneously the results were erratic, but generally followed the same pattern. The carbonate esters (**14**) were less active than the corresponding acylates (**13**) in these assays (Morozowich *et al.*, 1969b).

CHART I

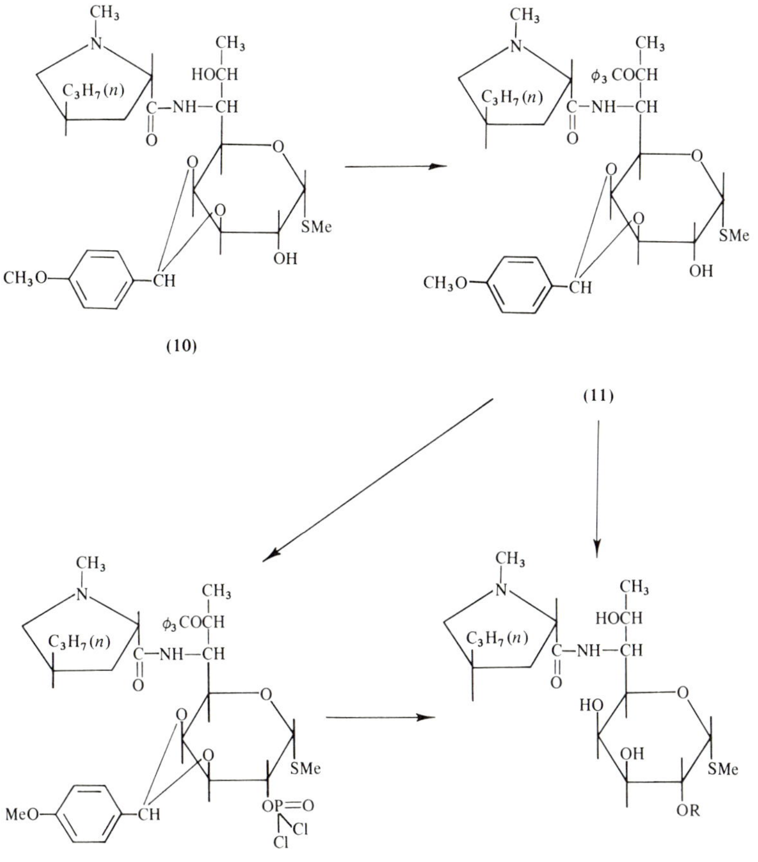

(10) → (11)

↓   ↓

(12) →

(9) R = P(=O)—(OH)$_2$
(13) R = C(=O)—Alkyl
(14) R = C(=O)—O—Alkyl

TABLE III
In Vitro AND In Vivo Antibacterial Assay of Lincomycin Monoesters

Ester	Standard curve assay[a] Sarcina lutea ATCC 9341 Lincomycin = 1	Mouse protection assay[b] Staphylococcus aureus (OSU 284) Lincomycin = 1	
		Subcutaneous	Oral
A. 2-Acylates and 2-Carbonates[b]			
Acetate	0.2	0.20	0.18
n-Propionate	0.40	0.77	1.20
n-Butyrate	0.78	1.03	0.80
n-Valerate	1.6	0.82	1.43
n-Hexanoate	1.05 (1.46)	0.71	1.37
n-Heptanoate	0.98		
n-Octanoate	1.41	1.00	1.40
n-Decanoate	0.22	1.14	1.60
n-Laurate	0.02	0.22	0.90
n-Myristate	0.07	1.80	1.23
n-Palmitate	0.02	0.22	0.79
n-Stearate	0.01	0.07	0.51
n-Butylcarbonate	0.5	0.57	0.69
n-Hexylcarbonate	0.82	0.72	1.07
n-Hexadecylcarbonate	0.02	0.80	0.80
B. 3-Acylates[c]			
n-Propionate	0.03	0.29	0.29
n-Hexanoate	0.49	0.50	1.27
C. 4-Hexanoate[d]			
n-Hexanoate	0.86	0.24	0.15
D. 7-Acylates and 7-Carbonates[e]			
Acetate	0.01	0.08	
n-Butyrate	0.03	0.32	0.36
n-Valerate	0.30	0.55	0.52
n-Octanoate	0.04	0.59	0.55
n-Laurate	0.059	1.06	0.55
n-Palmitate	0.02	0.01	0.05
n-Stearate	0.009	0.01	0.1
Methylcarbonate	0.004		
Hexylcarbonate	0.032	0.71	0.64
Hexadecylcarbonate	0.042	0.86	0.28
E. Lincomycin Hexanoates[d]			
4-Hexanoate	0.86	0.24	0.15

TABLE III (Continued)

7-Hexanoate	0.59	0.59	0.55
3-Hexanoate	4.90	0.50	1.27
2-Hexanoate	1.05	0.71	1.37

[a] Hanka et al. (1963).
[b] Lewis et al. (1963).
[c] Morozowich et al. (1970b).
[d] Morozowich et al. (1970c).
[e] Sinkula et al. (1969).

3. Lincomycin 3-Acylates

Acylation of lincomycin with an acid anhydride in the presence of a sterically hindered strongly basic tertiary amine selectively occurred at the 3-hydroxyl in good yield, This selectivity is attributed to the fact that the 3-hydroxyl is the least hindered of the two equatorial hydroxyls of lincomycin.

The antibacterial assays of lincomycin 3-monoesters are given in Table III,B. The 3-hexanoate was about equal to lincomycin in the mouse protection assay versus *Staphylococcus aureus* when administered orally, but less active when administered subcutaneously (Morozowich *et al.*, 1970b).

4. Lincomycin 4-Hexanoate

Lincomycin 4-hexanoate (16) was prepared by the acid or base catalyzed migration of the 3-hexanoate (15) to the 4-position (Chart II). The migration of the 3-hexanoate to the more stable axial 4-position is favored by virtue of the *cis* relationship of the positions involved. Lincomycin 4-hexanoate (Table III,C) was less active than the other monoesters in both *in vivo* and *in vitro* testing (Morozowich *et al.*, 1970c).

CHART II

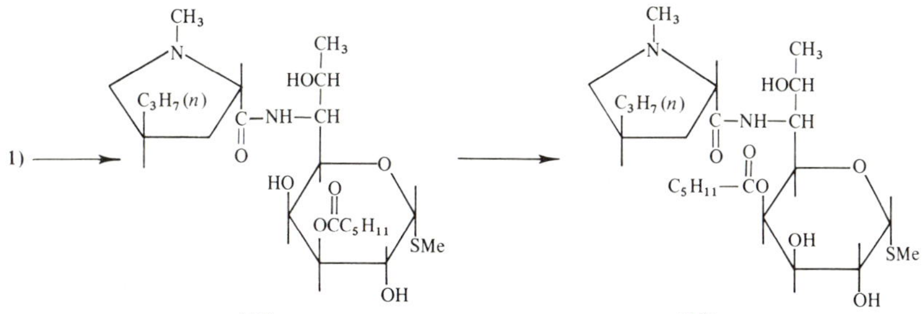

5. Lincomycin 7-Acylates and 7-Carbonates

A series of lincomycin 7-monoesters was described by Sinkula and colleagues (1969). Lincomycin tetra-trimethylsilyl ester (17) on mild acid hydrolysis afforded tris ether (**18**) (Houtman *et al.*, 1968). Acylation followed by removal of the silyl ethers gave lincomycin 7-acylates (**20**) and carbonates (**21**).

In vitro and *in vivo* antibacterial assays of lincomycin 7-esters are summarized in Table III,D. The 7-esters all showed a low order of *in vitro* activity in the standard curve plate assay. The activity which was observed was most likely due to lincomycin formed by hydrolysis of the ester on the plate.

In the mouse protection assay none of the 7-esters were as active as lincomycin when administered orally, although several were competitive with lincomycin when administered subcutaneously.

The data also show that the antibacterial potency of the 7-esters when administered subcutaneously varied with the number of carbon atoms in the acid reaching a maximum at C-12.

CHART III

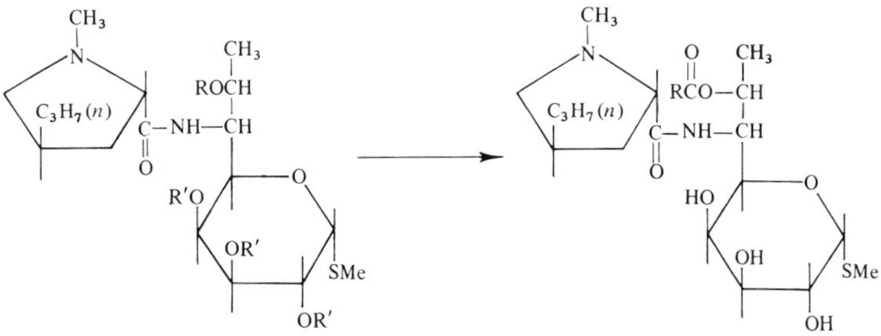

(17) R = R' = Si(CH$_3$)$_3$
(18) R = H; R' = Si(CH$_3$)$_3$
(19) R = C—alkyl or C—O—alkyl; R' = Si(CH$_3$)$_3$

(20) R = alkyl
(21) R = O—alkyl

6. *Comparative Evaluation of Lincomycin Hexanoates*

Antibacterial assays of the four monohexanoates of lincomycin are summarized in Table III,E. These data show that the 4-hexanoate possessed the least antibacterial potency of the positionally isomeric esters. The 2-hexanoate ester was somewhat superior in both *in vivo* and *in vitro* tests. The antibacterial potency of the 3- and 7-hexanoates was intermediate (Morozowich *et al.*, 1970c).

Oesterling (1970b) and Oesterling and Metzler (1970) reported in some detail on the acidic and alkaline isomerization and hydrolysis of lincomycin monohexanoates. They found that in an alkaline medium the 2-, 3-, and 4-hexanoates isomerize to an equilibrium mixture accompanied by hydrolysis to lincomycin. At a low pH the 3- and 4-

hexanoates isomerize to a mixture containing a trace of 2-hexanoate accompanied by hydrolysis to lincomycin. The 7-hexanoate at both high and low pH and the 2-hexanoate at high pH underwent strong forward hydrolysis to lincomycin and hexanoic acid.

B. Synthetic Lincomycins

Synthetic lincomycins, sometimes referred to as lincomycin analogs, possess the lincomycin skeleton, but have variation in one or more substituents. Antibacterial activity associated with such derived antibiotics is intrinsic, rather than due to regeneration of lincomycin.

Two different approaches have been exploited to prepare lincomycin analogs (Magerlein et al., 1967a). In the first method the intact molecule served as the substrate for chemical modification. Modification at C-7 or at the methylthiol group is most readily available by this type of approach. In the first step of a more versatile synthesis of analogs, lincomycin was cleaved by hydrazine hydrate into a sugar fragment, methyl 6-amino-6,8-dideoxy-1-thio-D-*erythro*-α-D-*galacto*-octopyranoside, given the name methyl thiolincosaminide (MTL) (**22**), and after acid hydrolysis, to amino acid, *trans*-4-propylhygric acid hydrochloride (PHA) (**24**) (Schroeder et al., 1967). The resulting fragments were modified by chemical reaction, or alternately similar entities prepared by total synthesis. Recombination of the appropriate fragments permitted the synthesis of analogs modified in either the sugar or amino acid moieties or both. The chemical synthesis of lincomycin and the preparation of 1'-demethyl-4'-depropyl-4'-alkyl analogs described in later sections are excellent applications of this method.

1. Modification of Sugar Moiety

a. 7-Substituted Lincomycins

Birkenmeyer et al. (1965) reported that when lincomycin (**1**) was treated with thionyl chloride in carbon tetrachloride, replacement of the 7-hydroxyl by chlorine occurred to yield a highly active antibiotic. They postulated that the selectivity of the reaction was due to the stepwise formation of a 3,4-cyclic sulfite-2-linear sulfite (**25**) which was hydrolyzed under the conditions of the workup yielding 7(S)-chloro-7-deoxylincomycin hydrochloride (**26**). This antibiotic as the free base form was given the generic name clindamycin.[2] An improved synthesis of clindamycin hydrochloride (**26**) from lincomycin (**1**) using triphenylphosphine dichloride or triphenylphosphine-carbon tetrachloride subsequently appeared (Birkenmeyer and Kagan, 1970).

Another method of synthesis was reported by Magerlein and Kagan (1969a) who converted MTL (**22**) to methyl 7(S)-chloro-7-deoxythiolincosaminide (**27**) which when recombined with PHA (**24**) afforded clindamycin hydrochloride (**26**). These authors presented substantial data to show that the 7-chloro group was in the 7(S)-configuration. This point was conclusively established by conversion of methyl 7(S)-chloro-7-deoxythiolincosaminide (**27**) to dimethylmercaptal (**28**) followed by oxidation to L-chloropropionic acid (**29**) (Birkenmeyer and Kagan, 1970). Confirmation of the structure of clindamycin by X-ray crystallography was reported by Duchamp (1967).

---

[2]Cleocin® is the Upjohn Company registered trademark for clindamycin.

$$(26) \longrightarrow \left[R-C\begin{smallmatrix} O-CH-CH_3 \\ + \\ HN-CH \end{smallmatrix} \right] Cl^- \longrightarrow (1)$$

(30)

Clindamycin shows its maximum stability in aqueous solution at pH 3–5. At a lower pH range the thioglycosidic linkage was slowly hydrolyzed (Oesterling, 1970a). At high pH values clindamycin was solvolized, slowly forming lincomycin. Magerlein and Kagan (1969a) postulated the formation of an intermediate oxazolonium ion (30) in this solvolysis, a hypothesis supported by Oesterling (1970a).

The 7(S)-chloro group of clindamycin is only slowly attacked by nucleophilic reagents as is discussed in a later section.

A comparison of the *in vitro* antibacterial potency of clindamycin (26) and lincomycin (1) is given in Table IV. These data show that clindamycin is significantly more active than lincomycin toward the test organisms shown. Garrison and co-workers (1968) determined the minimal inhibitory concentrations of clindamycin, lincomycin, and

TABLE IV
In Vitro ANTIBACTERIAL ACTIVITY OF CLINDAMYCIN HYDROCHLORIDE

Test Organism	MIC (μg/ml)	
	Clindamycin·HCl	Lincomycin·HCl
Staphylococcus aureus UC 76	0.125	0.5
Staphylococcus aureus UC 70	0.064	0.25
Staphylococcus aureus UC 3216[a]	500	> 1000
Staphylococcus aureus UC 749[b]	0.125	1.0
Streptococcus pyogenes UC 152	0.125	0.25
Streptococcus faecalis UC 157	0.064	0.5
Streptococcus viridans UC 155	0.064	0.25
Bacillus subtilis UC 564	1	32
Escherichia coli UC 51	64	1000
Proteus vulgaris UC 93	250	1000
Klebsiella pneumoniae UC 57	8	125
Salmonella schottmuelleri UC 126	64	1000
Pseudomonas aeruginosa UC 95	1000	> 1000

[a] Lincomycin-resistant *Staphylococcus* strain.
[b] Penicillin-, tetracycline-, and erythromycin-resistant *Staphylococcus* strain.

TABLE V
In Vivo Activity of Clindamycin Hydrochloride[a]

Test organism	LD_{50}	Clindamycin·HCl CD_{50}		Lincomycin·HCl CD_{50}	
		Subcutaneous	Oral	Subcutaneous	Oral
Staphylococcus aureus OSU 284	316	—	20 (13–27)	—	20 (13–27)
Staphylococcus aureus OSU 284	302	21 (16–26)	22 (17–27)	16 (12–20)	23 (13–33)
Staphylococcus aureus OSU 284	126	14 (8.8–22)	28 (19–43)	12 (7.6–19)	35 (23–52)
Streptococcus pyogenes UC 152	347	1.1 (0.7–1.5)	3.2 (2.3–3.9)	0.9 (0.6–1.2)	6.1 (5.1–7.1)
Streptococcus viridans UC 871[c]	10	7.6 (5.3–10.7)	17.7 (13.6–23.0)	10 (6.6–15.1)	25 (17.7–35.2)
Diplococcus pneumoniae I UC 41[b]	50	24 (17–36)	30 (21–43)	21 (15–29)	57 (40–80)
Diplococcus pneumoniae III UC 3214	50	19 (13–27)	22 (17–28)	28 (18–44)	66 (49–89)
Klebsiella pneumoniae UC 58	40	149 (95–234)	566 (389–823)	> 320	> 800
Escherichia coli UC 311	794	> 320	> 800	> 320	> 800

[a] Data supplied by C. Lewis.
[b] Animals infected with D. pneumoniae I were challenged subcutaneously. All other animals were challenged intraperitoneally. Mucin suspensions were employed only with the staphylococcal strains and with the strain of E. coli.
[c] This strain of Streptococcus viridans is unique in that it is pathogenic for mice by the intraperitoneal route.

TABLE VI
BLOOD LEVELS OF CLINDAMYCIN IN MICE AFTER THE ORAL
ADMINISTRATION OF DOSES OF 100 MG/KG OF BODY WEIGHT

Time after dose (minutes)	Clindamycin · HCl average of 17 mice (μg/ml)	Lincomycin · HCl average of 17 mice (μg/ml)
5	4.0	0.4
10	4.5	1.3
20	4.6	0.8
30	2.9	1.9
45	2.1	1.1
60	1.4	1.5
90	0.8	1.6
120	0.6	1.8
150	0.4	0.8
180	0.3	0.4

erythromycin for a number of strains of several genera of clinical isolates of antibiotics. They reported that for the majority of bacteria examined, clindamycin was inhibitory at a lower concentration than was lincomycin. The minimal inhibitory concentration of clindamycin for several strains of coagulase-positive staphylococci was lower than that for erythromycin.

In the mouse-protection assay this differential was not as great as expected from the *in vitro* data (Table V). Table VI shows the concentration of clindamycin found in the blood of mice dosed with the antibiotic. These data indicate that clindamycin is very rapidly and efficiently absorbed following oral administration in the mouse (Magerlein *et al.*, 1967a).

The antiplasmodial activity of clindamycin is discussed in a later section.

Clindamycin is relatively nontoxic as shown by the data of Table VII. The LD_{50} of clindamycin administered orally in rats is $> 5,000$ mg/kg (Magerlein *et al.*, 1967a).

Wagner and co-workers (1968) showed that clindamycin was absorbed extremely rapidly in man following oral administration. Peak serum activities were noted within 24 minutes. The data further indicated that essentially the entire dose was absorbed following oral administration. In another study Wagner *et al.* (1968) showed that in contrast to other antibiotics the absorption efficiency of clindamycin was relatively unaffected by the presence of food in the gastrointestinal tract.

TABLE VII
Acute Toxicity of Clindamycin Hydrochloride[a]

Species	Route[b]	Dose (mg/kg)	Mortality	Time of death	LD_{50}
Mouse	ip	630	10/10	8–29 min	262 (237–290)[c]
		500	10/10	2–4 hr	
		400	8/10	2 hr–4.5 days	
		320	9/10	2 hr–7 days	
		250	7/10	5–6 days	
		200	0/10		
		160	0/10		
Mouse	iv	200	10/10	10–20 sec	143 (143–152)
		160	9/10	10–15 sec	
		125	1/10	1 min	
		100	0/10		
Rat	Oral	5000	2/5	1–2 days	> 5000
		1000	0/5		

[a] Data supplied by J. E. Gray.
[b] Intraperitoneal, ip; Intravenous, iv.
[c] Spearman Karber method of calculating the LD_{50} with 95% confidence limits.

Examination of the biological fluids obtained from the test subjects in the above studies led Brodasky and colleagues (1968) to recognize a highly antibacterial metabolic product. They showed that this material was identical with 1'-demethylclindamycin (**31**) prepared by chemical synthesis (Magerlein and Kagan, 1969a).

Using radioactive tracer techniques, Sun (1970) showed that the metabolic products of clindamycin in the rat were the sulfoxide (**32**) and 1'-demethyl-clindamycin (**31**), while in the dog considerable clindamycin glucuronide (**33**) was also formed.

In order to improve the muscle tolerance of clindamycin when administered intramuscularly, the antibiotic was converted to its 2-phosphate (**34**). The 2-phosphate ester of clindamycin may be prepared by the method outlined above for the synthesis of lincomycin 2-phosphate. Clindamycin 2-phosphate (**34**) was antibacterially inactive *in vitro*, but rapid hydrolysis occurred *in vivo* to give acceptable blood levels of clindamycin. In human single and multiple dose tolerance studies the 2-phosphate was found to have much improved muscle tolerance over the parent antibiotic, clindamycin (Morozowich *et al.*, 1970a).

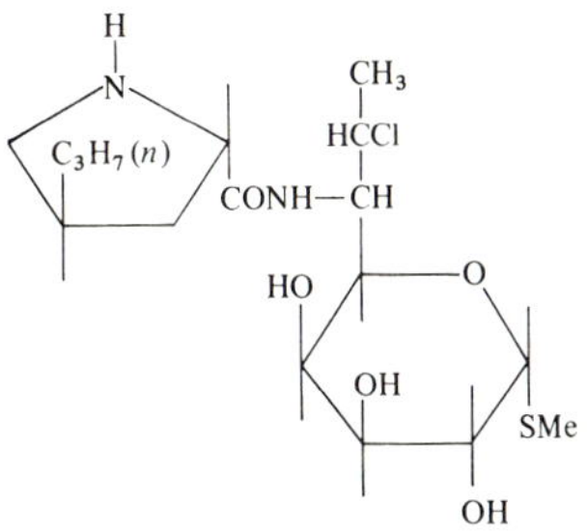

(31)

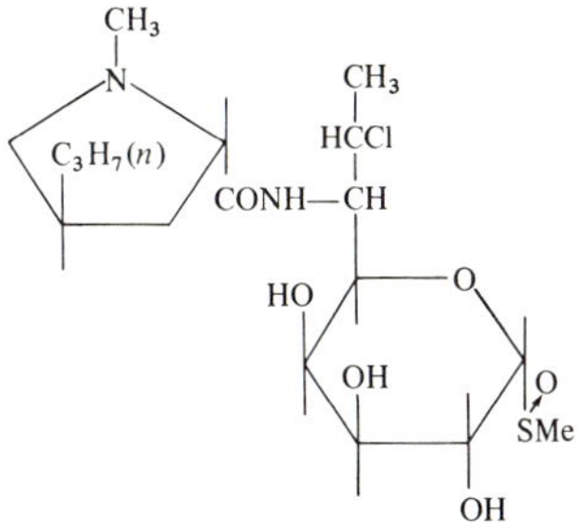

(32)

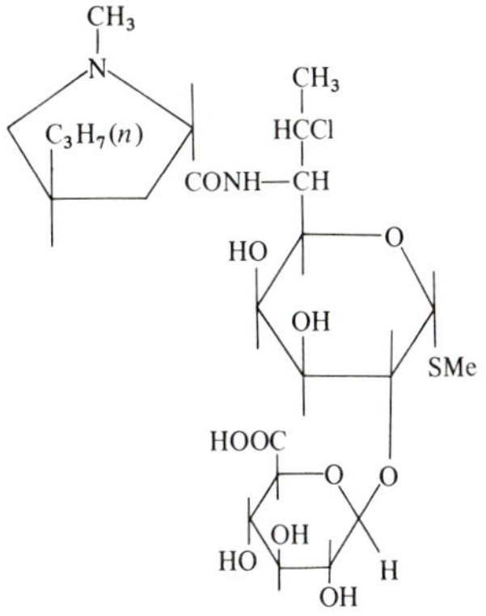

(33)

Aqueous stability studies with clindamycin 2-phosphate indicated that at pH less than 6, phosphate ester and thioglycoside hydrolysis predominate while at pH greater than 6, solvolysis of the 7(S)Cl to 7(R)OH is favored (Oesterling and Rowe, 1970).

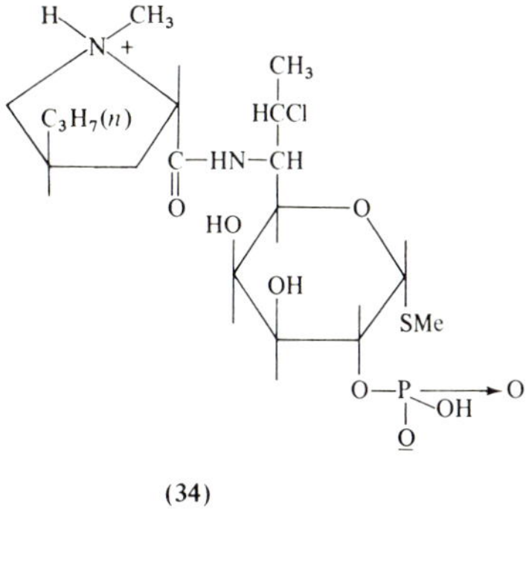

(34)

TABLE VIII
ANTIBACTERIAL ACTIVITIES OF 7-SUBSTITUTED-7-DEOXYLINCOMYCINS

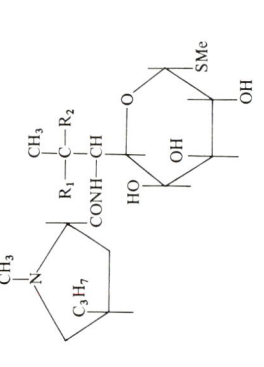

Compound	Standard curve assay with Sarcina lutea ATCC 9341 UC 130	Serial dilution minimal inhibitory concentration						Mouse protection assay Staphylococcus aureus OSU 284 – UC 76	
		Staphylococcus aureus OSU 284 UC 76	Staphylococcus aureus UC 552	Streptococcus faecalis UC 3235	Escherichia coli ATCC 26 UC 51	Proteus vulgaris ATCC 8427 UC 93	Salmonella schottmuelleri ATCC 9149 UC 126	Subcutaneous	Oral
1. $R_1 = OH; R_2 = H$	1	0.4	0.8	12.5	400	800	4000	1.0	1.0
26. $R_1 = H; R_2 = Cl$	4	0.1	0.1	6.2	25	250	64	1.0	1.3
35. $R_1 = H; R_2 = Br$	4	0.05	0.05	6.2	100	100	50	0.9	1.5
36. $R_1 = H; R_2 = I$	4–5	0.05	0.05	3.2	12.5	50	50	—	—
37. $R_1 = R_2 = H$	0.3	1.6	1.6	6.4	>200	>200	>200	—	—
38. $R_1 = H; R_2 = N_3$	0.8	0.4	0.4	25	>200	>200	>200	—	—
39. $R_1 = H; R_2 = NH_2$	0.01	50	50	100	>200	>200	>200	—	—
40. $R_1 = H; R_2 = CN$	0.8	0.8	0.4	25	>200	>200	>200	0.9	—
41. $R_2 = O$	0.02	50	50	50	>200	>200	>200	—	—
42. $R_1 = H; R_2 = OH$	0.5	1.6	1.6	25	>200	>200	>200	0.4	0.8
43. $R_1 = Cl; R_2 = H$	2	0.8	0.8	12.5	>200	>200	>200	0.4	0.5
44. $R_1 = SH; R_2 = H$	0.1	3.2	3.2	200	>200	>200	>200	1.5	—
45. $R_1 = H; R_2 = SH$	0.3	1.6	1.6	100	>200	>200	>200	0.8	—

MODIFICATION OF LINCOMYCIN 203

7(S)-Bromo and 7(S)-iodo-7-deoxylincomycin (**35** and **36**), were prepared from lincomycin by treatment with triphenylphosphine and the appropriate carbon tetrahalide. As shown in Table VIII, these halolincomycins are potent antibiotics, both being somewhat more active than clindamycin on *in vitro* testing (Birkenmeyer and Kagan, 1970).

Reduction of clindamycin with zinc in aqueous solution led to the isolation of 7-deoxylincomycin (**37**). This compound was significantly less active than the halolincomycins (Table VIII) (Birkenmeyer, 1969; Magerlein *et al.*, 1967a).

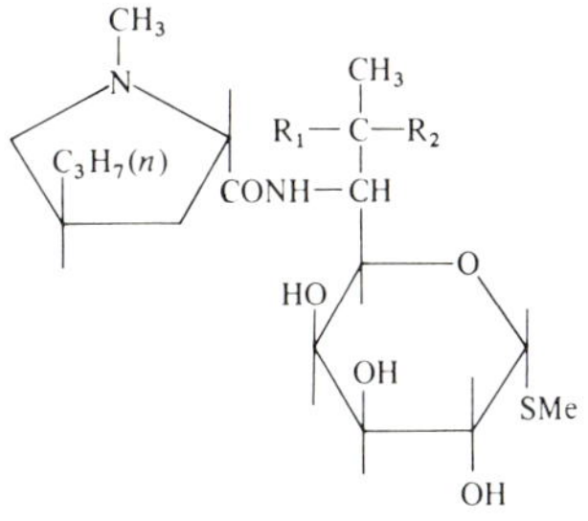

(35) $R_1 = H, R_2 = Br$
(36) $R_1 = H, R_2 = I$
(37) $R_1 = R_2 = H$
(38) $R_1 = H, R_2 = N_3$
(39) $R_1 = H, R_2 = CN$

(40) $R_1 = H, R_2 = NH_2$
(41) $R_1, R_2 = O$
(42) $R_1 = H, R_2 = OH$
(43) $R_1 = Cl, R_2 = H$

When treated with certain nucleophilic reagents in dimethylformamide the 7-chloro group of clindamycin was replaced, most probably with inversion. In this manner 7-azido and 7-cyano-7-deoxylincomycin (**38** and **39**), were prepared. Hydrogenation of 7-azido-7-deoxylincomycin afforded 7-amino-7-deoxylincomycin (**40**) (Bannister, 1969; Birkenmeyer, 1969).

Hoeksema (1965) described the synthesis of 7-deoxy-7-ketolincomycin (**41**) by careful oxidation of 2,3-isopropylidene lincomycin followed by hydrolysis. Reduction of 2,3-isopropylidene 7-deoxy-7-ketolincomycin followed by hydrolysis led to the isolation of 7-deoxy-7(S)-hydroxylincomycin (**42**). When treated with triphenylphosphine and carbon tetrachloride the latter was converted to 7-deoxy-7(R)-chlorolincomycin (**43**) (Birkenmeyer and Kagan, 1970). It is noteworthy that just as the 7-epimer of lincomycin was about one-half as active as lincomycin so was the 7-epimer (**43**) of clindamycin about one-half as active as clindamycin.

204   BARNEY J. MAGERLEIN

The synthesis of 7-deoxy-7(R) and 7(S)-thiolincomycin (**44** and **45**) outlined in Chart IV were reported by Magerlein and Kagan (1969b). Both thio analogs were about as active as lincomycin in the mouse protection assay when administered subcutaneously, but less active on *in vitro* assay.

The antibacterial activities of 7-deoxy-7-substituted lincomycins are compiled in Table VIII.

CHART IV
PREPARATION OF 7-DEOXY-7(R) AND 7(S)-THIOLINCOSAMINIDE

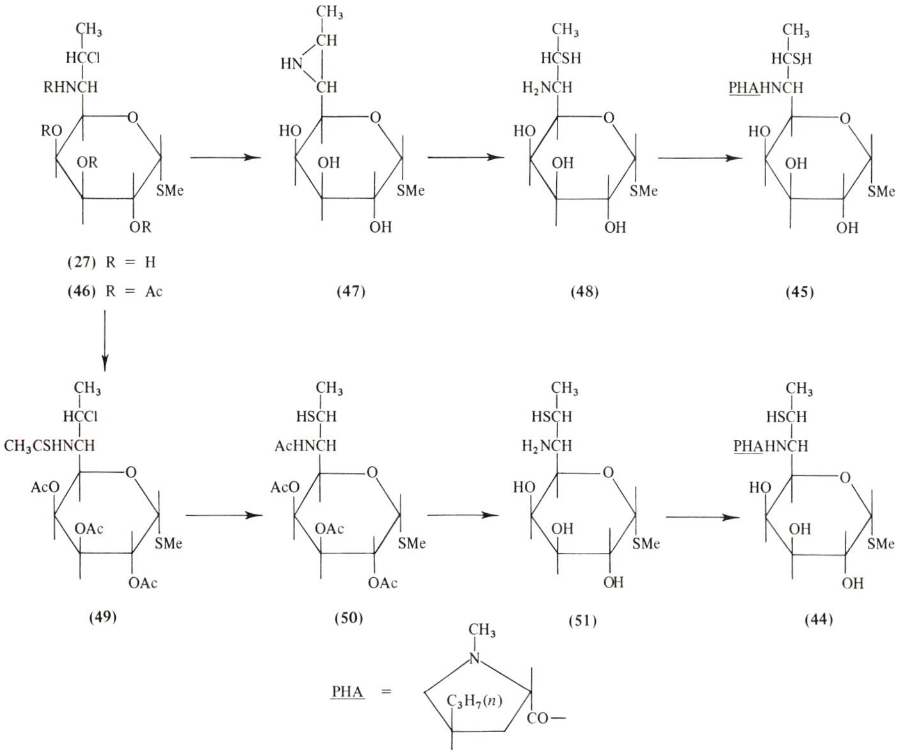

b. *1-Substituted Lincomycins*

In an earlier section of this review lincomycin-related antibiotics containing groups at C-1 other than methylthio were shown to possess significant antibacterial activity. Thus, celesticetin (**2**), which possesses a β-hydroxy-ethyl group as the salicylate ester at C-1 and 1-demethylthio-1-ethylthiolincomycin (**5**), which contains a 1-ethylthio group were isolated from fermentation sources.

CHART V
SYNTHESIS OF 1-DEMETHYLTHIO-1-ALKYLTHIOLINCOMYCINS

(53a) R = R' = Alkyl
(53b) R = Alkyl, R' = Methyl
(53c) R = R' = Methyl

(52) R = Alkyl

(54)

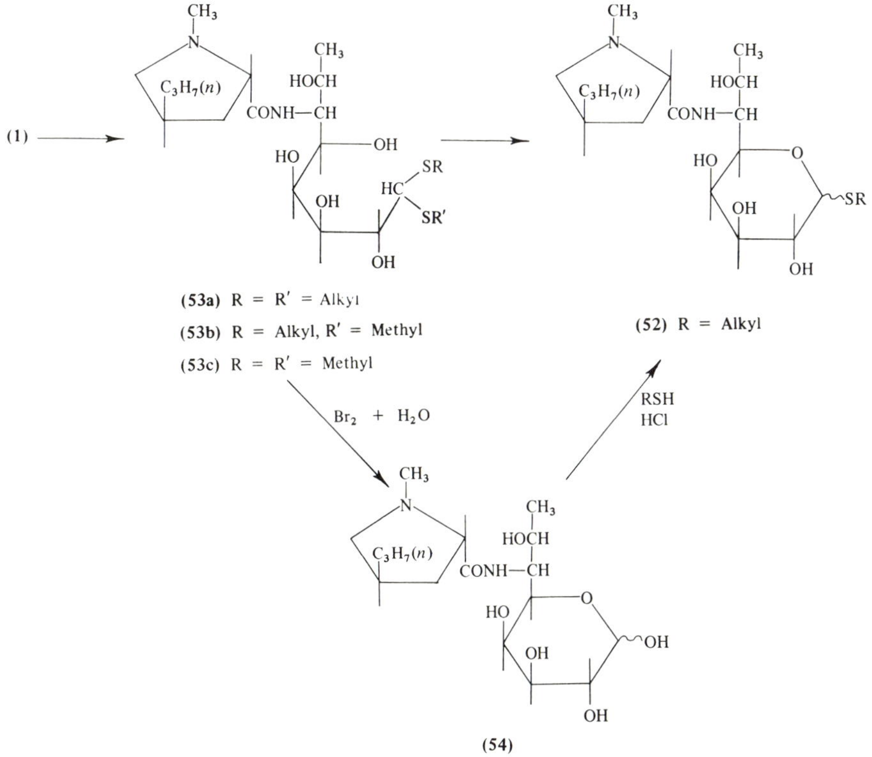

Birkenmeyer et al. (1965) described a series of 1-demethylthio-1-alkylthiolincomycins (52) prepared from lincomycin as shown in Chart V. In one process intermediate dialkyldithioacetals (53) prepared from lincomycin (1) were cyclized by treatment with acid to afford the given analogs. Although this sequence was used to prepare several homologs of lincomycin, the process was complicated by the necessity of separating the three possible dialkyldithioacetals (53a, b, c) before cyclization. To circumvent this problem, lincomycin (1) was brominated in aqueous solution to form lincomycose (54) which was treated *in situ* with alkanethiol and acid. A mixture of the epimeric 1-demethylthio-1-alkanethiolincomycins (52) and the dialkylthioglycoside (53) were isolated by chromatography.

The antibacterial activities of 1-demethylthio-1-alkylthiolincomycins prepared by these processes are given in Table IX. These data indicate that both the substituent and its configuration affect the bioactivity. 1-Demethylthio-1-ethylthiolincomycin (52, R = α-ethyl) was about as potent as lincomycin in *in vitro* and *in vivo* testing. The 1-β

TABLE IX
ANTIBACTERIAL ACTIVITIES OF 1-DEMETHYLTHIO-1-ALKYLTHIOLINCOMYCINS

Compound	Standard curve assay with Sarcina lutea ATCC 9341 UC 130	Serial dilution minimal inhibitory concentration						Mouse protection assay Staphylococcus aureus OSU 284 – UC 76	
		Staphylococcus aureus OSU 284 UC 76	Staphylococcus aureus UC 552	Streptococcus faecalis UC 3235	Escherichia coli ATCC 26 UC 51	Proteus vulgaris ATCC 8427 UC 93	Salmonella schottmuelleri ATCC 9149 UC 126	Subcutanous	Oral
α-CH$_3$	1.0	0.4	1.6	25	>200	>200	>200	1	1
α-C$_2$H$_5$	1.4	0.2	0.8	25	>200	>200	>200	1	1
β-C$_2$H$_5$	0.15	3.2	12.5	200	>200	>200	>200	–	0.16
α-CH(CH$_3$)$_2$	0.7	0.8	1.6	25	>200	>200	>200	0.3	0.2
α-C$_4$H$_9$(n)	2.2	0.4	0.8	50	>200	>200	>200	0.8	0.3
α-C$_{11}$H$_{23}$	<0.01	200	100	200	200	>200	>200	Inactive 320 mg/kg	–
β-C$_{11}$H$_{23}$	0.01	100	200	200	>200	>200	>200	Inactive 160 mg/kg	–
αC$_6$H$_{11}$	1.6	0.4	0.8	25	>200	>200	>200	–	Inactive 80 mg/kg

anomer (**52**, R = β-ethyl) was only about one-tenth as active as the α-anomer. The 1-α-butylthio analog (**52**, R = α-butyl), was also interesting, being quite active on *in vitro* testing, particularly in the standard curve lincomycin assay. *In vivo* testing was disappointing, particularly when the compound was administered orally.

c. *Miscellaneous Variation in the Sugar Moiety*

Methyl thiolincosaminide (**22**) was converted to methyl N-methyl-thiolincosaminide (**58**) as outlined in Chart VI (Magerlein, 1971). Attempts to couple either methyl N-methyl or N-benzylthiolincosaminide (**58** or **56**) with *trans*-4-propylhygric acid to form lincomycin were unsuccessful.

Variations in the stereochemistry and substituents at C-2 and C-3 indicated that modification of the ring hydroxyls was relatively unrewarding (Bannister, 1969).

SYNTHESIS OF 1-DEMETHYLTHIO-1-ALKYLTHIOLINCOMYCINS
CHART VI

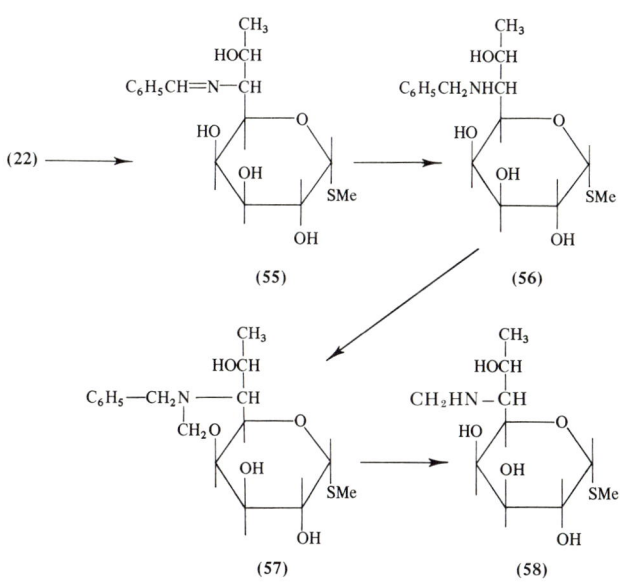

2. *Modification of Proline Moiety*

a. *4'-Alkyl-4'-depropyllincomycins and Related Compounds*

A series of 4'-alkyl-4'-depropyllincomycins (**59**), prepared as outlined in Chart VII, was reported by Magerlein and colleagues (1967c). 1-Carbobenzoxy-4-keto-L-proline (**60**) was treated with a triphenylalkylidene phosphorane to form unsaturated acid (**61**). Catalytic reduction of (**61**) afforded the saturated acids (**62**) as a mixture of epimers at C-4. The epimers were not separated, but condensed with MTL (**22**) to afford (**63**). Decarbobenzoxylation of (**63**) gave 1'-demethyl-4'-alkyl-4'-depropyllincomycins (**64**). These compounds possessed significant antibacterial activity and were also useful intermediates for the preparation of other active agents, as is described later. Reductive alkylation of (**64**) with either formaldehyde or acetaldehyde, followed by chromatographic separation of the epimers, led to the isolation of derived antibiotics (**59**, R_3 = Me or Et). The *trans* isomers possess the same stereochemistry at C-4' as lincomycin.

Antibacterial testing of 4'-alkyl-4'-depropyllincomycins (**59**) is summarized in Table X. The lower members of the 1'-demethyl series showed only slight *in vitro* activity; however, as the homologous series was ascended, *in vitro* potency increased until the 4'-octyl

CHART VII
SYNTHESIS OF 4'-ALKYL-4'-DEPROPYLLINCOMYCINS

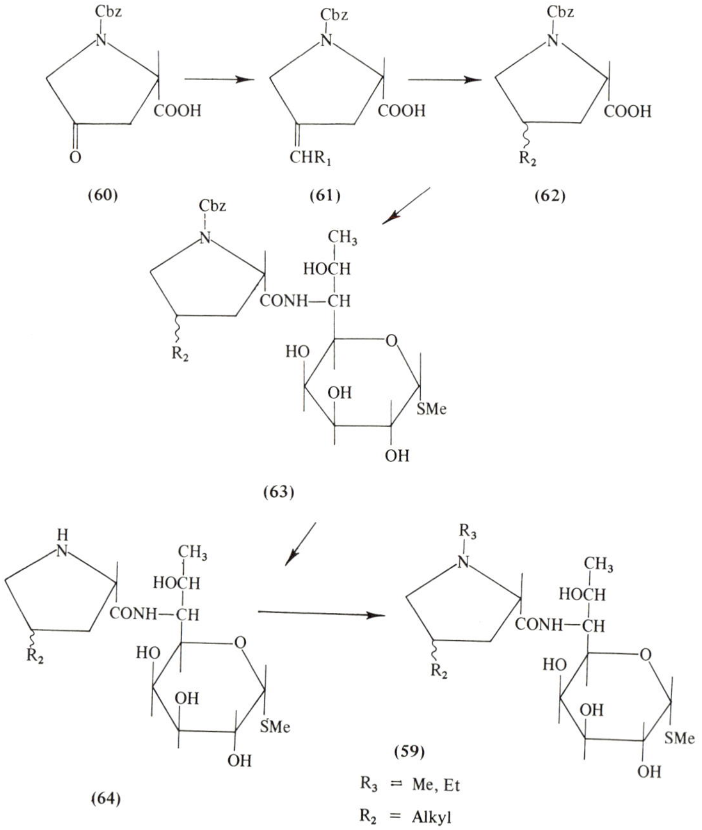

(65) R_1 = HO; R_2 = H; R_3 = OC_2H_5 (trans)

(66) R_1 = H; R_2 = Cl; R_3 = OC_2H_5 (trans)

(67) R_1 = H; R_2 = Cl; R_3 = >CH—CH_2—CH_3

(68a) R_1 = HO; R_2 = H; R_3 = ~CH_2—$COOC_2H_5$

(68b) R_1 = HO; R_2 = H; R_4 = —OTs

(70) R_1 = HO; R_2 = H
(71) R_1 = H; R_2 = Cl

homolog possessed 60% the activity of lincomycin in the standard curve assay. The 1'-demethyllincomycins were more potent toward streptococci bacteria than would be expected from their lincomycin standard curve assay. *In vivo* potency was not influenced by the substituent at C-4' in this series.

In the lincomycin series the standard curve assay indicated gradual increase in potency as the number of carbon atoms at C-4' was increased. A maximum was noted when pentyl or hexyl was present at C-4'. The broth dilution assays also indicated increased potency for these analogs. *In vivo* testing showed less dramatic but consistent increase in potency with a maximum of activity noted for a pentyl substituent. The 4'-*cis* counterparts of the analogs described above were about one-half as active as the *trans* compounds.

TABLE X
ANTIBACTERIAL ASSAY OF 4'-ALKYL-4'-DEPROPYLLINCOMYCINS

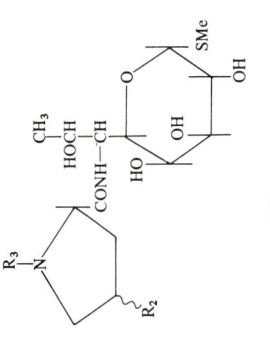

(59)

Compound	Standard curve assay with Sarcina lutea ATCC 9341 UC 130	Serial dilution minimal inhibitory concentration						Mouse protection assay Staphylococcus aureus OSU 284–UC 76	
		Staphylococcus aureus OSU 284 UC 76	Staphylococcus aureus UC 552	Streptococcus faecalis UC 157	Escherichia coli ATCC 26 UC 51	Proteus vulgaris ATCC 8427 UC 93	Salmonella schottmuelleri ATCC 9149 UC 126	Subcutaneous	Oral
I. 1'-Demethyl Series, $R_3 = H$									
$R_2 = H$	0.02	>200	>200	>200	>200	>200	>200		
n-C_3H_7 trans	0.06	6.0	3.2	1.6	>100	>100	>100	0.2	0.1
n-C_4H_9 cis-trans	0.01	3.2	6.4	1.6	>200	>200	>200	0.3	
n-C_5H_{11} cis-trans	0.15	3.2	6.4	0.8	>200	>200	>200	0.25	0.2
n-C_6H_{13} cis-trans	0.33	1.6	6.4	3.2	>200	>200	>200		
n-C_7H_{13} cis-trans	0.4	0.8	3.2	0.2	200	>200	>200		
n-C_8H_{17} cis-trans	0.6	1.6	6.4	0.4	>200	>200	>200	0.1	—
n-$C_{18}H_{37}$ cis-trans	<0.01	>200	>200	>200	>200	>200	>200		

MODIFICATION OF LINCOMYCIN 211

II. 1'-Methyl Series, $R_3 = CH_3$

$R_2 = H$		0.25	50	>200	>200	>200	>200	>200	0.6	
$-C_2H_5$										
$n-C_3H_7$	trans	1	0.4	0.8	0.4	>200	>200	>200		
$n-C_4H_9$	trans	2.1	0.2	0.2	0.4	>200	>200	>200	1.2	1.2(2.8)
$n-C_5H_{11}$	trans	3.5	0.2	0.2	0.2	200	>200	>200	1.5	1.3
$n-C_6H_{13}$	trans	3.6	0.2	0.2	0.2	100	>200	>200		
$n-C_7H_{15}$	trans	1.5	0.4	0.4	0.4	200	>200	>200	0.51	0.5
$n-C_8H_{17}$	trans	1.0	0.4	0.8	0.4	>200	>200	>200	0.3	
$n-C_3H_7$	cis	0.5				>200	>200	>200	0.03	
$n-C_4H_9$	cis	1.4	0.4	0.8	0.4	>200	>200	>200		
$n-C_5H_{11}$	cis	1.8	0.4	0.2	0.2	200	>200	>200	0.8	
$n-C_6H_{13}$	cis	2.1	0.8	0.4	0.4	200	>200	>200	—[a]	
$n-C_7H_{15}$	cis	0.9	1.6	1.6	0.8	>200	>200	>200	—[a]	
$n-C_8H_{17}$	cis	0.6								

III. 1'-Ethyl Series, $R_3 = C_2H_5$

$R_2 = H$		0.02	>200	>200	>200	>200	>200	>200		
$n-C_3H_7$	trans	1	0.4	0.2	0.4	>200	>200	>200	1	—
$n-C_4H_9$	trans	1.2	0.2	0.2	0.2	50	>200	>100	1.6	1.5
$n-C_5H_{11}$	trans	3.0	0.2	0.2	0.1	25	100	25	0.9	—
$n-C_6H_{13}$	trans	2.0	0.4	0.4	0.2	50	200	50	0.5	—
$n-C_8H_{17}$	trans	0.4	0.8	3.2	0.8	200	>200	>200	0.1	0.2
$n-C_3H_7$	cis	0.5				>200	>200	>200	0.5	—
$n-C_4H_9$	cis	0.7	0.4	0.4	0.4	100	>200	200	1	1
$n-C_5H_{11}$	cis	1.2	0.4	0.4	0.2	200	>200	>200	0.8	—
$n-C_6H_{13}$	cis	1.2	0.8	0.8	0.4	>200	>200	>200	0.3	—
$n-C_8H_{17}$	cis	0.4	1.6	1.6	0.8	>200	>200	>200		

[a] Inactive at dose tested.

Somewhat the same pattern was found in the 1'-ethyl series for both the *trans* and *cis* isomers. The 4'-pentyl or 4'-butyl homologs were more potent than higher or lower members of the series. The 1'-ethyl analogs in this series as well as in other series showed noticeably more activity toward gram-negative bacteria, such as *Escherichia coli* and *Klebsiella pneumoniae*, than the 1'-methyl analogs.

The concentration of antibiotic found in the blood of mice after oral dosing with members of a series of 4'-alkyl-4'-depropyllincomycins is given in Table XI. When the number of carbon atoms in the 4'-substituent was increased, the concentration of the analog found in the blood was found to be greater than that observed for lincomycin.

TABLE XI
CONCENTRATION OF 4'-ALKYL-4'-DEPROPYLLINCOMYCINS IN THE BLOOD OF MICE AFTER ORAL DOSES OF 100 MG/KG OF BODY WEIGHT

Substituent at C-4	Area under blood level curve relative to that of lincomycin
C_2H_5	0.6
n-C_3H_7	1.0
n-C_4H_9	1.6
n-C_5H_{11}	1.4

b. Other 4'-Substituents

4'-Depropyl-4' ethoxylincomycin (65), a bioisoster of lincomycin, was prepared in both the 4'-*trans* and 4'-*cis* series (Magerlein, 1967). Both compounds were only about 2% as potent as lincomycin on *in vitro* testing and inactive *in vivo* at the levels tested. 4'-Depropyl-4'-ethoxyclindamycin (66), was found to be more potent on *in vitro* testing than its lincomycin analog (Magerlein, 1968).

4'-Depropyl-4'-propylidene clindamycin (67), in which the three carbon substituents at 4' lie in the same plane as the proline ring, was found to possess about 0.3 the potency of lincomycin on *in vitro* testing (Magerlein, 1969).

4'-Depropyl-4'-carboethoxymethyl and 4'-tosyloxylincomycins (68a and 68b) were essentially inactive on antibacterial testing (Hoeksema, 1968a).

c. *5'-Alkyl-4'-depropyllincomycins*

4'-Depropyl-5'-propyllincomycin (**70**) prepared from 1-methyl-5-propyl proline was reported to possess a low order of antibacterial activity (Herr, 1967). A more recent synthesis of 4'-depropyl-5'-propylclindamycin (**71**) by Dolak (1969) indicated modest antibacterial potency for this 4'-propyl analog.

d. *3'-Alkyl-4'-depropyllincomycin*

Racemic 1-methyl-3-propylproline (**72**) was reported not to form lincomycin analogs when treated with MTL (**22**) (Magerlein *et al.*, 1967b).

e. D-*Amino acids*

In some experiments the 1-methyl-4-propylproline obtained by the hydrazinolysis of lincomycin was almost completely racemized (Magerlein *et al.*, 1967b). When this material was condensed with MTL (**22**) or methyl 7(S)-chloro-7−deoxythiolincosaminide (**27**) the resulting product possessed only 50% of the expected antibacterial activity, indicating that the D isomer possed a low order of activity (Schroeder and Birkenmeyer, 1967).

In the synthesis of 1'-demethyl-4'-depropyl-4'-pentylclindamycin (**73**), a highly potent derived antibiotic discussed in greater detail below, the D isomer at 2'- was also prepared and found to be essentially inactive (Magerlein, 1968).

f. *1'-Alkyl-1'-demethyllincomycins*

In Section 2,a the reductive alkylation of 1'-demethyllincomycins with formaldehyde or acetaldehyde was reported to yield the 1'-methyl and 1'-ethyl compounds, respectively. The 1'-ethyl analogs were about as active as the 1'-methyl compounds, but possessed greater *in vitro* activity toward gram-negative bacteria. Reductive alkylation with higher aldehydes led to decreased antibacterial potency, though the 1'-propyl and 1'-isopropyl homologs still possessed significant activity (Magerlein *et al.*, 1967a; Birkenmeyer, 1969; Magerlein, 1969).

3. *Modification in Both the Amino Acid and Sugar Moieties*

Modification of the sugar portion of lincomycin, particularly at C-7, was shown in a preceding section to lead to antibiotics of increased potency. Also a variation in the C-4' and N-1' substituent of the proline fragment produced enhanced and varied activity. To determine

CHART VIII

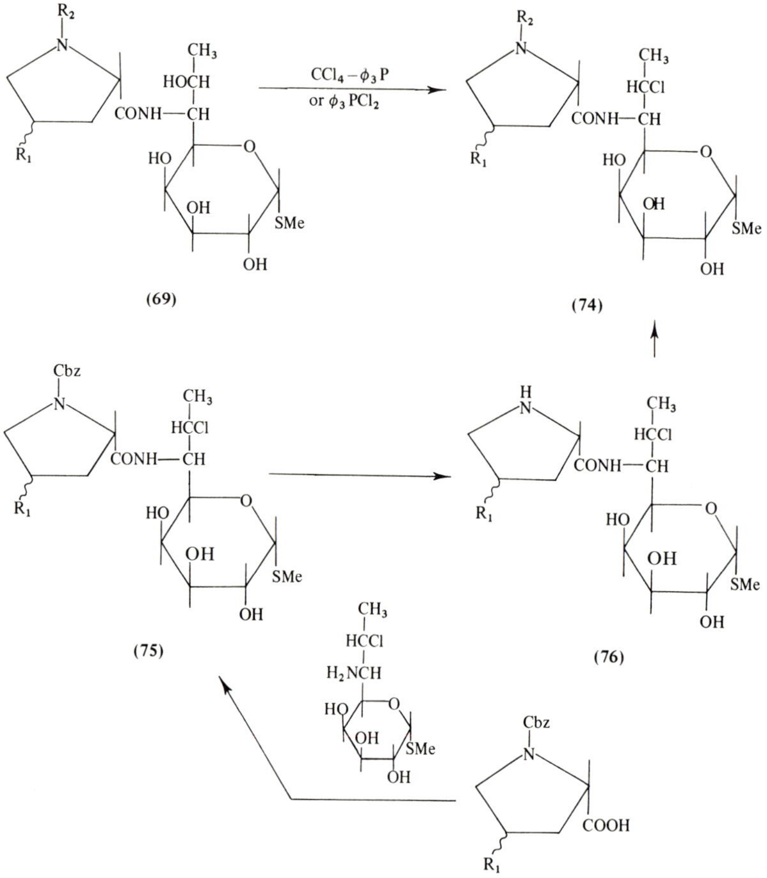

R_1 = Alkyl *cis* or *trans*
R_2 = H, Me, Et

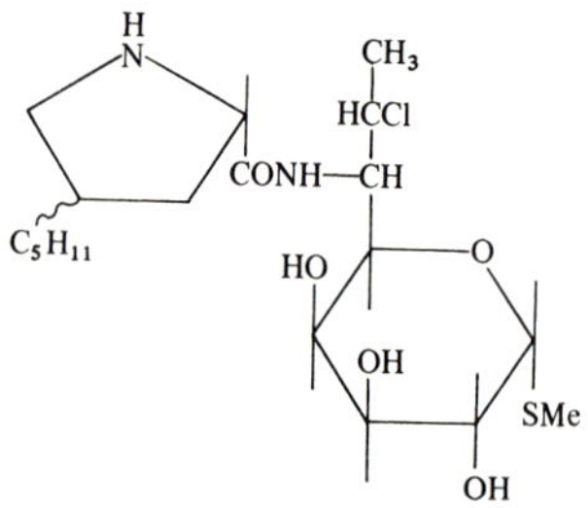

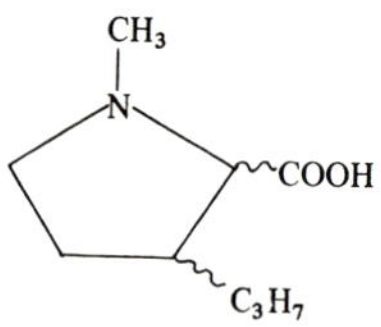

whether these effects were cumulative became of prime importance. This led to the synthesis and testing of lincomycin analogs substituted in both the amino acid and sugar moieties. Some of the more interesting multiple variations are discussed in the following paragraphs.

a. 4'-Alkyl-4'-depropylclindamycins

The general methods were developed by Magerlein and colleagues (1967a, 1969a) for the synthesis of 4'-alkyl-4'-depropylclindamycins (74). These processes are generically summarized in Chart VIII. The initial method used to prepared these analogs was simply to chlorinate the corresponding lincomycin analog whose synthesis was described in Section 2,a. Greater flexibility in analog synthesis was offered by the second method exemplified by (62) → (75) → (76) → (74). The 1'-demethylclindamycins (76), intermediates in this process, proved to be highly potent antibiotics whose antimicrobial activity is described in the following section. Reductive alkylation of (76) permitted the introduction of various alkyl groups at N-1', though in practice the substituent at this position was generally limited to methyl or ethyl.

The antibacterial activities of 4'-alkyl-4'-depropylclindamycins are summarized in Table XII (Magerlein and Kagan, 1969a). The structure-activity relationships noted in this series were very similar to those observed in the 4'-alkyl-4'-depropyllincomycin series. Once again, both *in vitro* and *in vivo* potency increased as the number of carbon atoms in the C-4' substituent increased, reaching a maximum with the pentyl group. The *cis* isomers were generally less active than the *trans* isomers. The pentyl analogs in both the 1'-methyl or 1'-ethyl series [(74), $R_1 = n\text{-}C_5H_{11}$; $R_2 = CH_3$ and (74), $R_1 = n\text{-}C_5H_{11}$; $R_2 = C_2H_5$], were the outstanding compounds in this group.

b. 4'-Alkyl-1'-demethyl-4'-depropylclindamycins

As noted above these compounds were isolated as intermediates in the synthesis of 4'-alkyl-4'-depropylclindamycins (Magerlein and Kagan, 1969a). Except for 1'-demethylclindamycin [(76), $R_1 = n\text{-}C_3H_7$], which was prepared by chlorination of 1'-demethyllincomycin, the 4'-alkyl-1'-demethyl-4'-depropylclindamycins (76) were isolated as a mixture of epimers at C-4'. Separation of epimers by chromatography proved to be unrewarding so that initial antimicrobial testing was performed on the epimeric mixtures. Indirect methods to separate epimers and to prepare pure *trans* and *cis* isomers of 4'-alkyl-1'-demethyl-4'-depropylclindamycins (76) were recently developed (Magerlein, 1969).

TABLE XII
ANTIBACTERIAL ACTIVITIES OF 4'-ALKYL-4'-DEPROPYLCLINDAMYCIN HYDROCHLORIDES

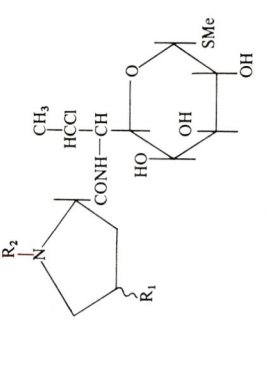

(74)

Compound[a]	Standard curve assay with *Sarcina lutea* ATCC 9341 UC 130	Serial dilution minimal inhibitory concentration						Mouse protection assay *Staphylococcus aureus* OSU 284 – UC 76	
		Staphylococcus aureus OSU 284 UC 76	*Staphylococcus aureus* UC 552	*Streptococcus faecalis* UC 157	*Escherichia coli* ATCC 26 UC 51	*Proteus vulgaris* ATCC 8427 UC 93	*Salmonella schottmuelleri* ATCC 9149 UC 126	Subcutaneous	Oral
Lincomycin	1	0.4	0.8	12.5	400	800	4000	1	1
$R_1 = C_2H_5^t$; $R_2 = CH_3$	0.5	0.8	0.4	6.4	200	200	200	0.3	0.7
$R_1 = C_3H_7^t$; $R_2 = CH_3$	4	0.125	0.125	6.2	25	250	64	1	1.3

MODIFICATION OF LINCOMYCIN

$R_1 = C_4H_9{}^t$; $R_2 = CH_3$	3.6–4	0.1	0.05	1.6	100	50	50	1	1.4
$R_1 = C_4H_{11}{}^c$; $R_2 = CH_3$	1.7	0.1	0.1	1.6	100	50	100	—	—
$R_1 = C_5H_{11}{}^t$; $R_2 = CH_3$	5–9	0.05	0.05	0.4	50	100	50	1.75	1.88
$R_1 = C_5H_{11}{}^c$; $R_2 = CH_3$	1	0.4	0.8	3.2	>200	>200	>200		0.3
$R_1 = C_3H_7{}^t$; $R_2 = C_2H_5$	2	0.4	0.4	25	50	100	100	0.6	1.4
$R_1 = C_4H_9{}^t$; $R_2 = C_2H_5$	4.7	0.01	0.1	0.4	25	50	25	—	1.5
$R_1 = C_4H_9{}^c$; $R_2 = C_2H_5$	1.2	0.2	0.2	1.6	100	200	200	—	0.6
$R_1 = C_5H_{11}{}^t$; $R_2 = C_2H_5$	4.1	0.1	0.1	0.4	50	50	50		
$R_1 = C_5H_{11}{}^c$; $R_2 = C_2H_5$	2.0	0.2	0.2	0.8	100	>200	100	—	—
$R_1 = C_2H_5{}^{c-t}$; $R_2 = H$	0.9	0.2	0.2	3.2	200	200	>200	2.8	1.0
$R_1 = C_3H_7{}^t$; $R_2 = H$	4.6	0.05	0.1	3.2	25	50	50	5.8	3.5
$R_1 = C_3H_7{}^c$; $R_2 = H$	3.6	0.05	0.1	1.6	100	100	100	4.5	
$R_1 = C_3H_7{}^{c-t}$; $R_2 = H$	4.2	0.05	0.05	1.6	50	100	100	4.0	2.3
$R_1 = C_4H_9{}^{c-t}$; $R_2 = H$	5	0.05	0.05	1.6	12.5	12.5	12.5	4.0	2.3
$R_1 = C_5H_{11}{}^{c-t}$; $R_2 = H$	4	0.025	0.025	0.05	12.5	50	12.5	4.0	3.0
$R_1 = C_6H_{13}{}^{c-t}$; $R_2 = H$	1.8	0.025	0.025	0.05	25	200	50	4.0	2.5
$R_1 = C_8H_{17}{}^{c-t}$; $R_2 = H$	0.1	0.05	0.025	3.2	12.5	>200	25	3.7	0.2

[a] t = *trans* isomer; c-t = *cis-trans* epimers at C-4'; c = *cis* isomer.

A process for the synthesis of 1'-demethyl-4'-depropyl-4'-pentylclindamycin [(**76**), $R_1 = n\text{-}C_5H_{11}$] (U-24,729) from racemic 4-pentylproline which embodies a resolution of the isomers at C-2' was recently described (Magerlein, 1971). The antibacterial activity of 1'-demethyllincomycin was markedly lower than that of its 1'-methyl cogenors, e.g., 1'-demethyllincomycin possesses only 2% the antibacterial potency of lincomycin in the standard curve assay. In the clindamycin series the 1'-demethyl analogs exhibited unexpectedly high *in vitro* and *in vivo* antibacterial activities which are summarized in Table XII. The most active compounds in this series, such as U-24,729 [(**76**), $R = n\text{-}C_5H_{11}$], exhibit *in vitro* antibacterial potency to both gram-positive and gram-negative bacteria comparable to cephalothin and ampicillin (Table XIII) (Mason, 1968).

TABLE XIII
In Vitro COMPARISON OF 1'-DEMETHYL-4'-DEPROPYL-4'-PENTYLCLINDAMYCIN WITH AMPICILLIN AND CEPHALOTHIN

	Minimum inhibitory concentrations, μg/ml		
Organism	Ampicillin	Cephalothin	1'-Demethyl-4'-depropyl-4'-pentylclindamycin
Staphylococcus aureus UC 76	0.1	0.2	0.025
Staphylococcus aureus UC 76	0.1	0.2	0.0125
Staphylococcus aureus UC 552	25	0.4	0.025
Staphylococcus aureus UC 3216	0.1	0.2	> 6.4
Streptococcus pyogenes UC 147	< 0.05	0.1	0.006
Streptococcus viridans UC 153	0.1	0.4	0.003
Streptococcus faecalis UC 694	1.6	50	0.1
Diplococcus pneumoniae UC 41	0.1	0.1	0.006
Escherichia coli UC 51	8	16	16
Klebsiella pneumoniae UC 55	8	32	8
Proteus vulgaris UC 93	64	250	64
Pseudomonas aeruginosa UC 95	95	250	250
Salmonella gallinarum UC 265	4	8	4
Clostridium perfringens UC 247	0.05	0.4	0.0125

c. Antimalarial Activity of 4'-Alkyl-4'-depropyl-1'-demethylclindamycins and Related Analogs

Lewis (1968a) observed that clindamycin when administered either orally or subcutaneously to *Plasmodium berghei*-infected mice possessed marked antimalarial activity not possessed by lincomycin. Subsequent studies showed that antimalarial activity was largely confined to the analogs of lincomycin that were intrinsically active as antibacterial agents and were halogenated at C-7. 4'-Alkyl-1'-demethyl-4'-depropylclindamycins (**76**) proved to be the most active group of analogs tested. Of this group 1'-demethyl-4'-depropyl-4'-pentylclindamycin [(**76**), R = n-C_5H_{11}] (U-24,729) was recognized as being particularly effective. Antimalarial testing data for these compounds as well as for chloroquine and diaminodiphenyl sulfone are summarized in Table XIII (Lewis, 1968b). The data indicate that U-24,729 is comparable to or better than chloroquine in this assay. Furthermore, the clindamycin analogs were not cross-resistant with either chloroquine, or diaminodiphenyl sulfone, indicating an inhibition mechanism different from that of existing antimalarials. Lewis (1968b) postulated that the antiplasmodial action of clindamycin analogs was due to inhibition of protein synthesis at the ribosomal level.

Initial studies evaluating clindamycin analogs against blood-induced *Plasmodium cynomolgi* infection in rhesus monkeys, the simian counterpart of *Plasmodium vivax* in man, was reported by Powers (1969). He showed that 1'-demethyl-4'-depropyl-4'-pentylclindamycin (U-24,729), 1'-demethylclindamycin (U-26,285), and clindamycin (**26**) effected cures in the animals when the antibiotics were administered orally or subcutaneously. Schmidt and co-workers (1970) studied in greater detail the effectiveness of U-24,729 and clindamycin in rhesus monkeys infected with *Plasmodium cynomolgi*. They found that both compounds effected marked delays in the onset of patency when administered at the time of sporozoite challenge and throughout the prepatent period. In a significant number of instances U-24,729 prevented the evolution of infections, thus functioning as a casual prophylactic. Radical cures were observed in a significant number of animals treated with U-24,729.

Of particular interest in this area of sensitivity to drug-resistant strains of *Plasmodium* was Schmidt's findings that both U-24,729 and clindamycin (**26**) given in prophylactic regimens were as effective against challenges with a pyrimethamine-resistant strain of *Plasmodium cynomolgi* as against challenges with the parent susceptible strain. Since a need exists for improved casual prophylactic and radical curative agents in human malaria, particularly in infections caused by strains possessing resistance to existing drugs, prompt testing in man was recommended.

4. Miscellaneous Modifications

Treatment of lincomycin and clindamycin tetraacetate with phosphorus pentasulfide followed by hydrolysis led to the isolation of thiamido analogs (**77**) and (**78**) (Magerlein et al., 1969b). These compounds were about one-quarter as active as their parent antibiotics in standard tests.

(77) $R_1 = OH; R_2 = H$
(78) $R_1 = H; R_2 = Cl$

1'-Demethyllincomycins undergo dihydroimidazole formation in the presence of an aldehyde and a base. (Argoudelis et al., 1971). Of particular interest were dihydroimidazoles (**79**) and (**80**) formed from 1'-demethyl-7-deoxy-7-halolincomycins and formaldehyde. These compounds were about as active as their parents *in vitro* but less active *in vivo*.

(79) $R_1 = Cl$
(80) $R_2 = Br$

An interesting method of 1'-demethylation of lincomycin and various of its analogs was recently reported by Birkenmeyer and Dolak (1970). These investigators found that when oxygen was bubbled through an aqueous solution of lincomycin or its analogs in the presence of platinum catalyst, highly satisfactory yields of demethylated products were obtained. This method offers a preparative process for the conversion of lincomycin (**1**) to 1'-demethyl-lincomycin (**7**) and clindamycin (**26**) to 1'-demethylclindamycin (**74**, $R_1 = C_3H_7^t$; $R_2 = H$).

5. Total Synthesis of Lincomycin

While a number of methods for chemical modification of lincomycin were studied with considerable success over an extended period of time, only recently has a total synthesis of the antibiotic been attained.

Within a few months several research groups, working independently, reported the synthesis of methyl α-thiolincosaminide (**22**) from D-galactose and its coupling with *trans*-1-methyl-4-propyl-L-proline hydrochloride (**24**) to form lincomycin (**1**). Since the synthesis of *trans*-1-methyl-4-propyl-L-proline hydrochloride (**24**) from 4-hydroxy-L-proline was previously reported (Magerlein *et al.*, 1967b), the total synthesis of lincomycin was realized. The synthesis of lincomycin was an interesting scientific accomplishment and may offer further analogs. For this reason, a summary of the various syntheses of methyl α-thiolincosaminide (**22**) appears justified.

a. *Lincomycin from Methylthio-α-D-galacto-pyranoside*

The synthesis of lincomycin from methylthio-α-D-*galacto*-pyranoside (**81**) reported by Magerlein (1970) is outlined in Chart IX. The starting thiogalactoside (**81**), formed when D-galactose was treated with methanethiol, was also available in high yield by the periodate oxidation of methyl α-thiolincosaminide (**22**) (Magerlein, 1969). Conversion to the 6-nitrosugar (**82**), as indicated, followed by condensation with acetaldehyde yielded nitroalcohols (**83**) as a mixture of epimers. Reduction of (**83**) followed by coupling with *trans*-1-methyl-4-propyl-L-proline (**24**) led to the isolation of lincomycin (**1**) and its 7-epimer (**42**).

This method offered the advantage that the methylthio group was introduced early in the synthesis so that the often tedious introduction of this function could be performed on a readily available intermediate.

CHART IX

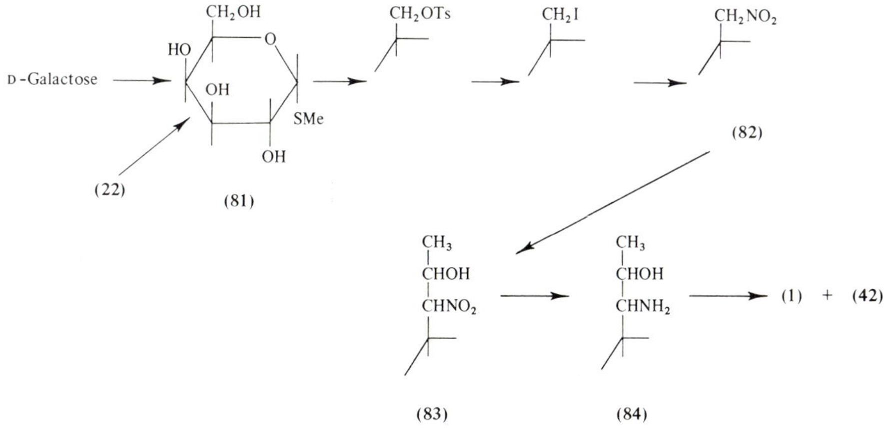

b. Lincomycin from 1,2:3,4-Di-O-isopropylidene-α-D-galacto-pyranose

Methods reported for the synthesis of lincomycin from 1,2:3,4-di-O-isopropylidene-α-D-*galacto*-hexodialdo-1,5-pyranose (**85**) utilized the common intermediate 1,2:3,4-di-O-isopropylidene-α-D-*galacto*-hexodialdo-1,5-pyranose (**86**). This aldehyde was obtained by Pfitzner-Moffett oxidation of diisopropylidene sugar (**85**) (Horton et al., 1968).

Szarek and colleagues utilized aldehyde (**86**) to elaborate the sugar side chain by two elegant methods outlined in Charts X and XI (Howarth et al., 1969a,b, 1970). In the first method, aldehyde (**86**) was treated with ethylidene triphenylphosphorane to give the Wittig product predominantly as the *cis* isomer (**87**). Hydroxylation followed by selective benzoylation and oxidation afforded keto sugar (**90**). The amino group was introduced by conversion to the oxime (**91**) followed by reduction to give after N-acylation 6-acetamido-6,8-dideoxy-1,2:3,4:di-O-isopropylidene-D-*erythro*-α-D-*galacto*-octopyranose (**92**) as well as its D-*threo* epimer.

In an alternate method, key intermediate (**92**) was prepared as shown in Chart XI. Aldehyde (**86**) was treated with nitromethane to form nitroalcohols (**93**) which were dehydrated via the 6-acetate to form unsaturated sugar (**94**) (Howarth et al., 1969b). Addition of ammonia to (**94**) followed by acylation afforded (**95**). Oxidative denitration of (**95**) gave the epimeric 7-ketones from which the D-*glycero* epimers (**96**) were isolated. Reduction of 7-ketone (**96**) gave (**92**) and also its 7-epimer, identical with that prepared by the previous process. Removal of the blocking groups in (**92**) and introduction of the 1-methylthiol group afforded methyl α-thiolincosaminide (**22**) whose conversion to lincomycin (**1**) was previously reported (Magerlein et al., 1967b).

Saeki and Ohki (1969, 1970a,b) introduced the substituted three carbon side chain onto aldehyde (**86**) as outlined in Chart XII. The addition of hydrogen cyanide to aldehyde (**86**) followed by tosylation yielded a mixture of epimers from which L-*glycero* epimer (**98**) was isolated. Metal hydride reduction of (**98**) led to aziridine (**99**). This aziridine (**99**) was opened and resulting alcohol (**100**) oxidized to aldehyde (**101**). When treated with Grignard Reagent, (**102**) was obtained though in the undesired configuration at C-7. Therefore (**102**) was oxidized at C-7, and reduced (Hoeksema, 1965) to give (**92**) and its 7-epimer.

Removal of the ketal groups afforded N-acetyllincosaminide (**92a**) identical with a sample prepared from methyl α-thiolincosaminide (**22**).

CHART X

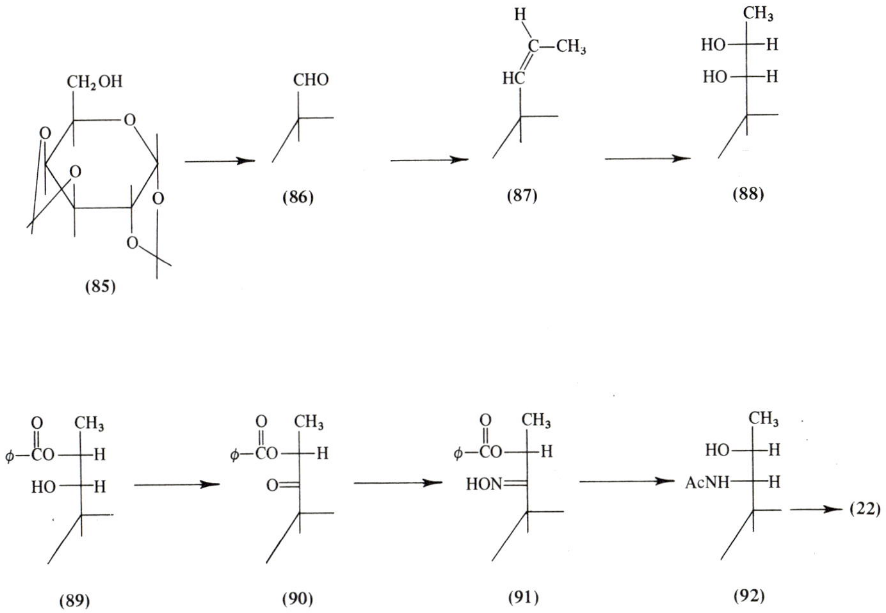

(85) → (86) → (87) → (88)

(89) → (90) → (91) → (92) → (22)

CHART XI

(86) → (93) → (94) → (95)

(95) → (96) → (92) → (92a)

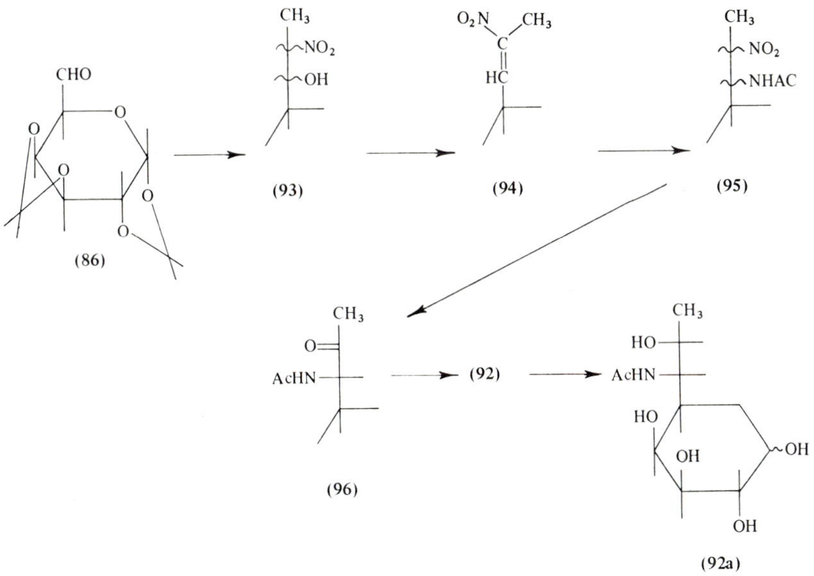

CHART XII

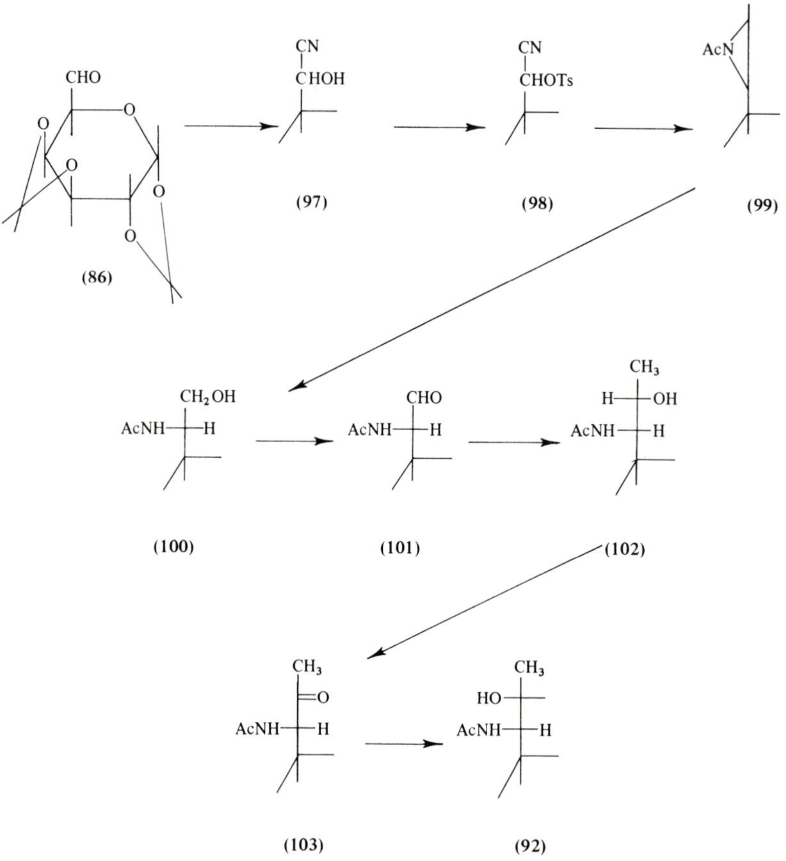

Horton *et al.* (1968) reported still another method for elaborating the α-aminoalcohol sugar side chain. When treated with acetylene magnesium bromide, aldehyde (**86**) gave a mixture of the expected acetylene alcohols from which epimer (**104**) was separated (Chart XIII). This alcohol was tosylated and oxidized to form (**106**). Replacement of the 6-tosylate by azide and then reduction led to aminoalcohol (**108**).

IV. Microbial Modification of Lincomycin and Clindamycin

Microbial modification of lincomycin and clindamycin does not lend itself to the formation of the number and diversity of transformation products encountered in chemical modification. However, several types of interesting microbial transformations have been reported.

CHART XIII

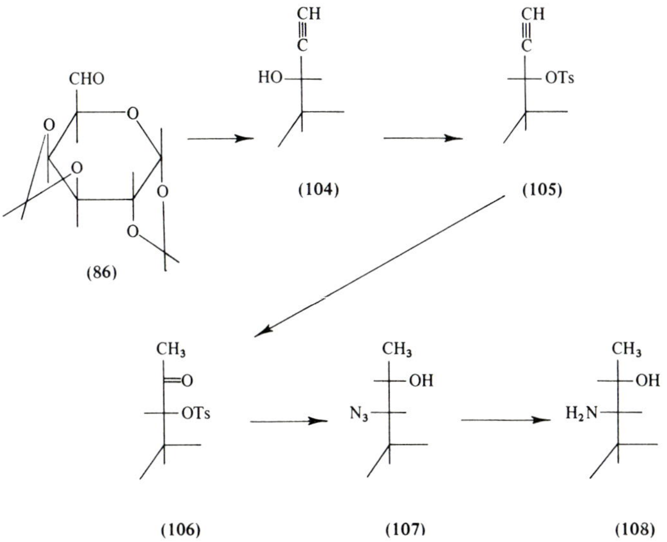

(86) → (104) → (105) → (106) → (107) → (108)

A. SULFOXIDE FORMATION

The addition of lincomycin to a fermentation of *Streptomyces lincolnensis* var. *lincolnensis* resulted in the formation of lincomycin sulfoxide (109) (Argoudelis and Mason, 1969). This sulfoxide was also prepared by oxidation of lincomycin hydrochloride with hydrogen peroxide (Birkenmeyer, 1969). It is about 0.01 as active as lincomycin against *Sarcina lutea*. A lesser yield of 1-demethylthio-1-hydroxylincomycin (110) was also isolated from the fermentation beer.

Similarly, clindamycin sulfoxide (111) was the major transformation product when clindamycin (26) was added to fermentations of *Streptomyces armentosus* (Argoudelis *et al.*, 1969). Trace amounts of sul-

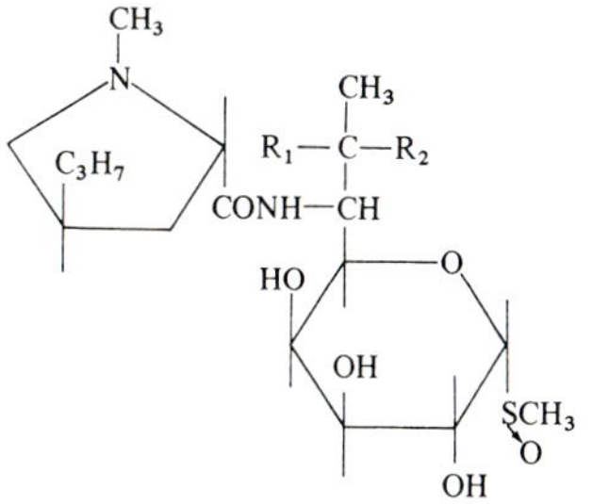

(109) $R_1 = OH; R_2 = H$
(111) $R_1 = H; R_2 = Cl$

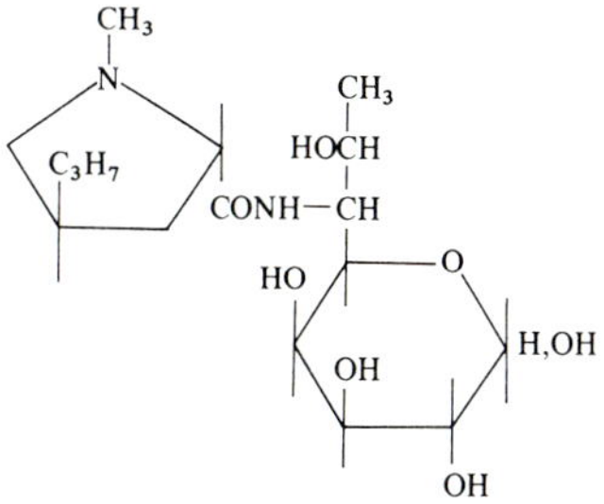

(110)

TABLE XIV
Antimalarial Activity of Lincomycin and Clindamycin Analogs in *Plasmodium Berghei*-Infected Mice

	MED (mg/kg)[a]		CD_{50} (mg/kg)[b]		CD_{100} (mg/kg)[c]	
	Subcutaneous	Oral	Subcutaneous	Oral	Subcutaneous	Oral
1'-Demethyl-4'-depropyl-4'-pentylclindamycin (U-24,729)	0.31	1.6	4.7	12	10	25
1'-Demethyl-4'-depropyl-4'-hexylclindamycin	0.6	1.6	6.6	14	20	50
1'-Demethylclindamycin	1.2	3.1	19.0	37	40	50
1'-Demethyl-4'-depropyl-4'-butyl-1'-ethylclindamycin	8.0	15	57	62	100	150
Clindamycin	5.0	12	113	141	160	250
1'-Demethyl-1'-ethylclindamycin	> 100	250	> 100	> 250	> 160	> 250
1'-Demethyl-4'-depropyl-4'-pentyllincomycin	50	125	> 100	> 250	> 100	> 250
Lincomycin	> 160	> 400	> 160	> 400	> 160	> 400
Chloroquine	10	12.5	8.1	14	20	25
Diaminodiphenylsulfone	3.12	6.5	25	38	50	100

[a] Dosage at which median survival time (ST_{50}) was increased significantly ($p = 0.05$) over ST_{50} of untreated controls.
[b] CD_{50} is median protective dose in mg/kg (95% limits).
[c] Dosage that protected 100% of animals; no parasites could be demonstrated by staining or subinfecting normal mice.

foxide were detected in similar fermentations involving other species of streptomyces, particularly *Streptomcyes punipalus*. Clindamycin sulfoxide is about equal in potency to lincomycin when assayed with the test organism, *S. lutea*.

B. 1'-Demethylation

In addition to sulfoxide formation, several species of streptomcyetes including those mentioned above effect partial 1-demethylation (Argoudelis *et al.*, 1969). *Streptomyces punipalus* was particularly successful in demethylating clindamycin (**26**) to 1'-demethylclindamycin [(**74**), $R_1 = n\text{-}C_3H_7$] which was previously prepared by chemical modification.

C. Phosphorylation

Argoudelis and Coats (1969) observed that *Streptomyces rochei* grown in synthetic medium converted lincomycin and lincomycin-related antibiotics to the 3-phosphate esters. Lincomycin 3-phosphate was inactive *in vitro* against several organisms. It did protect *Staphylococcus aureus*-infected mice with a CD_{50} of about 30 mg/kg when administered subcutaneously.

V. Summary

Chemical and microbial modifications of lincomycin have succeeded in producing derived antibiotics possessing greater potency, extended spectrum including antiparasitic activity, improved taste, and varied absorption characteristics. Several total syntheses of lincomycin have been achieved. These syntheses offer opportunities for further modification of the lincomycin molecule.

The direction and extent to which these new opportunities, as well as the older synthetic patterns, are developed and extended, make modification of lincomycin an exciting area of future research.

References

Argoudelis, A. D., and Coats, J. H. (1969). *J. Antibiot.* **22**, 341–343.
Argoudelis, A. D., and Mason, D. J. (1965). *Biochemistry* **4**, 704–709.
Argoudelis, A. D., and Mason, D. J. (1969). *J. Antibiot.* **22**, 289–291.
Argoudelis, A. D., Fox, J. A., Mason, D. J., and Eble, T. E. (1964). *J. Amer. Chem. Soc.* **86**, 5044–5045.
Argoudelis, A. D., Fox, J. A., and Eble, T. E. (1965a). *Biochemistry* **4**, 698–703.
Argoudelis, A. D., Fox, J. A., and Mason, D. J. (1965b). *Biochemistry* **4**, 710–713.
Argoudelis, A. D., Coats, J. H , Mason, D. J., and Sebek, O. K (1969). *J. Antibiot.* **22**, 309–314.
Argoudelis, A. D., Eble, T. E., and Mason, D. J. (1970a). *J. Antibiot.* **23**, 1–8.

Argoudelis, A. D., Coates, J. H., Birkenmeyer, R. D., and Magerlein, B. J. (1972). *Abstr. 11th Intersci. Conf. Antimicrob. Ag. Chemother.*
Bannister, B. (1969). Personal communication.
Birkenmeyer, R. D. (1965). *Abstr. 5th Intersci. Conf. Antimicrob. Ag. Chemother.* p. 18.
Birkenmeyer, R. D. (1969). Personal communication.
Birkenmeyer, R. D., and Dolak, L. (1970). *Tetrahedron Lett.* pp. 5049–5051.
Birkenmeyer, R. D., and Kagan, F. (1970). *J. Med. Chem.* **13**.
Birkenmeyer, R. D., Magerlein, B. J., and Kagan, F. (1965). *Abstr. 5th Intersci. Conf. Antimicrob. Ag. Chemother.* p. 17.
Brodasky, T. F., Argoudelis, A. D., and Eble, T. E. (1968). *J. Antibiot.* **21**, 327–333.
DeBoer, C., Dietz, A., Wilkins, J. R., Lewis, C. N., and Savage, G. M (1955). *In* "Antibiotics Annual 1954–1955" (H. Welch and F. Marti-Ibanez, eds.), pp. 831–836. Med. Encycl., New York.
Dolak, L. (1969). Personal communication.
Duchamp, D. J. (1967). *Amer. Crystallogr. Ass. Program, Abstr. Summer Meet., Minneapolis* Paper D-5.
Garrison, D. W., DeHaan, R. M., and Lawson, J. B. (1968). *Antimicrob. Ag. Chemother.* 1967, pp. 397–400.
Hanka, L. J., Mason, D. J., Burch, M. R., and Treick, R. W. (1963). *Antimicrob. Ag. Chemother.* 1962, p. 565.
Herr, R. R. (1967). Personal communication.
Herr, R. R., and Bergy, M. E. (1963). *Antimicrob. Ag. Chemother.* 1962, p. 560.
Herr, R. R., and Slomp, G. (1967). *J. Amer. Chem. Soc.* **89**, 2444–2447.
Herrell, W. E. (1969). "Lincomycin." Mod. Sci. Publ., Chicago, Illinois.
Hoeksema, H. (1965). *Abstr. Papers, 149th Meet. Amer. Chem. Soc., Detroit* p. 9-C
Hoeksema, H. (1968a). Personal communication.
Hoeksema, H. (1968b). *J. Amer. Chem. Soc.* **90**, 755–757.
Hoeksema, H., Crum, G. F., and DeVries, W. H. (1955). *In* "Antibiotics Annual 1954–1955" (H. Welch and F. Marti-Ibanez, eds.), pp. 837–841, Med. Encycl., New York.
Hoeksema, H., Bannister, B., Birkenmeyer, R. D., Kagan, F., Magerlein, B. J., MacKellar, F. A., Schroeder, W., Slomp, G., and Herr, R. R. (1964). *J. Amer. Chem. Soc.* **86**, 4223–4224.
Hoeksema, H., Bannister, B., and Herr, R. R. (1967). U.S. Patent No. 3,326,891.
Horton, D., Nakadate, M., and Tronchet, J. M. J. (1968). *Carbohyd. Res.* **7**, 56–65.
Houtman, R. L., Kaiser, D. G., and Tarazka, A. J. (1968). *J. Pharm. Sci.* **57**, 693–695.
Howarth, G. B., Lance, D. G., Szarek, W. A., and Jones, J. K. N. (1969a). *Can. J. Chem.* **47**, 75–79.
Howarth, G. B., Szarek, W. A., and Jones, J. K. N. (1969b). *Chem. Commun.* pp. 2218–2224.
Howarth, G. B., Szarek, W. A., and Jones, J. K. N. (1970). *J. Chem. Soc. London* pp. 2218–2224.
Lewis, C. (1968a). *J. Parasitol.* **54**, 169–170.
Lewis, C. (1968b). *Antimicrob. Ag. Chemother.* 1967, pp. 537–542.
Lewis, C., Clapp, H. W., and Grady, J. E. (1963). *Antimicrob. Ag. Chemother.* 1962, p. 570.
Magerlein, B. J. (1967). *J. Med. Chem.* **10**, 1161–1163.
Magerlein, B. J. (1968). Unpublished observations.
Magerlein, B. J. (1971). *J. Org. Chem.* **36**, 596.
Magerlein, B. J. (1970). *Tetrahedron Lett.* pp. 33–36.
Magerlein, B. J. (1971). *Abstr. Papers 161st. Meet. Amer. Chem. Soc. Los Angeles*, Med. Div., Abstr. No. 38.

Magerlein, B. J., and Kagan, F. (1969a). *J. Med. Chem.* **12**, 780–784.
Magerlein, B. J., and Kagan, F. (1969b). *J. Med. Chem.* **12**, 974–977.
Magerlein, B. J., Birkenmeyer, R. D., and Kagan, F. (1967a). *Antimicrob. Ag. Chemother.* 1966, p. 727.
Magerlein, B. J., Birkenmeyer, R. D., Herr, R. R., and Kagan, F. (1967b). *J. Amer. Chem. Soc.* **89**, 2459–2462.
Magerlein, B. J., Birkenmeyer, R. D., and Kagan, F. (1967c). *J. Med. Chem.* **10**, 355–359.
Mason, D. J. (1968). Personal communication.
Mason, D. J., and Lewis, C. (1965). *Antimicrob. Ag. Chemother.* 1964, pp. 7–12.
Mason, D. J., Dietz, A., and DeBoer, C. (1963). *Antimicrob. Ag. Chemother.* 1962, p. 554.
Morozowich, W., Lamb, D. J., Karnes, H. A., MacKellar, F. A., Lewis, C., Stern, K. F., and Rowe, E. L. (1969a). *J. Pharm. Sci.* **58**, 1485–1489.
Morozowich, W., Sinkula, A. A., MacKellar, F. A., and Lewis, C. (1969b). *Abstr. Papers, Amer. Pharm. Ass. Meet., Montreal* p. 57.
Morozowich, W., Lamb, D. J., DeHaan, R. M., and Gray, J. E. (1970a). *Abstr. Papers, Amer. Pharm. Ass. Meet., Washington, D.C.*, p. 63.
Morozowich, W., MacKellar, F. A., and Lewis, C. (1970b). *Abstr. Papers, Amer. Pharm. Ass. Meet., Washington, D.C.*, p, 62.
Morozowich, W., MacKellar, F. A., and Lewis, C. (1970c). *Abstr. Papers, Amer. Pharm. Ass. Meet., Washington, D.C.*, p. 63.
Oesterling, T. O. (1970a). *J. Pharm. Sci.* **59**, 63–67.
Oesterling, T. O. (1970b). *Carbohydr. Res.* **15**, 285–290.
Oesterling, T. O., and Metzler, C. M. (1970c). Personal communication.
Oesterling, T. O., and Rowe, E. L. (1970). *J. Pharm. Sci.* **59**, 175–179.
Powers, K. G. (1969). *Amer. J. Trop. Med. Hyg.* **18**, 485–490.
Saeki, H., and Ohki, E. (1969). *Chem. Pharm. Bull.* **17**, 1974–1976.
Saeki, H., and Ohki, E. (1970a). *Chem. Pharm. Bull.* **18**, 412–413.
Saeki, H., and Ohki, E. (1970b). *Chem. Pharm. Bull.* **18**, 789–802.
Schmidt, L. H., Harrison, J., Ellison, R., and Worcester, P. (1970). *Amer. J. Trop. Med. Hyg.* **19**, 1–11.
Schroeder, W., and Birkenmeyer, R. D. (1967). Personal communication.
Schroeder, W., Bannister, B., and Hoeksema, H (1967). *J. Amer. Chem. Soc.* **89**, 2448–2453.
Sinkula, A. A., Morozowich, W., Lewis, C., and MacKellar, F. A. (1969). *J. Pharm. Sci.* **58**, 1389–1392.
Slomp, G., and MacKellar, F. A. (1967). *J. Amer. Chem. Soc.* **89**, 2454–2459.
Sun, F. F. (1970). *Fed. Proc. Fed. Amer. Soc. Exp. Biol.* **29**, 2429.
Taraszka, M. J., and Morozowich, W. (1968). *J. Org. Chem.* **33**, 2349–2354.
Wagner, J. G., Novak, E., Patel, N. C., Chidester, C. G., and Lummis, W. L. (1968). *Amer. J. Med. Sci.* **256**, 25–37.

Fermentation Equipment

G. L. SOLOMONS

*The Lord Rank Research Centre,
High Wycombe, Buckinghamshire, England*

I.	Introduction	231
II.	Auxiliary Equipment	233
	A. Mechanical Seals	233
	B. Aseptic Metering Pumps	234
III.	Process Control Instruments	235
	A. Temperature Control	236
	B. Temperature Indication	237
IV.	Sterilization of Culture Medium	238
V.	Analytical Instrumentation	240
	A. Rheology	240
	B. Gas Analysis	240
VI.	Biochemical Analysis of Fermentation Cultures	244
	References	246
	Commercial References	246

I. Introduction

Fermentation studies are conducted either to produce products such as antibiotics, enzymes, and biomass, or to study the kinetics or metabolic pathways, etc., of microorganisms. In either case, fermentation research, in common with every other branch of science, calls for ever more refined techniques and apparatus to control and analyze the systems.

It is the purpose of this article to indicate some of the newer equipment now in use and often to speculate upon some of the equipment and techniques that may become commonplace or necessary in the future. Often the more reliable and accurate instruments become the the more they cost. There are exceptions to this trend, e.g., deviation-dependent sensitivity (DDS) electronic equipment is approximately 10% of the cost of three-term (P.I.D.)[1] control equipment but is claimed to provide equal performance. Nevertheless, fermentation equipment is likely to become more expensive, and various users of equipment will be faced with having to decide how best to solve their problems. For example, it is still the custom of some producers of antibiotics to use large numbers of very simply equipped fermentor vessels. By using a large number of vessels they generate the information required to increase productivity. Because they rely, in the main, on culture mutation to obtain high yields (some basic parameters having been established in some detail beforehand) this approach is perhaps the

---
[1] P.I.D. (Proportional and Integral + Derivative).

best economic one in their circumstances. On the other hand, those workers concerned with, for example, the production of biomass do not rely on mutation to anything like the same extent and they can obtain more information relevant to their needs by carrying out fewer fermentations but analyzing them in more detail.

Fundamental studies will certainly require the use of better control of variables, for example, it is now considered adequate to control fermentation temperatures to within ± 0.1 to $0.25°C$. In some recent experiments on heat balance, it was found necessary to reduce these levels to $\pm 0.05°C$ (Flynn et al., 1971).

Allied to the closer control of operational variables such as temperature, pH, redox and dissolved gas concentration, the requirements of better control of medium constituents will become apparent. It was shown many years ago by Pfeifer and Vojnovich (1952) that the same culture medium sterilized by continuous flow methods involving high temperature–short time treatment, provided better yields of a number of metabolites than batch sterilized medium. Large industrial fermentors are often supplied with culture medium sterilized by the continuous flow method, but small laboratory fermentors must often still employ batch sterilization. To partially overcome this problem, more rigorous control of the heating and cooling cycles is now provided by automatic sterilization equipment on some of the commercial fermentors available (Solomons, 1968). However, small-scale continuous sterilizers will need to be studied and applied.

It is, however, in the fields of analysis and uses of the computer that perhaps the greatest advances will lie. The use of autoanalyzers coupled with fermentors is, of course, well established, but these are usually linked by a discrete sampling system and for very detailed information may well be insufficient. Moreover, small fermentors often cannot provide a sufficient number of samples without affecting the agitation pattern; as a rule no more than 10% of the vessel content can be removed without effect. In addition, since the time required to sample and either separate or inactivate can be too long, more methods of "instant" analysis are required.

Coupled with advances in better control and more detailed analytical information, computer techniques should enable us to obtain much more useful assessments of the course of fermentation processes. Computers have been used to provide the process control systems for fermentation for some time (Grayson, 1969), and some work has been carried out using computers to assess and simulate fermentations. We feel, however, that a major role can be played by the combination of computer process control and biochemical analysis.

II. Auxiliary Equipment

Although the use of more sophisticated instrumentation will be stressed in this article, it should not be assumed that this is the only area in need of better equipment. Many items of equipment used at present are not totally satisfactory and some are downright unsatisfactory. Of course, this is not a universal statement of fact. While one well-equipped laboratory can successfully use a given item of equipment, another, less well provided, can use it only with a great deal of inconvenience.

A. Mechanical Seals

An example of this type of equipment is the shaft seal. Originally stuffing boxes were used, but because of their many inherent disadvantages they have largely been replaced by mechanical seals. Various arrangements of these have been described (Solomons, 1969), but the more usual designs call for two seals, back to back, with a lubricating and cooling fluid surrounding them. This fluid is usually sterile water provided by condensing steam and fed through the seal chamber at the rate of 50–100 ml per minute. Unfortunately, water possesses poor lubricant properties and sterile condensate is often not available in laboratories on a 24-hour basis. Another requirement for mechanical seal application is that the manufacturers specify that one atmosphere overpressure should be applied to the seal chamber in order to force the lubricant across the very narrow gap between the seal faces. Perhaps an alternative could be offered on the basis of the mechanical seal arrangement illustrated in Fig. 1. In this there are two chambers, completely isolated from one another. The inner one is in contact with the seals containing the silicone oil lubricant and the outer one allows for either steam or cold water. The steam is used for heating the inner chamber during sterilization and the cold water for cooling the silicone oil during running. The replacement of water (usually as steam condensate) by silicone oil should greatly improve the service life of the seals faces, as water is a poor lubricant compared to oil. Provisions must be made for pressurizing the chamber and this could be done with either a conventional diaphragm or perhaps by using a "Bellofram" seal (Geo. Angus & Co., Ltd.). The actual pressure could be applied with a spring or by a static air pressure on the nonsterile side. Because a certain amount of lubricant must pass across the face of the seal the loss of volume must be compensated for by the pressurization device. The presence of very small traces of silicone oil would not usually affect fermentations.

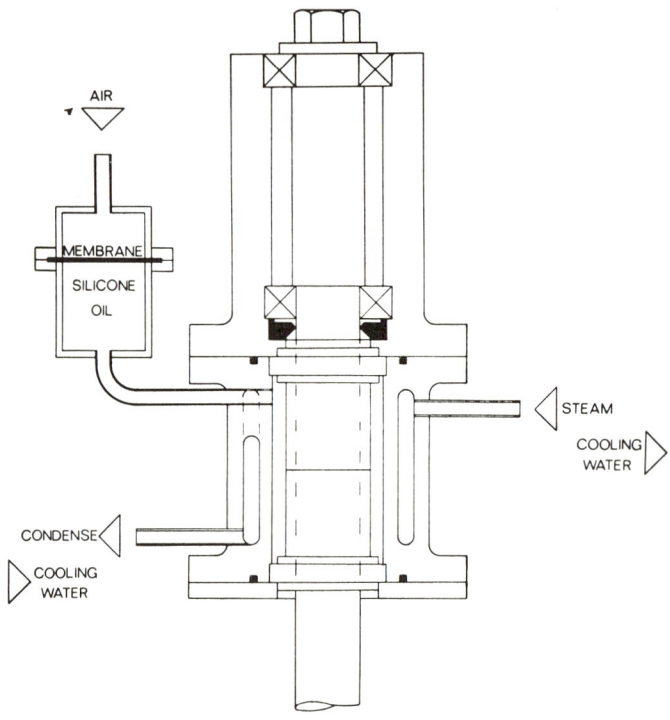

Fig. 1. Mechanical seal system for stirrer shaft.

It should be pointed out that at least one company concerned with agitation (Mixing Equipment Co., Inc.) does offer larger scale equipment similar in principle to that outlined. However, it is not designed with aseptic operation in mind. Another approach to aseptic lubrication of seals was described by Reisman *et al.* (1968) who used hot lubricating oil in place of normal cool condensate.

B. Aseptic Metering Pumps

Long-term aseptic metering still presents many problems. While clear fluids handled with conventionally valved piston pumps can be metered successfully when a final filter is included in the line immediately prior to the pump, fluids containing solids cannot be successfully dealt with. Diaphragm pumps fitted with PTFE (polytetrafluoroethylene) or similar diaphragms are not very satisfactory as metering pumps due to the distortion of the diaphragm on sterilization with steam. We have also been unsuccessful with the replacement of plastic diaphragms by Viton rubber. The peristaltic pump and flow inducer can cope with fluids containing particles, but for long-term duties complications arise due to dependence on the "memory" of rubber tubing. When the pump is used to supply fermentors operating under pressure, the position is aggravated.

There would seem to be two alternative types of pump that could solve this problem; either valved pumps, with the valves operated by cams and springs, or desdromonic (mechanically linked to open and close) gear, or rotary sleeve valves. A pump of the first type using spring-loaded valves (QP-Shirley Metering Pump, Model 3004) similar to those used in motor car engines is produced by Ormerod Engineering Ltd., but it is of relatively large capacity and is not suitable unmodified for aseptic operation. For a piston pump to be suitable for aseptic use over a long period of time, it is necessary to incorporate either a "steam lock" on the piston so that the ingress of microorganisms is prevented or, alternatively, the piston can be housed in a bellows so that once sterilized it is not subject to exposure to microorganisms from the surroundings. Moving poppet valves and their associated springs and cams are perhaps inherently difficult to operate aseptically, and the alternative, that is rotary valves, is used by New Brunswick Scientific Inc. in their latest design. The turning action of the sleeve is powerful enough to cut through any particles which are present in the fluid and the pump has been used for handling diced vegetable soup, a severe challenge for a metering pump. For really precise metering it should not be forgotten that the prime mover, the electric motor, is subject to mains fluctuation and these can be considerable in some areas. To guard against variation from this source, ideally the necessary voltage regulators should be incorporated.

III. Process Control Instruments

Undoubtedly, the present most sophisticated method of process control is the use of a computer, but it seems unlikely that computers will become inexpensive enough to be used on small laboratory applications. Therefore, improvements in or applications of conventional process instruments are of considerable importance to the better control of fermentation installations. An application example is a sequence control system of 2500 wire-spring relays operating over 1000 on-off valves and 15 motors in a Japanese monosodium glutamate plant (Mori and Yamashita, 1967).

All the major companies marketing process-control instruments have recently introduced marked improvements in the control operation of their equipment. For example, the change-over from manual to automatic operation has now been made "bumpless," that is a smooth change-over control can be made, instead of having a period of instability which was at one time common.

A. Temperature Control

Small fermentors are usually heated by means of electric heaters placed either in a jacket within the vessel or in a circulating water line to the vessel. Simple on-off regulators can often prove suitable, but changes in load can often produce substandard control. A newer type of controller is available from Ether Controls Ltd.—type 17-90B/1 Series—in which, an adjustable time proportional circuit is fitted, the degree of feedback being adjustable. A solid state controller with nonlinear mode control has been marketed by S K Instruments Ltd. The nonlinear characteristics are described as DDS (deviation-dependent sensitivity) and Fig. 2 illustrates this characteristic com-

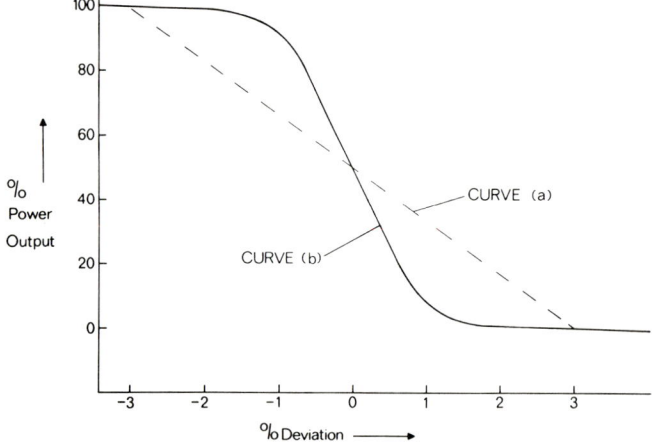

FIG. 2. Controller characteristics: curve (a) conventional linear proportional mode; curve (b) deviation dependent sensitivity mode.

paring it to conventional linear proportional control. Considerable claims are being made for its performance, including a four-to-one reduction in offset without resorting to a reset mode, a more than two-to-one improvement in stability without the need for a rate mode and stable straight line control almost down to ambient temperature. In short, the claim is made that in many applications, control is equivalent to that usually obtained with a three-term (P.I.D.) controller. We cannot yet offer an opinion as to the performance of these controllers since we have not yet made operational the fermentors for which we have installed them as temperature regulators, but if we are able to substantiate them it will be of great interest since the cost of them is approximately 10% of equivalent three-term control and less than some simple on-off regulators that we have used.

B. Temperature Indication

If we have available a sensitive accurate controller, we then have the problem of accurately measuring the temperature. This is necessary since the normal controller adjustment dial cannot be read with sufficient accuracy. If we require an instrument to measure temperatures just for fermentation operation, say 20 to 70°C (to include thermophilic fermentations), we require a range of 50°C. At the same time, it would be very useful to have at least a measurement of the sterilization cycle — say at 120–130°C. In all then, the instrument should have a range of 20–130°C or 110°C indicated. The instrument we have found most useful for this duty is a Comark Electronics Ltd. electronic thermometer. This instrument is available for operating both with thermocouples and resistance thermometers. These indicators have multiranges, and we use one covering $-60°C$ to $+170°C$ in 23 steps of 10°C, which are indicated on a 120-mm scale with an overlap of 0.5°C at the top end of the scale. Resolution on the scale is in steps of 0.1°C which should enable 0.05°C to be estimated. The switching steps are, of course, manual so that during sterilization it would be necessary to either switch to the appropriate range in a slow sweep or more probably just use the upper temperature which is required. In conjunction with this indicator it is more satisfactory to use a deviation recorder, rather than a wide range model. Thus, the installation of a simple Rustrack Instrument Co., Inc. recorder, giving a full span of $\pm 2.5°C$ (5°C overall) can provide resolution of 0.1°C (50 lines on the chart).

Almost all the control instruments in current use are closed loop feedback systems. By implication, no corrective measures are taken by the regulators until a deviation from the set point occurs. If good control is to be effected, much depends, therefore, on the rate of change from the set point and the rate of response of the corrective measure from the control system. Fortunately, most fermentation changes are slow in comparison to many chemical reactions and feedback control, well tuned, is capable of providing adequate accuracy. The one major area where that is possibly not true is in the continuous sterilization of culture medium.

IV. Sterilization of Culture Medium

There are some combinations of microorganisms and culture medium in which the latter can be heated over a wide range without affecting the growth and metabolic processes of the former. By and large, these are a small minority of fermentations and the majority can be affected either partially or wholly by the "quality" of the culture medium. It is, therefore, our contention that it is at least equally important to control the quality of the culture medium as it is to provide adequate control or analytical systems. The effects of overheating culture medium on scale-up have been dealt with by Deindoerfer and Humphrey (1961). Batch sterilizations are comparatively straightforward to control since, provided the vessel is adequately stirred, there should be good reproducibility between batches. Even so, Sikyta and Mastner (1967) have found it of value to provide an automatic control of sterilization which determines the exact amount of heat energy supplied during the cycle. The problem of reduced medium quality due to degradation of vitamins, or undesirable reactions between, say, glucose and amino acids, remains. If we are to use culture medium with the absolute minimum of heat damage, then we shall be forced to use continuous sterilization. An exception would be in the use of cold sterile filtration, not a real alternative for many fermentation processes due to the use of a complex medium with suspended solids and prohibitive costs on a large scale.

The basic design features of continuous sterilizers have been dealt with by Aiba *et al.* (1965). In order to provide a medium with minimum overheating, the need is for high temperatures and short heating times. The problem therefore is to ensure that there is little or no deviation from the temperature, as even a small amount of incompletely heated medium can lead to lack of sterilization and consequent infection. As very close control is required, the use of feed-forward feedback control seems attractive. In this, a change in controlled condition is *predicted* from knowledge of other process parameters which can be measured or whose variation causes upset in the primary control parameter.

In continuous sterilization the two main problems are variation of medium flow and of steam flow into the sterilizer. The former is more of a problem than the latter since, provided good engineering practice has been maintained, steam flow should be very steady. Minor variations can be caused by changes in the temperature of the input medium. From a practical point of view it is changes in the flow rate of medium which present the greatest hazard. In a conventional feedback closed loop system (Fig. 3), a temperature sensor would detect

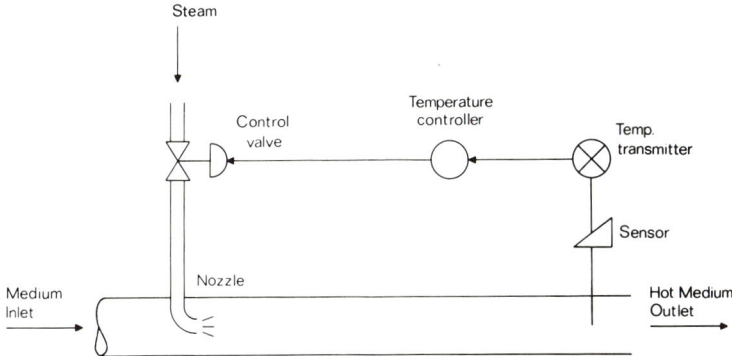

FIG. 3. Closed loop temperature control system for heater section of a continuous medium sterilizer.

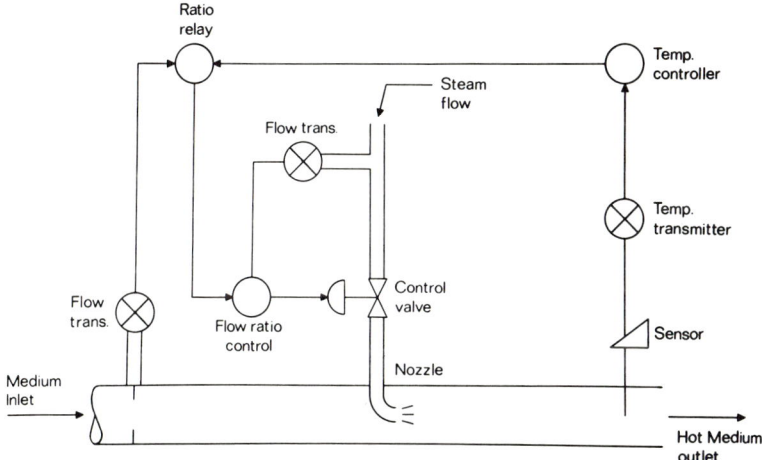

FIG. 4. Feedforward feedback control system for heater section of a continuous medium sterilizer.

changes and operate the steam supply valve accordingly. Sudden changes of flow would, however, cause considerable variations in heating. By monitoring the flow of medium *before* the heater, the steam valve can be regulated to cope with variations. In practice a feedback loop would be provided to take account of small variations in inlet medium temperatures, steam quality, etc. (Fig. 4). This example is taken from Brown (1970), who was actually describing a water heating system, but it obviously holds just as well for the heating stage of continuous sterilization.

V. Analytical Instrumentation

A. RHEOLOGY

The most obvious difference between submerged cultures of micro-fungi and actinomycetes or bacteria and yeasts is the "thickness" of the former. While some bacteria produce exocellular polysaccharides, and are therefore viscous, most fungal cultures have the appearance of porridge. It is therefore surprising to find that scarcely a dozen or so papers have appeared in the literature which have dealt with the measurement of these characteristic rheological properties (Deindorfer and West, 1960). These properties however dominate the fermentation system of most filamentous cultures by their effects on mass and heat transfer. The study of fluid behavior is in fact a difficult field, particularly in the case of micro-fungal cultures as the texture of them is seldom "smooth" or "even" but tends to be "lumpy." It is this "lumpiness" which makes rheological measurements difficult. A usual method of measurement is with a "cup and bob" type of viscometer (Solomons, 1969) where reproducibility of the readings is sometimes seriously affected by clumping of mycelium in the annulus of the instrument.

Moreover, the determination of rheological properties of samples taken from a fermentor do not necessarily provide a true picture of what is happening within the vessel. What is required therefore is an "in-situ" measurement system. Because non-Newtoniam fluids do not have a fixed viscosity, it would be necessary to have measurements in various positions within a fermentor in order to evaluate the effects of agitation on the flow of the culture.

Due to the effects of air bubbles present in a fermentation culture, "cup and bob" viscometers would not appear to offer a practical possibility of measurement within a fermentor. Instead, if a paddle shaft is mounted through a seal which has a very low resistance to turning (to obviate very high blank values) a torque transducer fitted as an integral part of this paddle would provide an electrical signal of the torque required to turn the paddle at a series of selected speeds (Brading, 1970). With this information the K (flow consistency index) and n (flow behavior index) values of the whole broth could perhaps be estimated.

B. GAS ANALYSIS

1. Dissolved Gases

The importance of dissolved oxygen (DO) has been recognized as affecting the course of microbial metabolism. In recent years much attention has been paid to the measurement of DO and there are now commercial electrodes available, some of which can be sterilized by autoclaving. While in no way trying to underestimate the importance of the work done on oxygen, we are of the opinion that a more integrated form of gas analysis is necessary. There are, for example, no sterilizable forms of carbon dioxide or ammonia electrodes and no electrodes at all for methane and only one for hydrogen. A different probe for each gas can also present considerable space problems in small laboratory fermentors. We also consider that "active-probes" (that is a probe in which some electrochemical reaction occurs) are on the whole undesirable, especially in very long term continuous culture experiments and perhaps on industrial scale equipment. Admittedly, pH electrodes may be considered to come into this category, but they are a special case. Some disadvantages of "active probes" are (a) as they are electrochemical in reaction, part of the system is consumed during operation. In a batch fermentation system this is of no great importance, but for long term continuous culture, it is possible for the electrolyte to be exhausted before the termination of the fermentation. (b) Unless each electrode is mounted in a self-sterilizable chamber (an expensive and bulky undertaking) the electrode cannot be replaced if it fails without breaking sterility and therefore usually terminating the fermentation. (c) Once the fermentation has started, it is usually not possible to restandardize the probe.

The system which we are investigating is based on a "passive-probe," originally used by Phillips and Johnson (1961) only for DO and which they termed the "coil probe." Basically the idea is to have a probe, either with a flat membrane or a coil, immersed in the fermentor and through which an inert carrier gas is passed. Gases in solution and permeable to the membrane will pass through the membrane in an equilibrium between the level of dissolved gas and carrier gas. The carrier gas and sample gas can then be passed to an analyzer which does not need to be aseptic in operation.

In the original method and adaptations of it (Solomons, 1966; Roberts and Shepherd, 1968), the coil used was of PTFE. This material has the great advantage of being inert and mechanically quite strong, its disadvantage is the poor permeability as shown in Fig. 5 and it can be seen that silicone rubber is approximately four or five orders of magnitude more permeable. This effect can be used to either greatly reduce the size of membrane surface or make the system more sensitive.

The probe must be designed with the viscosity of the fermentation in mind; Aiba and Huang (1969) have shown the importance of this factor. Because of the low permeability of oxygen through PTFE an

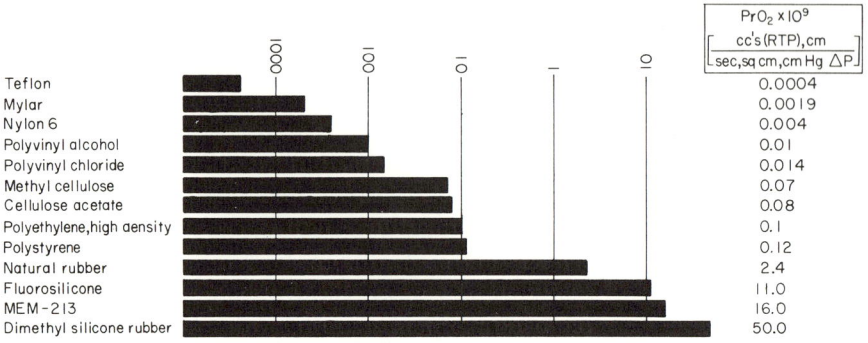

FIG. 5. Oxygen permeabilities in various polymers.

expensive analyzer is required to determine the level of oxygen, typically a Hersch meter (Engelhard Industries Ltd.) has been used. An alternative, but still expensive analyzer is produced by Analytical Instruments Ltd. The principle of this is an electron capture detector, which is very sensitive to small quantities of oxygen. Present analyzers are calibrated 0–5, 0–10, etc., up to 0–250 ppm oxygen in nitrogen. Unfortunately, the equipment is not yet ready in the form of a continuous duty process instrument. By using a thin silicone rubber membrane (approximately 0.001 inch thickness) and using oxygen-free nitrogen as carrier gas, we have made an electrode which is both passive and easy to mount. In place of an expensive Hersch meter, we have used a Rank Bros. electrode, which is a simple polarographic cell; by replacing the plastic membrane with silicone rubber, we can obtain fast response times and good sensitivity (Fig. 6). Of course, conventional oxygen electrodes could be used to detect the oxygen in the nitrogen stream, such as the MacKereth type, which is not normally heat sterilizable. As an alternative to the conventional electrolyte-filled oxygen electrode, Bergman (1968) has described metallized membrane electrodes. He has described three electrodes for measuring oxygen, hydrogen, and carbon dioxide. Three such electrodes fed from a "passive-probe" would be of great interest in the study of *Hydrogenomonas*. If a methane probe could be developed, the requirements of almost all fermentation studies would be met.

Another approach to the problem of overcoming the requirement to sterilize "active-probes" was described by Ring *et al.* (1969). Their method was simply to divide the electrode into two parts, an outer sleeve terminating in a membrane which was autoclaved with the vessel and the inner electrode proper which was inserted afterward.

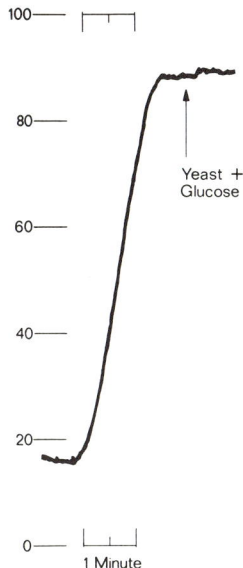

FIG. 6. Response of dissolved oxygen probe.

2. Fermentor Gas Effluent

The analysis of fermentor gas-effluent presents unique problems of its own. As well as the very severe problem of the cost of the instruments, the major problem to date has been the necessity for most analyzers to be presented with moisture-free samples. One problem, at least, is missing; the need for aseptic operation.

The gases most commonly measured are oxygen and carbon dioxide, the first usually by a paramagnetic analyzer and the second by infrared analysis. In order to dry the gas stream we have used a system described by Fiechter and von Meyenburg (1968), first a refrigerant/condenser system to remove "water mist" and desaturate and then a column of anhydrous sodium sulfate Drierite (W. A. Hammond & Co.). We have found this to least affect the CO_2 content of the gas, which is the component most likely to be absorbed by some desiccants. Even so, losses of gas can occur in the condensate from the refrigerant condenser and certainly the desiccant column needs constant attention to ensure it is not depleted. Consequently an analytical system that would not require a predrying stage would be extremely advantageous. For oxygen measurement there are now two instruments which meet this demand; both work on a similar principle, basically a high temperature fuel cell, and are available from the

Westinghouse Electric Company and the Bailey Meter Co., who designate their instrument OE. Neither water vapor nor CO_2 influence the reading which has an extremely fast response and covers a wide range of oxygen concentrations and is accurate to approximately $\pm 2\%$. Because of the temperature of operation, the presence of hydrogen or methane is detrimental!

Unfortunately, no similar system is yet available for carbon dioxide. Methane, however, in the range of 0.1–10% can be determined, again without interference from water vapor or CO_2 by the inexpensive VQ1 instrument produced by the English Electric Valve Co.

VI. Biochemical Analysis of Fermentation Cultures

The usual system currently employed is to arrange for samples to be taken aseptically from fermentors at fixed time intervals; the samples are often frozen at $-20°C$ until a sufficient number have been accumulated and these can then be analyzed (Maddix et al., 1970). The sampling process can of course be automated (Eisler, 1968), but the devices work best with nonviscous broths and for thick mold fermentations no automatic sampling device is yet fully satisfactory.

There can be several drawbacks to manual sampling. Most nonindustrial researchers do not have shift workers available to sample during the night. The necessity to take large numbers of samples can often lead to poor sampling technique, with consequences of nonrepresentative samples or infection problems. When studying transient states moreover, discrete sampling may not provide sufficient information. For these and other reasons, attention is increasingly focusing upon methods which can analyze directly in the fermentor, or methods which can measure continuously without withdrawing samples from the culture vessels.

An excellent example of the first type mentioned is the fluorimetric technique used by Harrison and Chance (1970) for monitoring levels of NADH by recording the light absorption at 340 nm. As they state, the oxygen electrode they used in their experiment had a 90% response time of approximately 1 minute, whereas for their fluorimeter it was 1 second, enabling small fluctuations or changes to be more easily detected.

Changes in the concentration of inorganic ions can be followed by the use of ion-electrodes; these, however, are not heat sterilizable and therefore always entail risk or very elaborate precautions.

The second type of continuous analysis system is more generally applicable, since it depends upon a stream of dialyzate from a culture broth. Ferrari *et al.* (1965) have described a system which recycles broth from a fermentor through a dialyzer and returns the stream to the vessel. External loop systems, however, suffer from the disadvantage of delay time in the piping and values of limiting substrate in a chemostat could, therefore, be altered appreciably. An internal system could consist of a metering pump passing a stream of fluid, usually water, through a semipermeable membrane coil or probe positioned inside a culture vessel. By selecting a suitable flow rate and a suitable thickness and area of dialysis membrane, a dialysis rate will be set up; in the case of continuous culture this should be a steady state. Conventional dialysis tubing passes molecules with a molecular weight below approximately 1–2000, allowing oligosaccharides, small peptides, inorganic ions, etc., to be determined. Naturally, diffusion rates vary between compounds and one rate would not be likely to allow all desired compounds to be analyzed from one such stream. Although the concentration of compounds in a dialysis stream would be quite low, modern analytical techniques, used for example in the Technicon system, are usually operated at no more than a few parts per million and many can be made more sensitive if required.

A few points to mention about the positioning of such a system are that the mounting within the culture vessel requires a volume of turbulent flow to overcome stagnant film conditions, also an adjustable nonreturn valve on the outlet of the stream would enable a suitable pressure to be built up inside the tubing to balance the pressure within the fermentor.

Perhaps the major limitation of any dialysis system is the danger of growth of microorganisms on the tubing or film, thus restricting dialysis rates. One possible solution is to heat the fluid so that the film surface is just above the temperature at which the organism in question will grow.

For macromolecules, a probe utilizing an ultra membrane filter disk could perhaps be used. As there would be at most a few pounds per square inch driving pressure available, filtration rates would be very slow, but perhaps sufficient for sampling. Careful positioning of the probe would be required in order to minimize the build-up of cells, etc., on the culture broth side of the membrane. Perhaps the best position would be on top of the impeller discharge flow, so that a fast moving jet of fluid was passing across the face of the film.

Acknowledgments

We wish to thank the following for their kind permission in allowing us the use of their material: SK Instruments Ltd., for Fig. 2; The Editor, Instrument Practice, for Figs. 3 and 4; General Electric Co., for Fig. 5.

REFERENCES

Aiba, S., and Huang, S. Y. (1969). *Hakko Kogaku Zasshi* **47**, 372.
Aiba, S., Humphrey, A. E., and Millis, N. F. (1965). "Biochemical Engineering." Academic Press, New York.
Bergman, I. (1968). *Nature (London)* **218**, 266.
Brading, D. J. (1970). *Process Eng.* Aug. 4, p. 31.
Brown, G. (1970). *Instrum. Pract.* **24**, 239.
Deindoerfer, F. H., and Humphrey, A. E. (1961). *Appl. Microbiol.* **9**, 134.
Deindoerfer, F. H., and West, J. M. (1960). *J. Biochem. Microbiol. Technol. Eng.* **2**, 165.
Eisler, W. J., Jr. (1968). *Appl. Microbiol.* **16**, 1381.
Ferrari, A., Gerke, J. R., Watson, R. W., and Umbreit, W. W. (1965). *Ann. N.Y. Acad. Sci.* **130**, 704.
Fiechter, A., and von Meyenburg, K. (1968). *Biotechnol. Bioeng.* **10**, 535.
Flynn, D. S., Corso, V., and Gaden, E. L., Jr. (1971). Unpublished data.
Grayson, P. (1969). *Process Biochem.* 4(3), 44.
Harrison, D. E. F., and Chance, B. (1970). *Appl. Microbiol.* **19**, 446.
Maddix, C., Norton, R. L., and Nicolson, N. J. (1970). *Analyst* **95**, 738.
Mori, M., and Yamashita, S. (1967). *Contr. Eng.* July, p. 66.
Pfeifer, V. F., and Vojnovich, C. (1952). *Ind. Eng. Chem.* **44**, 1940.
Phillips, D. H., and Johnson, M. J. (1961). *J. Biochem. Microbiol. Technol. Eng.* **3**, 261.
Reisman, H. B., Gore, J. H., and Day, J. T. (1968). *Chem. Eng. Progr., Symp. Ser.* No. 86, **64**, 26.
Ring, K., Schlecht, S., Eschweiler, W., and Kutscher, J. (1969). *Arch. Mikrobiol.* **65**, 48.
Roberts, A. N., and Shepherd, P. G. (1968). *Process Biochem.* 3(2), 23.
Sikyta, B., and Mastner, J. (1967). *Biotechnol. Bioeng.* **9**, 575.
Solomons, G. L. (1966). *Process Biochem.* 1(6), 307.
Solomons, G. L. (1968). *Process Biochem.* 3(8), 17.
Solomons, G. L. (1969). "Materials and Methods in Fermentation." Academic Press, New York.

COMMERICAL REFERENCES

Analytical Instruments Ltd., Fowlmere, Cambridge, U.K.
Bailey Meter Co., Wickliffe, Ohio 44092, U.S.A.
Comark Electronics Ltd., Brookside Avenue, Rustington, Littlehampton, Sussex, U.K.
Engelhard Industries Ltd., Valley Road, Cinderford, Gloucestershire, U.K.
English Electric Valve Co., Ltd., Chelmsford, Essex, U.K.
Ether Controls, Ltd., Caxton Way, Stevenage, Hertfordshire, U.K.
Geo. Angus & Co., Ltd., Coast Road, Wallsend, Northumberland, U K.
Mixing Equipment Co., Inc., 200 Mt. Read Blvd., Rochester 3, N.Y., U S.A.
New Brunswick Scientific Co., Inc., Box 606, 1130 Somerset St., New Brunswick, New Jersey 08903, U.S.A.
Ormerod Engineering Ltd., Shawclough, Rochdale, Lancashire, U.K
Rank Bros., High Street, Bottisham, Cambridgeshire, U.K.
Rustrak Instrument Co., Inc., 130 Silver Street, Manchester, N.H., U.S.A.

S K Instruments Ltd., Greenhey Place, East Gillibrands, Skelmersdale, Lancashire; U.K.
Technicon Instruments Corp., Research Park, Chauncey, N.Y. U.S.A.
W. A. Hammond & Co., Xenia, Ohio, U.S.A.
Westinghouse Electric Corp., P.O. Box 8606, Pittsburgh, Pa. 15221, U.S.A.

The Extracellular Accumulation of Metabolic Products by Hydrocarbon-Degrading Microorganisms

BERNARD J. ABBOTT

*Esso Research and Engineering Company,
Linden, New Jersey*

AND

WILLIAM E. GLEDHILL

*Monsanto Company,
St. Louis, Missouri*

I.	Introduction	249
II.	Products from Aliphatic Hydrocarbons	252
	A. Alkane Transformation Products	252
	B. Amino Acids	272
	C. Organic Acids	290
	D. Nucleic Acid-Related Products	298
	E. Vitamins and Pigments	303
	F. Sugars and Polysaccharides	311
	G. Other Products	314
	H. Cell Yields, Oxygen Requirement, and Heat Evolution	314
III.	Products from Cyclic Hydrocarbons	318
	A. Biosynthetic Products	319
	B. Transformation Products	320
IV.	Concluding Remarks	375
	References	378

I. Introduction

The intent of the present review is to consolidate the current knowledge of microbial hydrocarbon fermentations in order to demonstrate the types, uses, and potential of microbial hydrocarbon products, mechanisms and concepts of product formation, and methods for increasing product yields. The review will be limited to discussion of the accumulation of extracellular products resulting from hydrocarbon metabolism. The term hydrocarbon will be employed in its strict sense (i.e., a compound containing only carbon and hydrogen). Since detailed previous reviews have not appeared, except for a few reports dealing with products from specific types of hydrocarbon substrates (53, 132, 259, 265), we have attempted to be as comprehensive as possible. Since numerous patents are surveyed in the current review the reader must be cautioned that yields, taxonomy of organisms, or possibly even the existence of particular products claimed might not always be accurate. Many of the fermentations discussed may have no commercial practicality; however, concepts involved in these fermentations can often be extended to other hydrocarbon processes.

The limitation of this review to hydrocarbon products will necessitate omission of a great deal of work in other areas of hydrocarbon microbiology. Publications, including some comprehensive reviews, have appeared in the following areas: (a) single cell protein (35, 236, 257); (b) petroleum dewaxing (177, 281) and desulfurization (233); (c) petroleum microbiology (52, 56, 120, 244, 276, 323, 332); (d) metabolism (287, 316, 320, 373, 386); (e) ecology (103, 259, 405); (f) oil contamination of aqueous environments (259); (g) corrosion and fuel tank contamination (110, 111); (h) waste disposal (293); (i) engineering and continuous culture (6, 26, 92, 232, 317). In addition, three books concerned with hydrocarbon microbiology have been published (18, 53, 298).

Three-quarters of a century have elapsed between the initial observation of microbial paraffin degradation (242) and the commercialization of hydrocarbon processes by the British and Japanese. The realization of these processes can be attributed to several unique advantages afforded by the hydrocarbon system. During the past two decades hydrocarbon production has expanded tremendously. Consequently, many types of hydrocarbons have become commercially available in large quantity and in relatively high purity as substrates for microbial fermentations. The availability, purity, high carbon content and low density of many hydrocarbons have contributed to considerably lower product and transportation costs per unit of available carbon than conventional carbon sources such as sugar or grain. Because hydrocarbon fermentation systems consist primarily of cells, substrate, product, and inorganic salts and not the excessive amount of extraneous material in many sugar- and grain-based media, product recovery after fermentation is facilitated greatly. In addition, the solubility properties of the majority of hydrocarbon substrates facilitate product recovery. Since the substrate is removed readily following fermentation, water soluble and hydrocarbon soluble products are readily separated. Perhaps most important, the multiple and varied types of hydrocarbon substrates have made possible the formation of a vast number of unique and functional products.

In another light, the major disadvantages of hydrocarbon fermentation processes, the excessive heat evolution and oxygen demand, may hinder adoption of these fermentations. The conventional fermentors currently used in many production plants are inadequate for many hydrocarbon processes. Consequently, a large capital outlay would be required to convert outdated fermentors or to build new facilities specifically for hydrocarbon processes.

From a historical viewpoint, initial observations of a microbial hydrocarbon metabolite were made by Sohngen (309) and Kaserer (162) who demonstrated methane utilization by *Methanomonas methanica* with the release of CO_2. Subsequent reports revealed, in addition to CO_2, fatty acid formation from benzene (310, 389) and from various petroleum fractions (310). *Actinomyces elastica* and *A. fuscus* were shown to split methane from rubber and to utilize this product as a carbon source for growth (311). Tausson (362) reported the complete oxidation of paraffin to CO_2 and H_2O by *Penicillium*, *Aspergillus*, and *Pseudomonas* species. He doubted the intermediate formation of fatty acids since no pH drop was seen during culture growth. Jacobs (152), in studies of plant growth stimulation by partial soil sterilization with aromatic compounds, demonstrated the formation of phthalic acid as an intermediate degradation product of naphthalene metabolism. Tausson (362), however, had previously reported that this product was not formed from naphthalene. The production of CO_2 and acetaldehyde as growth products of *Mycobacterium lacticola* on acetylene was reported (23). In studies of bacterial contamination of gasoline storage tanks Thaysen (365) demonstrated the formation of both methane and ethane. Further studies by Thaysen (366) revealed the production of acetaldehyde, acetic acid, and lactic acid from kerosene.

Bushnell and Haas (31) demonstrated that growth of *Pseudomonas* species on petroleum fractions resulted in the production of CO_2 and fatty acids. In addition, they presented evidence for the appearance of unsaturated hydrocarbons. Subsequent studies by these workers indicated that several carotenoid pigments (astacin and β-carotene plus others) were produced during growth of mycobacteria, corynebacteria, and fungi on petroleum fractions (105, 106). These studies also revealed ergosterol and cholesterol formation by these organisms (106). In Stone's laboratory, Strawinski (324) demonstrated the formation of salicylic acid during growth of a *Pseudomonas* species on naphthalene. This study might be considered the forerunner of current hydrocarbon fermentations because it was the first instance in which parameters for increasing product yields were investigated. Shortly afterward, Taggert (333) discovered the formation of alcohols, fatty acids, and esters from gaseous hydrocarbons and found that metal salts favored product formation.

With the advent of increased production of varied and relatively pure hydrocarbon substrates, curiosity, demand for new products, improved methods for product analysis and recovery, and development of microbiological techniques such as sequential induction and co-oxidation, research into microbial hydrocarbon fermentations accelerated. Today it is recognized that hydrocarbon fermentations offer possibilities for development of a multitude of novel fermentation products. Economic advantages of certain hydrocarbon fermentations over their sugar- or grain-based counterparts may eventually result in the elimination of the latter.

II. Products from Aliphatic Hydrocarbons

A. ALKANE TRANSFORMATION PRODUCTS

1. Introduction

Transformation products, for the ensuing presentation, are defined as catabolic products in which most of the carbon skeleton is derived intact from the alkane substrate. Earlier reports of products included in this category described classes of compounds such as waxes, alcohols, esters, and acids without complete identification (333, 359, 360, 364). These reports, and possibly others, are subject to the caveat of Bird and Molton (25), who made a comparison of the metabolic products of five *Pseudomonas* species during growth on alkanes or glucose (24, 25). Many of the products (including fatty acids, hydroxy fatty acids, dicarboxylic acids, and amides) were found regardless of the growth substrate. Structural relationships of these products with alkane substrates have often been used to establish pathways of alkane degradation. It is apparent, however, that these products may be of synthetic rather than of degradative origin. The following can be used to distinguish between these possibilities: (a) isotopic labeling, (b) metabolic inhibitors, (c) sequential induction experiments, and (d) mutation. Additional evidence could be obtained by demonstrating the dependence of product structure on alkane structure, the absence of the product when carbohydrate is substituted for hydrocarbon, and substantial increases in product quantity in the presence of hydrocarbons. Some of the products listed in Table I have not been subjected to these tests. They are included only on the basis of a structural similarity to the substrate or the implication of the investigator that they are of degradative origin. In the tables and the discussion which follow the reader should keep this in mind. Reports of the various carboxylic acid products are particularly subject to this precaution.

One of the earliest confirmed reports of an aliphatic hydrocarbon transformation product described the production of 1,2-hexadecanediol from 1-hexadecene (29). The yeast *Candida lipolytica* when cultivated on 1-hexadecene accumulated about 5% of the consumed hydrocarbon as the diol. Subsequent studies confirmed diol production and also indicated that the conversion of hydrocarbon to diol was appreciable (175, 176).

2. Fatty Acids

Fatty acids are the most commonly encountered product of alkane oxidation (Table I). They have been implicated widely as early intermediates of alkane degradation and like most other alkane degradation products rarely accumulate in substantial quantity. Most attempts to increase accumulation have been made to facilitate detection in order to establish a pathway of hydrocarbon degradation.

Heringa et al. (114) noted that the respiratory quotient of resting *Pseudomonas* cells oxidizing heptane was about 0.62, indicating that the hydrocarbon was completely oxidized to CO_2 and water. However, by reducing the oxygen supply to cells in the last few hours of growth, it was possible to induce accumulation of small amounts of fatty acids. A maximum of 0.36 mg/ml of heptanoic acid was produced from heptane.

Thijsse and Van der Linden (369) promoted fatty acid accumulation from alkanes by using chloramphenicol, an inhibitor of protein synthesis. A resting cell suspension (2.7 mg/ml) of hexane-grown *Pseudomonas* sp. KSLA 473 oxidized heptane to propionic acid in the presence of chloramphenicol. About 0.54 mg/ml of propionic acid accumulated after only 5 hours of incubation. This represented a 60% conversion on a molar basis. Using the same organism, 2-methylhexane was oxidized to isovaleric acid, and 1-heptene was oxidized to 4-pentenoic and 2,4-pentadienoic acids.

In a similar approach, acrylic acid, an inhibitor of β-oxidation, was used to induce fatty acid accumulation (367). Acrylate was believed to function by interacting with coenzyme A, thus inhibiting coenzyme-A-dependent reactions. In the procedure described by Thijsse (367) and in a patent granted to Shell International (299), heptane-grown resting cells in a concentration of 1 to 2 mg/ml were incubated with either hexane, heptane, or octane, and 0.1% acrylic acid. The resulting products (Table I) did not accumulate to concentrations greater than 0.8 mg/ml. Methyl ketones, one carbon shorter than the substrate, also accumulated. These products probably arose via decarboxylation of β-keto derivatives of the substrate.

The concentration of products obtained in these experiments approximately equaled the concentration of inhibitor used to induce the accumulation. This may indicate that acrylic acid functions as a competitive inhibitor. As such, it is doubtful that it could be employed economically to produce higher yields in a commercial fermentation.

TABLE I
ALKANE TRANSFORMATION PRODUCTS

Products	Substrate	Organism	Yield	Remarks	References
Acetaldehyde	Acetylene	*Mycobacterium lacticola*			23
Acetone	30% Propane–air mixture	*Mycobacterium smegmatis*	193 mg/liter	Employed nitrogen-deficient medium.	228
2-Butanone	30% Butane–air mixture	*Mycobacterium smegmatis*		Cells grown on propane and resuspended in nitrogen-free medium with butane.	228
2-Pentanone	0.33% v/v Pentane	*Mycobacterium smegmatis*	83 mg/liter	Cells grown on propane and resuspended in nitrogen-free medium with butane.	228
2-Hexanol and 2-hexanone	0.66% v/v Hexane	*Mycobacterium smegmatis*		Cells grown on propane and resuspended in nitrogen-free medium with butane.	228
2-Undecanoic, undecanoic, and 1,11-undecanedioic	Undecane	*Mycobacterium smegmatis*	Very small quantities	Cells grown on propane and resuspended in nitrogen-free medium with butane.	228
1,10-Decanedioic acid	5% v/v Decane	*Corynebacterium* 7EIC	800 mg/liter	Whole cultures were extracted and the amounts of dioic acids	114, 165

10-Hydroxydecanoic acid	Decane	*Corynebacterium* 7EIC	20 mg/liter	—	114, 165
1,14-Tetradecanedioic	Tetradecane	*Corynebacterium* 7EIC	—	—	114, 165
Propionic acid	Heptane	*Pseudomonas* KSLA 473	35 mg/liter	Oxygen supply reduced 25% toward the end of the fermentation. Fatty acids accumulated simultaneously.	114
Pentanoic acid	Heptane		145 mg/liter		
Heptanoic acid	Heptane		360 mg/liter		
Hexanoic	Hexane	*Pseudomonas* KSLA 473	204.5 mg/liter	Oxygen supply reduced. Only trace quantities of other fatty acids.	114
Suberic acid	5% v/v Octane	*Pseudomonas* sp.	2.5 mg/liter	Trace quantities of acetic and butyric acids also detected.	11, 12
Adipic acid			0.55 mg/liter		
Sebacic acid	2.5% Decane	*Pichia polymorpha*	25 mg/liter		15
Propionic acid	0.17% wt/v Heptane	*Pseudomonas aeruginosa* KSLA 473	0.54 mg/ml	Cells were grown on hexane and resuspended to 2.7 mg/ml in buffer. Chloramphenicol also added.	369

(Continued)

Table I (*Continued*)

Products	Substrate	Organism	Yield	Remarks	References
Isovaleric acid	0.13% wt/v Methylhexane	*Pseudomonas aeruginosa* KSLA 473	0.12 mg/ml	Used heptane-grown resting cells plus chloramphenicol.	369
4-Pentenoic	0.2% wt/v Heptene-1	*Pseudomonas aeruginosa* KSLA 473	0.25 mg/ml	Cells treated as above. Lesser quantities of other acids accumulated.	369
Hexanoic acid 2-Pentanone	1% v/v Hexane	*Pseudomonas aeruginosa* KSLA 473	0.68 mg/ml hexanoic and 0.68 mg/ml 2-pentanone	Heptane-grown resting cells (1–2 mg/ml) incubated with substrate and 0.1% acrylic acid.	299, 367
Heptanoic acid 2-Hexanone Pentanoic acid	1% v/v Heptane	*Pseudomonas aeruginosa* KSLA 473	0.78 mg/ml Heptanoic acid 0.24 mg/ml 2-Hexanone 0.14 mg/ml Pentanoic acid	Same as above. Products accumulated simultaneously.	299, 367
Octanoic acid 2-Heptanone 2-Pentanone	1% v/v Octane	*Pseudomonas aeruginosa*	0.04 mg/ml Octanoic acid 0.24 mg/ml 2-Heptanone 0.18 mg/ml 2-Pentanone	Same as above.	299, 367

Product	Substrate	Organism	Concentration	Notes	Ref.
1,2-Octene epoxide	1.5% v/v 1-Octene	*Pseudomonas* sp.	25 mg/liter	Employed resting heptane-grown cells suspended to 1.3 buffer.	300, 385
1,2-Hexadecanediol	3.0% v/v Hexadecene-1	*Candida lipolytica*	5% of consumed substrate	—	29, 150, 174, 175, 320
1-Decene	0.007% wt/v Decane	*Candida rugosa*	4 mg/liter	Used decane-grown resting cells we suspended to 0.7 mg/ml.	141
1-Decanol	1% v/v 1-Decene	*Candida rugosa*	0.648 mg/ml	Used decene-grown resting cells incubated anaerobically.	140
2-Hexadecanone 3-Hexadecanol 4-Hexadecanol	8% v/v Hexadecane	*Arthrobacter* sp.	1.92 mg/ml	These products produced in a ratio of 72:24:4.	171
2-Pentadecanone 3-Pentadecanone 4-Pentadecanone	8% v/v Pentadecane	*Arthrobacter* sp.	0.33 mg/ml	These products produced in a ratio of 80:14:16.	171
2-Hexadecanol	2% v/v Hexadecane	*Arthrobacter* sp.	0.016 mg/ml	Ketones also accumulated.	172
1-Decanol 1-Decanal 1-Decanoic acid 1-Octanoic acid Other carboxylic acids	2% v/v Decane	*Mycobacterium rhodochrous*	—	Resting cells grown on decane produce these products simultaneously in a ratio of 5:17:20:16:42.	86

(*Continued*)

Table I (Continued)

Products	Substrate	Organism	Yield	Remarks	References
1-, 2-, 3-Decanols 4- and/or 5-Decanols 2-Decanone 3-Decanone 4- and/or 5-Decanone Decanoic acid Nonanoic acid	2% v/v Decane	*Pseudomonas aeruginosa*	—	Resting cells grown on decane produced these products simultaneously in a ratio of: 21:7:20:14:20:6:10.	86
Fatty acids	Liquid paraffin	*Bacillus* sp.	2% of the hydrocarbon consumed	Assayed for just C_{14}, C_{16}, C_{18} fatty acids.	227
Fatty acids	Liquid paraffin	*Pseudomonas aeruginosa*	5.3% of the hydrocarbon consumed	Continuously grown cells.	227
Fatty acids	13% v/v Paraffin	*Candida* sp.	21 mg/ml	Employed nitrogen limited medium. Cells and supernatant extracted for acid determination. See ref. 280 for listing of other organisms producing fatty acids from alkanes.	279, 280
Cetylpalmitate	2% v/v Hexadecane	*Micrococcus cerificans*	4.3 mg/ml	Maximum yield obtained with strain H101N.	78, 322

Octadecylstearate Octadecylpalmitate	1% v/v Octadecane	*Micrococcus cerificans*	3.3 mg/ml	Equal quantities of both esters produced simultaneously.	321
Tetradecylpalmitate	1% v/v Tetradecane	*Micrococcus cerificans*	0.57 mg/ml	Esters also produced when decane or dodecane was used as a substrate.	321
15-Hexadecenyl-palmitate	1% v/v 1-Hexadecene	*Micrococcus cerificans*	2.6 mg/ml	Several esters also produced from 1-octadecene.	320
Heptadecyl pentadecanoate Heptadecyl palmitate Heptadecyl margarate	Heptadecane	*Micrococcus cerificans*	—	Esters produce in a ratio of: 12 : 10 : 78.	319
Esters	Hexadecane	*Micrococcus cerificans*	3.5 mg/ml	Production by continuous fermentation.	318
Esters	Various paraffins	*Cladosporium Hormodendrum hordei*	—	Ester obtained from pentadecane was unsaturated.	143, 305
Cetylpalmitate	Hexadecane	*Mycobacterium*	45 mg/liter	Larger quantities of the product were associated with the cells.	180
Hexadecylstearate	1% wt/v Kerosene	*Micrococcus* sp.	1 mg/ml	—	121

(*Continued*)

TABLE I (*Continued*)

Products	Substrate	Organism	Yield	Remarks	References
Hydroxy fatty acids (glycolypids)	2% wt/v Hexadecane	*Torulopsis apicola*	11 mg/ml	Hydroxy fatty acids were isolated as extracellular glycolipids. Similar yield obtained from octadecane, eicosane, and docosane.	314
Hydroxy fatty acids (glycolipids)	1% wt/v of an *n*-alkane	*Torulopsis gropengresseri*	Trace to 7 mg/ml	See ref. 157 for the identity of the various lipids found.	157
Hydroxy fatty acids (glycolipids)	1% wt/v of 1-hexadecane, 1-heptadecane, or 1-octa-decene	*Torulopsis gropengresseri*	Trace to 0.7 mg/ml	See ref. 157 for the identity of the various lipids found.	157
Hydroxy fatty acids and branched chain alcohols (glycolipids)	1% wt/v of a long-chain branched alkane	*Torulopsis gropengresseri*	Trace to 2.7 mg/ml	See ref. 157 for the identity of the various lipids found.	157
cis-7-Hexadecene	20% v/v Hexadecane	*Nocardia salmonicolor*	40 mg/ml	Cells were glucose-grown, concentrated, and used as a resting cell suspension.	1

Product	Substrate	Organism	Amount	Comments	Reference
cis-9-Octadecene	20% v/v Octadecane	Nocardia salmonicolor	15 mg/ml	Cells were glucose-grown, concentrated, and used as a resting cell suspension.	1
Heptanoic acid	1% wt/v Heptane	Pseudomonas aeruginosa	100 mg/liter	—	295
Ethanol	Ethane	Pseudomonas methanica	4.6 mg/liter ethanol	Cells co-oxidized substrate while growing on methane. Acid, alcohol, and ketones also produced from propane or butane.	220
Acetaldehyde			49 mg/liter acetaldehyde		
Acetate			300 mg/liter acetate		
1,10-Decanoic acid	3.1% v/v Decane	Candida rugosa JF 101	16.7 mg/liter	Smaller quantities of other dicarboxylic acids also found.	142
Trehalose glycolipid	10% wt/v C_{12}–C_{12} paraffins	Arthrobacter paraffineus KY 4303	1.3 mg/liter	Similar glycolipids produced by other bacteria.	205, 329

Dicarboxylic acids were isolated from the culture broth of the yeast *Pichia* growing on undecane as a sole carbon source (261). These acids were identified as 1,11-undecanedioic, azelaic, pimelic, and glutaric acids. When decane was supplied as a substrate 45 mg/liter of sebacic acid accumulated (15). In addition, resting cells transformed azelaic into pimelic and glutaric acids, and 1,11-undecanedioic into azelaic, pimelic, and glutaric acids.

A strain of *Candida rugosa* also accumulated various dicarboxylic acids during growth on decane (142). Acids with even-numbered carbon chains from sebacic (C_{10}) to succinic (C_4) were detected. Small amounts of 1-decanoic acid and 1-decanol were also found. Sebacic acid, which accumulated in greatest quantity, was present at only 16 mg/liter.

Foster, in a patent (85), described a procedure for producing dicarboxylic acids. The C_{10}, C_{12}, and C_{14} α-ω-dioic acids were formed from their analogous alkane substrates by a *Corynebacterium*. Lesser amounts of ω-hydroxy fatty acids were also produced (165). Several techniques were described for conducting the fermentation. For example, acids were produced by cells during growth on the substrate or by resting cell suspensions of alkane-grown cells. In another approach, the cells were grown on propane. After the desired population was obtained, the propane was vented and the substrate for acid production added. Incubation then continued in the absence of propane, and dicarboxylic acids accumulated. Maximum acid production in all of these experiments was less than 2% of the added hydrocarbon (85). Also, since the acids were extracted from whole culture broth, it is not known to what extent they accumulated extracellularly.

Linday and McDonald (227) were the first to make a detailed study of the factors influencing fatty acid production from hydrocarbons. They examined several *Nocardia* and *Pseudomonas* species, measuring the production of C_{14}, C_{16}, and C_{18} fatty acids in culture supernatants. Acid production was favored by nitrate as opposed to ammonia as a nitrogen source, and liquid as opposed to solid paraffins as a carbon source. In one batch experiment, about 11% of the hydrocarbon was recovered in the fatty acid fraction, of which about 1% was represented by the C_{14}, C_{16}, and C_{18} fatty acids. Significant increases in fatty acid production occurred during continuous cultivation. Residence time in these experiments was about 24 hours, and the C_{14}, C_{16}, and C_{18} fatty acids accumulated to the extent of about 5% of the hydrocarbon consumed. It was postulated that the higher conversion afforded by continuous culture was due to the prevention of degradation by washout of the acids.

Ratledge (279, 280) has made an extensive study of the commercial feasibility of fat production from hydrocarbons. His approach was to produce the shorter-chain fatty acids, such as dodecanoic, since these represent the principal components of the most expensive fatty acid mixtures (i.e., palm kernel oil and coconut oil). This approach was possible because selection of the alkane substrate can be used to direct the accumulation of specific fatty acids. The substrate selected was a C_{13} to C_{20} paraffin mixture containing a high proportion of the C_{12}, C_{13}, and C_{14} alkanes. Individual pure alkanes of low molecular weight are prohibitively expensive for this type of product.

The *Candida* sp. used in this study previously was reported to accumulate, extracellularly, lipids containing low molecular weight fatty acids from nonhydrocarbon substrates (299). When cultivated on the paraffin fraction, acid production was favored by increased aeration and nitrogen limitation. Under these conditions, after 108 hours of incubation, fatty acids accumulated to 21 mg/ml, representing a 24.8% conversion of the substrate, with about 54% of the acids shorter than C_{16}. After making several assumptions regarding potential improvements in this process, Ratledge concluded that it would not be economically competitive with conventional sources of fatty acids.

It is likely that substantial improvements can be made in this process. Because of the higher oxygen requirement and heat evolution associated with producing cells from hydrocarbon (see Section II, H) it would be desirable to produce cells from a carbohydrate, conserving the hydrocarbon exclusively for fatty acid synthesis. This might be accomplished by utilization of a mutant incapable of β-oxidation, two-stage continuous fermentation, or a combination of the two. Most organisms capable of degrading alkanes can also utilize various non-hydrocarbon substrates. It should be possible to isolate mutants of these organisms capable of growing on acetate but not on alkanes. Such a phenotype would arise by loss of an enzyme required for β-oxidation. These mutants then could be grown on acetate or sugar, and in the presence of alkanes co-oxidize the hydrocarbon to fatty acids.

One can also envision a two-stage continuous fermentation in which cells produced in the first stage enter a second stage where, in the presence of alkanes and a nitrogen limitation, fatty acids accumulate. This system possesses the advantages (and disadvantages) inherent in continuous fermentation while enabling production of cells from an inexpensive carbon source and conservation of alkanes for fatty acid synthesis.

An important and perhaps controlling factor which still remains in this fermentation is the toxicity or inhibitory effects of the product. This limitation is particularly important in view of the findings of Aida and Yamaguchi (7) who reported increased cell growth on alkanes when the medium was dialyzed during cultivation. The inhibitory diffusible product was tentatively identified as dodecanoic acid.

Product inhibition of cell growth would not be a factor in a two-stage fermentation where the second stage serves simply as a conversion step. The transformation step, however, may still be susceptible to such regulatory mechanisms as feedback inhibition, repression, or a mass action shift of reaction equilibria. A recent report on ketone production from hexadecane illustrates the last-mentioned type of inhibition. Henning et al. (113) found that a simple equation of first order kinetics would describe ketone production by resting cells oxidizing hexadecane. The relationship also contained an exponential decay function to account for enzyme inactivation. In addition, a term was included to account for an increased product concentration at the cell surface. Hydrocarbon attached to cells apparently contained a higher concentration of the product than the remainder of the hydrocarbon droplets in the medium. Vigorous agitation might reduce this gradient by increasing the rates of adsorption and desorption of cells to hydrocarbon droplets. Alternatively, the hydrocarbon solubility of products such as fatty acids can be reduced by increasing the pH of the medium. In this way, more product is transferred to the aqueous phase, consequently reducing the product concentration in the oil droplet at the cell surface.

Fatty acid inhibition might also be minimized by selecting fatty acid-tolerant strains of the production organism. However, there are few data available to suggest that such an approach would be fruitful. Other ways of preventing inhibition include using one of the following during production: (a) ion-exchange resins, (b) dialysis or ultrafiltration, or (c) continuous centrifugation to separate cells and product. However, these techniques are likely to be uneconomical for the commercial production of an inexpensive material such as fatty acids. Their applicability to the production of more expensive intermediates remains to be determined.

3. Glycolipids

During submerged, aerated cultivation of the osmophilic yeast *Torulopsis apicola* (previously designated *T. magnoliae*) an extracellular heavier-than-water oil was produced (314). This compound was identified as a glycoside of a partially acetylated sophorose and a 17-hydroxy C_{18} fatty acid. More precisely, the main component of the oil consisted of 2-0-β-D-glucopyranosyl-D-glucopyranose units linked β-glycosidically to 17-L-hydroxyoctadecanoic and 17-L-hydroxy-9-hydroxyoctadecanoic acid (100, 112). Glycolipid accumulation was favored by increased aeration, lower temperatures, and reduced nitrogen levels. Varying the growth substrate resulted in modification of the lipid moiety of the glycoside. With hydrocarbon substrates (C_{16}–C_{22}) the yield of hydroxy fatty acid increased three- to fourfold over carbohydrate-derived glycolipid and amounted to 50–60% of the hydrocarbon added. The principal fatty acids produced from hexadecane were 15-hydroxyhexadecanoic and 16-hydroxyhexadecanoic acids. From the C_{18}–C_{22} paraffins 17-hydroxyoctadecanoic acid was the primary lipid indicating that substrates of more than eighteen carbons are reduced to eighteen carbons in length before incorporation into the glycoside.

Spencer *et al.* (315) were granted a patent on this process as a means of producing hydroxy fatty acids. Long-chain hydroxy acids, except for ricinoleic, are neither readily available nor easily synthesized. Inexpensive unsaturated fatty acid substrates such as tall oil could be used to produce unsaturated hydroxy acids which might be useful in resin or polymer preparation.

The obvious attractive feature of this fermentation, other than high yields, is the ease of separation of the product from the culture broth. The heavy oil can be separated simply by decantation permitting conservation of substrate and recycling of cells. Although this was not attempted in the above studies, it may increase yields since glycolipids can exhibit antimicrobial activity (154).

Glycolipid formation appears to be widespread among hydrocarbon-utilizing bacteria. Many of the bacterial glycolipids contain trehalose as the sugar moiety, as opposed to the sophorose of *Torula*. The glycolipid of *Arthrobacter paraffineus* (205, 329) was characterized as an ester of α,α'-trehalose and two α-branched-β-hydroxy fatty acids. Glycolipid production by this organism paralleled growth and reached a maximum accumulation of 1.2 mg/ml.

Penicillin inhibited glycolipid formation, causing the release of free trehalose and fatty acids into the culture medium (328). The glycolipid was found in the emulsion layer of the culture and exhibited significant surfactant activity in water-hydrocarbon mixtures. The latter observation, together with the wide distribution of trehalose lipid among hydrocarbon utilizers, suggests that its natural function is to facilitate growth by emulsification of hydrophobic substrates.

Jones and Howe (155, 157), using a strain of *Torula gropengresseri* made a thorough study of glycolipid formation from various aliphatic substrates. After initial growth on glucose, during which the nitrogen source was depleted, an aliphatic substrate was added and converted into sophorosides during a subsequent 6 days incubation. The sophorosides isolated by Jones et al. were crystalline rather than oily and possessed a macrocyclic ring system. The principal component was identified by Tulloch et al. (381) as 17-L-(2'-O-β-D-glucopyranosyl)-oxyoctadecanoic acid 1-4" lactone 6',6"-diacetate (Fig. 1).

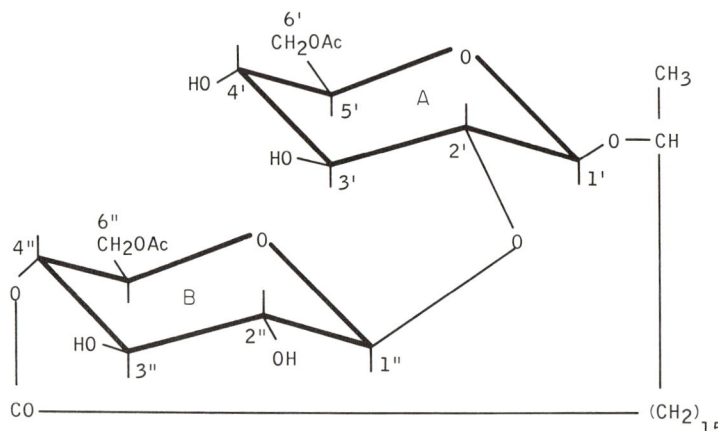

FIG. 1. Structure of glycolipid produced by *Torula gropengresseri*. From Tulloch, Hill, and Spencer (381).

The glycolipid formed from alkane substrates contained primarily ω or ω-1 hydroxylated fatty acids of the same chain length as the substrate (Table II). Alkenes also generated ω- and ω-1 hydroxy fatty acids; however, a significant portion of these acids were one carbon shorter than the alkene substrate (Table III). Jones and Howe (157) concluded that these shorter acids were probably formed by hydroxylation of fatty acids derived by oxidative fission of the double bond.

Qualitative and quantitative comparisons of the lipid component of the glycosides formed from various alkane and fatty acid substrates revealed a pattern of lipid synthesis. Fatty acids longer than C_{18} were not hydroxylated efficiently but were metabolized primarily by β-oxidation to C_{16}–C_{18} acids that were readily hydroxylated. Fatty acids shorter than C_{16} were not incorporated effectively into glycolipid; whereas, C_{15} and C_{14} ω-hydroxy acids were glycosidated readily. The lack of effective incorporation of the C_{15} and C_{14} alkanoic acids was attributed therefore to their low susceptibility to hydroxylation.

Numerous other aliphatic compounds including branched chain alkanes also were incorporated into glycolipids (156, 158, 159). Among these were various 1- and 2-alkanols, methyl-β-alkoxypropionates,

TABLE II
METHYL ESTER PRODUCTS OF THE GLYCOLIPIDS
PRODUCED FROM ALKANES BY *Torula gropengresseri*[a]

Substrate	Product 1	Product 2	Product 3
Dodecane	$CH_3 \cdot CH(OH) \cdot [CH_2]_9 \cdot CO_2 \cdot CH_3$ (trace)	$HO \cdot [CH_2]_{11} \cdot CO_2 \cdot CH_3$ (trace)	
Tridecane	$CH_3 \cdot CH(OH) \cdot [CH_2]_{10} \cdot CO_2 \cdot CH_3$ (0.6)	$HO \cdot [CH_2]_{12} \cdot CO_2 \cdot CH_3$ (0.4)	
Tetradecane	$CH_3 \cdot CH(OH) \cdot [CH_2]_{11} \cdot CO_2 \cdot CH_3$ (1.8)	$HO \cdot [CH_2]_{13} \cdot CO_2 \cdot CH_3$ (1.5)	$CH_3 \cdot O_2C \cdot [CH_2]_{12} \cdot CO_2 \cdot CH_3$ (trace)
Pentadecane	$CH_3 \cdot CH(OH) \cdot [CH_2]_{12} \cdot CO_2 \cdot CH_3$ (2)	$HO \cdot [CH_2]_{14} \cdot CO_2 \cdot CH_3$ (2)	$CH_3 \cdot O_2C \cdot [CH_2]_{13} \cdot CO_2 \cdot CH_3$ (0.3)
Hexadecane	$CH_3 \cdot CH(OH) \cdot [CH_2]_{13} \cdot CO_2 \cdot CH_3$ (27)	$HO \cdot [CH_2]_{15} \cdot CO_2 \cdot CH_3$ (34)	$CH_3 \cdot O_2C \cdot [CH_2]_{14} \cdot CO_2 \cdot CH_3$ (7)
Heptadecane	$CH_3 \cdot CH(OH) \cdot [CH_2]_{14} \cdot CO_2 \cdot CH_3$ (70)	$HO \cdot [CH_2]_{16} \cdot CO_2 \cdot CH_3$ (7)	$CH_3 \cdot O_2C \cdot [CH_2]_{15} \cdot CO_2 \cdot CH_3$ (1)
Octadecane	$CH_3 \cdot CH(OH) \cdot [CH_2]_{15} \cdot CO_2 \cdot CH_3$ (55)	$HO \cdot [CH_2]_{17} \cdot CO_2 \cdot CH_3$ (3)	$CH_2 \cdot O_2C \cdot [CH_2]_{16} \cdot CO_2 \cdot CH_3$ (0.2)
Nonadecane	$CH_3 \cdot CH(OH) \cdot [CH_2]_{16} \cdot CO_2 \cdot CH_3$ (16)		
Eicosane	$CH_3 \cdot CH(OH) \cdot [CH_2]_{14} \cdot CO_2 \cdot CH_3$ (34)	$HO \cdot [CH_2]_{16} \cdot CO_2 \cdot CH_3$ (2)	$CH_3 \cdot O_2C \cdot [CH_2]_{15} \cdot CO_2 \cdot CH_3$ (trace)
	$CH_3 \cdot CH(OH) \cdot [CH_2]_{17} \cdot CO_2 \cdot CH_3$ (7)		
	$CH_3 \cdot CH(OH) \cdot [CH_2]_{15} \cdot CO_2 \cdot CH_3$ (30)	$HO \cdot [CH_2]_{17} \cdot CO_2 \cdot CH_3$ (2)	$CH_3 \cdot O_2C \cdot [CH_2]_{16} \cdot CO_2 \cdot CH_3$ (trace)
Docosane	$CH_3 \cdot CH(OH) \cdot [CH_2]_{15} \cdot CO_2 \cdot CH_3$ (30)	$HO \cdot [CH_2]_{17} \cdot CO_2 \cdot CH_3$ (1)	$CH_2 \cdot O_2C \cdot [CH_2]_{16} \cdot CO_3 \cdot CH_3$ (trace)
Tetracosane	$CH_3 \cdot CH(OH) \cdot [CH_2]_{15} \cdot CO_2 \cdot CH_3$ (22)	$HO \cdot [CH_2]_{17} \cdot CO_2 \cdot CH_3$ (trace)	$CH_3 \cdot O_2C \cdot [CH_2]_{16} \cdot CO_2 \cdot CH_3$ (trace)

[a] Number in parenthesis is the lipid yield based on the substrate added. From Jones and Howe (157).

TABLE III
METHYL ESTER PRODUCTS OF THE GLYCOLIPIDS
PRODUCED FROM 1-ALKENES BY *Torula gropengresseri*[a]

1-Hexadecene	$CH_3 \cdot CH(OH) \cdot [CH_2]_{13} \cdot CO_2 \cdot CH_3$ (3)	$HO \cdot [CH_2]_{15} \cdot CO_2 \cdot CH_3$ (3)	$CH_3 \cdot O_2C \cdot [CH_2]_{14} \cdot CH_2 \cdot CH_3$ (4)
	$CH_3 \cdot CH(OH) \cdot [CH_2]_{12} \cdot CO_2 \cdot CH_3$ (trace)	$HO \cdot [CH_2]_{14} \cdot CO_2 \cdot CH_3$ (trace)	$CH_3 \cdot O_2C \cdot [CH_2]_{13} \cdot CO_2 \cdot CH_3$ (trace)
1-Heptadecene	$CH_3 \cdot CH(OH) \cdot [CH_2]_{14} \cdot CO_2 \cdot CH_3$ (6)	$HO \cdot [CH_2]_{16} \cdot CO_2 \cdot CH_3$ (0.3)	$CH_3 \cdot O_2C \cdot [CH_2]_{15} \cdot CO_2 \cdot CH_3$ (2)
	$CH_3 \cdot CH(OH) \cdot [CH_2]_{13} \cdot CO_2 \cdot CH_3$ (3)	$HO \cdot [CH_2]_{15} \cdot CO_2 \cdot CH_3$ (3)	$CH_3 \cdot O_2C \cdot [CH_2]_{14} \cdot CO_2 \cdot CH_3$ (0.5)
1-Octadecene	$CH_3 \cdot CH(OH) \cdot [CH_2]_{15} \cdot CO_2 \cdot CH_3$ (7)	$HO \cdot [CH_2]_{17} \cdot CO_2 \cdot CH_3$ (0.6)	$CH_3 \cdot O_2C \cdot [CH_2]_{16} \cdot CO_2 \cdot CH_3$ (1.1)
	$CH_3 \cdot CH(OH) \cdot [CH_2]_{14} \cdot CO_2 \cdot CH_3$ (3.5)	$HO \cdot [CH_2]_{16} \cdot CO_2 \cdot CH_3$ (0.4)	$CH_3 \cdot O_2C \cdot [CH_2]_{15} \cdot CO_2 \cdot CH_3$ (2.2)

[a] Numbers in parentheses indicate the lipid yield based on the substrate added. From Jones and Howe (157).

1-halogenalkanes, and alkanes having various nitrogen-, sulfur-, and cyanide-containing terminal functional groups. Glycolipid formation, therefore, is a relatively nonspecific enzymatic reaction. This characteristic and the ability to control product composition by substrate selection makes this fermentation remarkably versatile.

4. Esters

An important contribution toward establishing a pathway of alkane degradation resulted from the isolation and identification of cetylpalmitate in cultures of *Micrococcus cerificans* (78, 322). Ester formation from hexadecane reached its peak during the late logarithmic growth phase and declined thereafter. Various strains of *M. cerificans* differed in their ability to produce cetylpalmitate; however, maximum accumulation was about 4.3 mg/ml. Ester composition was governed by the substrate. The alcohol moiety was invariably the same chain length as the alkane. The fatty acid was often palmitic acid, but could also be two carbons longer or shorter than the substrate. 1-Alkenes, after hydroxylation of the terminal methyl group, were also incorporated in esters. The fatty acid component of these esters was saturated.

Finnerty *et al.* (79) examined 15 cultures of various *Pseudomonas*, *Nocardia*, *Mycobacterium*, and *Flavobacterium* species. Esters could not be isolated from any of these cultures. Shimakara and Yamashita (305) screened 57 hydrocarbon-utilizers for accumulation of metabolic products and found that lipid extracts of 8 of these organisms gave a positive hydroxamic acid test for esters. Ester formation by fungal metabolism of alkanes also has been reported. These esters were not of high molecular weight but contained about 17 carbons and a double bond. It appears, therefore, that ester production by hydrocarbon-assimilating organisms is not a commonly encountered characteristic. This is somewhat unexpected in view of the fact that many organisms produce the alcohol and acid moieties required for ester formation.

Patents have been issued describing ester production by microorganisms that metabolize hydrocarbons (121, 318). In one process *Micrococcus cerificans* was cultivated by continuous fermentation on *n*-hexadecane (318). Wax synthesis was stimulated by restricting various mineral salts and by the addition of azelaic acid. Ester production reached about 3.2 mg/ml in the presence of 7.6 mg/ml cells. These data must be interpreted cautiously because of the manner in which ester was quantitated. Ester yield was defined as the amount of substrate carbon which could not be accounted for as either cells or CO_2.

Raymond and Davis (281) described wax production by a *Nocardia* sp. Unlike *Micrococcus cerificans* esters, the nocardial esters were not found extracellularly even though the organism was grown in a fermentor with vigorous agitation (1750 rpm). During growth on hexadecane and octadecane, the lipid content of the *Nocardia* was as high as 70%, of which about 40% was wax. Davis (53) suggested that the wax was an intracellular storage material and that excretion by *M. cerificans* was due to cell lysis. However, Finnerty *et al.* (79) reported that *M. cerificans* did not lyse during growth, but the apparent lysis was an artifact of the cell mass determination. Although Stewart *et al.* (322) routinely extracted cells and culture broth for esters, they found cetylpalmitate only in culture filtrates. Small quantities of cetylpalmitate were also found in the culture filtrate of a *Mycobacterium* sp. (180). This organism, however, contained a larger quantity of the wax as a component of its cellular lipid. It appears, therefore, that the ability to "excrete" wax may vary from one organism to another.

5. Ketones

Ketones were among the earlier recognized products of hydrocarbon degradation. Lukins (228) described the production of methyl ketones from C_3 to C_6 alkanes. Ketone formation was promoted by growing cells on propane and resuspending them in nitrogen-free medium with the substrate for ketone formation. However, only small quantities of ketone accumulated under these conditions. For example, about 190 mg/liter of acetone was formed from a 30% propane-air mixture.

Leadbetter and Foster reported that methyl ketones were produced from propane and butane by *Pseudomonas methanica* (220, 221). This organism, an obligate methane utilizer, co-oxidized ethane, propane, and butane while growing on methane (Table I). Ketone formation from longer chain alkanes could not be demonstrated.

Fredericks (86) observed that *Pseudomonas aeruginosa* NCIB 9904 oxidized decane to a mixture of 2-, 3-, and 4- and/or 5-decanones and the corresponding alcohols. The ketones accumulated in greatest quantity, and the methyl ketone and 4- and/or 5-ketone represented the major ketonic products. It was postulated that the ketones arose by a nonspecific subterminal enzymatic attack of decane.

Klein *et al.* (171, 172) screened for organisms which would accumulate oxidation products from long-chain alkanes. Their approach was to isolate an organism which grew poorly on the alkanes, on the assumption that poor utilization indicated partial oxidation of the substrate. With this approach an *Arthrobacter* species was isolated that accumulated 2,3- and 4-ketones and the 2-, and 3-alcohols of hexadecane. The organism grew poorly on hexadecane as a sole carbon source but utilized the substrate when cornsteep liquor (CSL) or yeast extract was added to the medium. The ratio of the ketonic products did not vary as changes were made in the medium, the 2-ketone being produced in greatest quantity with lesser amounts of the 3- and 4-ketone. The ease of ketone formation decreased as the distance of the reactive carbon from the terminal end of the substrate increased. Clearly, a different mechanism of ketone formation occurs in *Arthrobacter* than in the *Pseudomonas* sp. described by Fredericks.

Klein *et al.* suggested that ketone formation by *Arthrobacter* was a type of co-oxidation where yeast extract or CSL serves as the growth substrate (171). It was found subsequently, however (170a), that yeast extract and CSL could be replaced by thiamine. Ketone formation, therefore, was probably due to suboptimal thiamine concentrations rather than to co-oxidation. It would not be surprising to find that ketone oxidation is a thiamine-requiring reaction quite analogous to the α-ketoglutarate dehydrogenase system. Regulation of the thiamine concentration may enable control of ketone formation in much the same way that thiamine regulation can be used to control α-ketoglutarate and glutamic acid production from alkanes (see Sections II,B,2 and II,C,1). If this relationship can be exploited, a large increase in ketone production may result.

6. Alkenes

Alkene formation from alkanes was first reported by Azoulay *et al.* (17, 37). They found that heptane-grown cells of *Pseudomonas aeruginosa* contained an enzyme which in the presence of heptane reduced pyocyanine. A small quantity of 1-heptane was isolated from the reaction mixture and identified by its infrared spectrum. Numerous attempts to demonstrate this reaction in other organisms were unsuccessful. Subsequently Wagner *et al.* (388) reported that 1-hexadecene could be isolated from a lipid extract of various microorganisms grown on hexadecane. Small amounts of alkene also were produced from hexadecane by a crude cell extract of a *Nocardia* sp. More recently, Iida and Iizuka (140, 141) reported that the anaerobic oxidation of decane by *Candida rugosa* yielded 1-decene. Maximum 1-decene accumulation was estimated at 4 mg/l.

Abbott and Casida (1) reported that large quantities of monointernal hexadecenes were accumulated by a resting cell suspension of *Nocardia* (Table I). This organism was grown on glucose and resuspended at about ten-fold its original concentration in phosphate buffer containing 25% v/v hexadecane. Alkene accumulated during a subsequent seven-day incubation period, after which as much as 40% of the substrate was unsaturated. Maximum alkene production occurred when cells were grown with both ferrous and ferric ion in the medium. When hexadecane replaced glucose as the growth substrate, only small quantities of alkene were detected during growth or during the subsequent resting cell incubation. Apparently, hexadecene induced alkene-degrading enzymes which prevented alkene accumulation.

This strain of *Nocardia* dehydrogenated alkanes from C_{14} to C_{18}. The susceptibility to dehydrogenation increased with increasing chain length. The double bonds were located primarily in the 9,10-position with smaller quantities of alkenes containing a double bond adjacent to the 9,10-position. The nuclear magnetic resonance spectrum revealed that the alkenes were exclusively of *cis*-configuration.

The location and configuration of the double bond makes these alkenes very difficult and expensive to synthesize chemically. Consequently, a microbial process would be quite attractive. However, as with many of the novel products of microbial metabolism, there is at present no market which specifically demands them.

7. *Epoxides*

Epoxides have been identified among the oxidation products of 1-alkenes. It is uncertain whether they represent an intermediate of alkane metabolism or a metabolic dead end formed by nonspecific hydroxylating enzymes. Regardless of their role, they can be accumulated to an appreciable extent by resting cells of a *Pseudomonas* sp. (300, 385). Heptane-grown cells resuspended in buffer to 1.3 mg/ml produced 25 mg/liter of 1,2-epoxyoctane from 1-octene in just 30 minutes. Epoxides were also formed from 1-heptene and 1-nonene; 2-alkenes, however, were not epoxidized. Oxidation of epoxides by this organism was either very slow or did not occur. Also, the lack of inhibition of epoxide formation by chloramphenicol indicated that continual protein synthesis was not required. Therefore, maximum epoxide yields may be determined only by enzyme decay and product inhibition.

B. AMINO ACIDS

1. *Introduction*

The successful commercialization of the glutamic acid fermentation provided an impetus for the development of high-yield fermentations for other amino acids. Whereas L-glutamic acid is valued for its flavor-enhancing characteristics, the natural L-isomers of other "essential" amino acids may be of value as dietary supplements. Efficient chemical syntheses exist for the production of most amino acids; however, the products are racemic mixtures that are expensive to resolve. The D-isomers usually do not exhibit biological activity and, consequently, do not satisfy nutritional requirements. Fermentation processes, however, lead almost exclusively to the L-isomers. Thus, by developing high-yield fermentation processes employing inexpensive substrates, it may become economically feasible to eliminate dietary deficiencies by supplementation with the proper amino acids.

Reports describing the use of hydrocarbons as substrates for extracellular amino acid accumulation began to appear in the early 1960s. Yamada et al. (399) screened 112 kerosene-utilizing organisms and found that 47 accumulated small quantities of amino acids extracellularly. Strain S10B1, subsequently identified as a *Corynebacterium* (402), accumulated 485 mg/liter of glutamic acid, 18.5 mg/liter of alanine, 7.5 mg/liter of lysine, and lesser amounts of glycine from a kerosene and liquid paraffin mixture. In a similar study, Shah et al. (297) isolated 91 organisms which accumulated small quantities of amino acids from kerosene and liquid paraffin. Maximum accumulation amounted to about 500 mg/liter of which glutamic acid and valine were the major constituents. A *Pseudomonas* sp. was identified as the most productive strain in this study (296).

The use of auxotrophic mutants provides a more effective method for increasing amino acid production. Ishii et al. (149) isolated a number of such mutants of *Corynebacterium hydrocarboclastus* and *Alcaligenes marshallii* and found that they accumulated various amino acids. The yields of many of the products were in excess of 1 mg/ml; however, no attempt was made to establish the optimum concentration of the required metabolite. It is likely that when this is done significant increases in yield will result.

Several patents were issued describing procedures for the extracellular production of various amino acid mixtures. Douros and Raymond (66) reported that *Mycobacterium tuberculosis* ATCC 15073 cultivated on decane accumulated 3.1 mg/ml of methionine and 0.6 mg/ml of lysine after 96 hours incubation (146). In another report (73) various organisms were found to accumulate 2–3 mg/ml of amino acids from a C_{17}–C_{19} olefin fraction. The predominant product of the mixtures was either lysine, glutamic acid, or arginine. *Brevibacterium ketoglutamicum* ATCC 15587 in the presence of gaseous hydrocarbons such as butane accumulated a mixture of amino acids containing 3.0 mg/ml of glutamic acid, 1.2 mg/ml of alanine, 0.5 mg/ml of valine, and 0.1 mg/ml of lysine. When penicillin was added to the medium, 1.0 mg/ml of trehalose and 4.5 mg/ml of glutamate accumulated (200). Yeast, as well as bacteria, have been reported to accumulate amino acids (8). The principal yeast product, glutamic acid, was produced at about 25 mg/liter.

A novel approach to amino acid production from alkanes was described in a recent patent (203). Two organisms were cultivated as a mixed culture in a medium containing a paraffin fraction as a substrate. One of the organisms was capable of utilizing paraffin as a sole carbon source, whereas the second organism was not. The nonhydrocarbon-utilizing organism was any of several strains selected for their ability to produce amino acids from carbohydrates. Inoculum for the production medium was obtained by cultivating each organism separately in a complex nonhydrocarbon medium. The production medium contained 10% C_9–C_{18} paraffin fraction and a small quantity of CSL. In this medium *Arthrobacter paraffineus* accumulated 15 mg/ml of glutamic acid; however, when cultivated with a glutamate-producing strain of *Corynebacterium glutamicum*, product yields exceeded 30 mg/ml. Some of this increase was probably due to increased production by *A. paraffineus*. Valine, lysine, homoserine, and ornithine were also produced by the above technique using the appropriate auxotrophic mutants of *C. glutamicum*. The latter products accumulated to about 2 mg/ml.

2. Glutamic Acid

The production of glutamic acid by microorganisms is one of the most extensively investigated product fermentations. Many of the techniques evolved to increase glutamate yield from carbohydrates have been successfully applied to alkane substrates (Table IV). Thus, to place the alkane fermentations in proper perspective, it is necessary to describe briefly the essential features of glutamate production from glucose.

The ability of *Corynebacterium glutamicum* (*Micrococcus glutamicus*) to produce large quantities of glutamic acid from carbohydrates can be atributed to two factors: (a) lack of the enzyme α-ketoglutarate dehydrogenase, and (b) a requirement for biotin. The absence of α-ketoglutarate dehydrogenase has the obvious effect of preventing the oxidation of α-ketoglutarate (the immediate precursor of glutamic acid) to succinate. The biotin requirement provides a means of controlling cellular permeability. Biotin is required by *C. glutamicum* for the CO_2-fixation reactions in fatty acid synthesis (58); thus, by controlling the biotin concentration, fatty acid synthesis can be restricted. This causes an increase in permeability of the cell membrane which

TABLE IV
GLUTAMIC ACID PRODUCTION BY MICROORGANISMS UTILIZING HYDROCARBON SUBSTRATES

Substrate	Organism	Yield (max)	Remarks	References
5% v/v C_{12}–C_{14} Paraffin mixture	Arthrobacter paraffineus ATCC 21167	25.3 mg/ml	Organism requires biotin and thiamine.	183
5% wt/v Undecane	Arthrobacter simplex ATCC 15799	4.6 mg/ml	Employed mineral salts medium containing 0.1% meat extract. Similar yields obtained with Corynebacterium hydrocarboclastus ATCC 15592.	186
1% wt/v Acetate	Arthrobacter simplex ATCC 15799	3.8 mg/ml		186
5% wt/v Undecane + 1% wt/v acetate	Arthrobacter simplex ATCC 15799	18.8 mg/ml		186
3% wt/v Decane	Nocardia globerula ATCC 15076	2.5 mg/ml	—	62
11% v/v Kerosene	Corynebacterium hydrocarboclastus M-104, ATCC 15110	2.1 mg/ml	30 Units/ml of penicillin and vitamin B_1 added to salts medium.	9
10% v/v Tetradecane	Corynebacterium hydrocarboclastus M-104, ATCC 15110	4.0 mg/ml	30 Units/ml of penicillin and vitamin B_1 added to salts medium.	9
5% v/v Undecane	Corynebacterium hydrocarboclastus No. 2438, ATCC 15592	15.5 mg/ml	Utilized salts medium containing 5 μg/liter of thiamine.	351
5% v/v Kerosene	Corynebacterium hydrocarboclastus No. 2438, ATCC 15592	3.0 mg/ml	Utilized salts medium containing 5 μg/liter of thiamine.	351

(Continued)

TABLE IV (Continued)

Substrate	Organism	Yield (max)	Remarks	References
10% v/v 180–350°C Gas-oil fraction	*Brevibacterium alkanolyticum* ATCC 21033	10 mg/ml	Medium contained 0.1% cornsteep liquor.	336
10% v/v Hexadecane	*Brevibacterium alkanolyticum* ATCC 21033	18 mg/ml	Medium contained 0.1% cornsteep liquor.	336
10% v/v C_{11}–C_{23} Paraffin fraction	*Brevibacterium alkanolyticum* ATCC 21033	16 mg/ml	Medium contained 0.1% cornsteep liquor.	336
10% v/v C_{11}–C_{23} Paraffin fraction	*Brevibacterium alkanolyticum* ATCC 21033	25 mg/ml	10 Units/ml of penicillin added to medium.	336
10% v/v C_{11}–C_{18}	*Corynebacterium hydrocarboclastus* No. 2438, ATCC 15592	75 mg/ml	Employed salts medium containing yeast extract, Tween 80, and penicillin.	182
5% v/v Kerosene	*Pseudomonas methanica*	7.9 mg/ml	Employed salts medium containing biotin and Tween 20. Other *Pseudomonas* sp. and *Serratia marcescens* also reported to produce glutamate from kerosene.	270

0.38 vvm 50% Butane + 50% air	*Corynebacterium alkanum* ATCC 21194	18 mg/ml	Salts medium contained 0.05% cornsteep liquor.	196
5% v/v Tetradecane	*Corynebacterium hydrocarboclastus* SB101	6.3 mg/ml	Thiamine incorporated in production medium.	146
5% v/v Nonane	*Corynebacterium hydrocarboclastus* SB101	1.3 mg/ml	Thiamine incorporated in production medium.	146
5% v/v Octadecene-1	*Corynebacterium hydrocarboclastus* SB101	4.3 mg/ml	Thiamine incorporated in production medium.	146
10% wt/v C$_{12}$–C$_{14}$ Paraffin fraction	*Arthrobacter simplex* ATCC 15799	39 mg/ml	Thiamine and mannitol (2%) added to medium. Lesser amount of α-ketoglutarate also accumulated.	353
5% wt/v Dodecane	*Arthrobacter paraffineus* ATCC 19064	20 mg/ml	Organism required adenine and medium also contained thiamine.	353
10% wt/v C$_{12}$–C$_{17}$ Paraffin mixture	*Corynebacterium alkanolyticum*	40 mg/ml	Glycerol required for growth.	248

facilitates the leakage of glutamate to the surrounding medium (58). The optimum biotin concentration for glutamate production is 2.5 µg/liter; whereas, maximum cell growth requires 50 µg/liter (350, 354).

Crude carbon sources, such as molasses, often contain biotin levels in excess of the optimum for glutamate production. This necessitates the use of additives (e.g., surfactants and antibiotics) to counteract the excessive biotin concentration. Since hydrocarbons can also reverse the effect of biotin (194) it would seem that permeability may not be a limiting factor for glutamate production from alkanes. Experimentally, however, it has been found that hydrocarbon substrates do not eliminate the need for other permeability-increasing agents (Table IV).

Otsuka et al. (268), noting previous reports that penicillin increased the permeability of cells producing glutamate from glucose (312), tested the effect of this antibiotic on a species of *Corynebacterium hydrocarboclastus* growing on tetradecane. The addition of penicillin (30 units/ml) increased glutamate production from 0.3 mg/ml to about 4.0 mg/ml (312).

Iguchi et al. (136) also reported increased glutamate yields from hydrocarbon in the presence of penicillin. In this study, it was found that the free intracellular glutamate levels of cells grown in the presence or absence of penicillin were about equal, but that only the penicillin-grown cells would release their free glutamic acid when washed with a buffer solution. In addition, cells grown in the absence of penicillin released glutamate only when incubated with penicillin. Thus, penicillin improves glutamate yields from both alkane and carbohydrate substrates through mechanisms involving changes in permeability.

The importance of permeability to glutamate production from alkanes is perhaps best exemplified in a patent granted to Kyowa Hakko Kogyo (182). *Corynebacterium hydrocarboclastus* ATCC 15592 accumulated 75 mg/ml of glutamic acid from 100 mg/ml of a C_{11} to C_{18} alkane fraction in the presence of penicillin and Tween 80. Deletion of the Tween 80 lowered the yield to 50 mg/ml while the absence of penicillin restricted accumulation to 30 mg/ml. In the absence of both Tween 80 and penicillin, glutamate levels reached only 15 mg/ml.

The "biotin effect" on glutamate production from carbohydrate has also been demonstrated with alkane substrates (183). *Arthrobacter paraffineus* ATCC 21167 accumulated glutamic acid when grown on a C_{12} to C_{14} paraffin mixture. Biotin-requiring mutants of this organism were isolated and found to accumulate greater quantities of glutamate than the parent organism. The response of *A. paraffineus* to biotin level was similar to that of *Corynebacterium glutamicum* (Fig. 2) except that the optimal biotin concentration for *A. paraffineus* (0.3 µg/liter) was only one-tenth of the optimum for *C. glutamicum*. Very low biotin optima also have been reported for some carbohydrate-utilizing organisms. For example, *A. globiformis* required only 0.3 µg/liter for maximum glutamate production from glucose; whereas, the optimum for *Brevibacterium flavum* decreased from 3.0 µg/liter to 0.3 µg/liter when glucose was replaced with acetate as a substrate. It is tempting, nevertheless, to speculate that the reduced requirement of *A. paraffineus* reflected a sparing effect of fatty acids generated via alkane degradation.

Recently, another way was found to control cellular permeability. Nakao *et al.* (248) isolated glycerol auxotrophs of *Corynebacterium alkanolyticum* that accumulated up to 40 mg/ml of glutamic acid from paraffin. The prototrophic strain produced similar quantities of glutamate only when penicillin was added to the culture. Penicillin increased cellular permeability by inhibiting phospholipid synthesis. The mutant strain also was unable to synthesize phospholipids since glycerol, a phospholipid precursor, was present in low concentration. The addition of penicillin to glycerol-limited cells did not increase yields beyond those obtained when penicillin was added to the prototrophic culture.

While cellular permeability is an important factor for glutamate accumulation, it is not the only requirement for high glutamate yields. An important contribution toward controlling glutamate synthesis from alkanes resulted from the observation that *Corynebacterium hydrocarboclastus* required thiamine for growth (335). Low thiamine concentrations restricted growth but greatly stimulated glutamate accumulation (Fig. 3). This phenomenon is remarkably similar to the effect of biotin on glutamate production from glucose (compare Figs. 2 and 3). The effect of thiamine, however, does not appear to be related to cellular permeability. Two key enzymes of carbon metabolism, pyruvate dehydrogenase and α-ketoglutarate dehydrogenase, require thiamine pyrophosphate as a cofactor. Thus, organisms requiring thiamine for growth should exhibit low α-ketoglutarate dehydrogenase activity under a thiamine limitation. This conserves α-ketoglutarate for the reductive amination to glutamate. The restriction of pyruvate dehydrogenase that also occurs during thiamine limitation, does not prevent the entry of carbon into the TCA cycle since the degradation of alkanes by β-oxidation generates acetate which can enter the cycle through oxalacetate or glyoxalate. Direct experimental evidence supporting this mechanism of "restricted thiamine-requiring

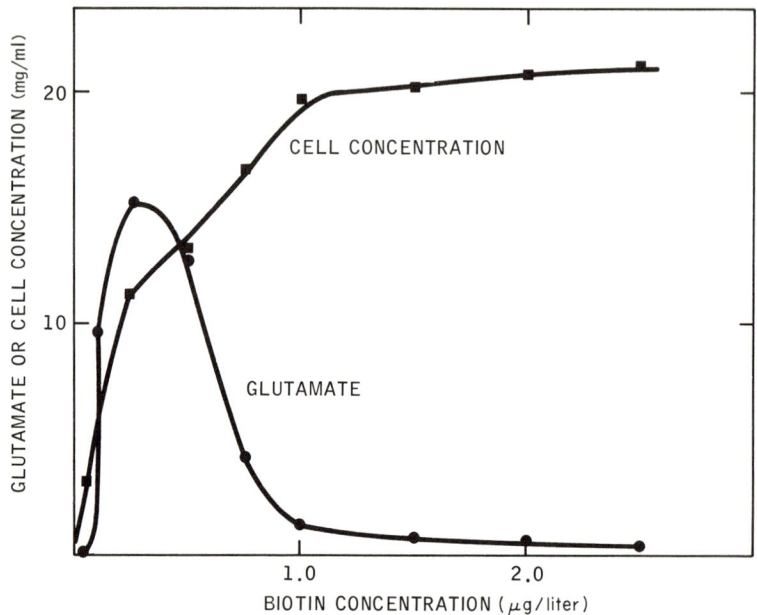

FIG. 2. Effect of biotin concentration on growth and glutamate production by a biotin auxotroph of *Arthrobacter paraffineus*. Drawn from data in Kyowa Hakko Kogyo patent (183).

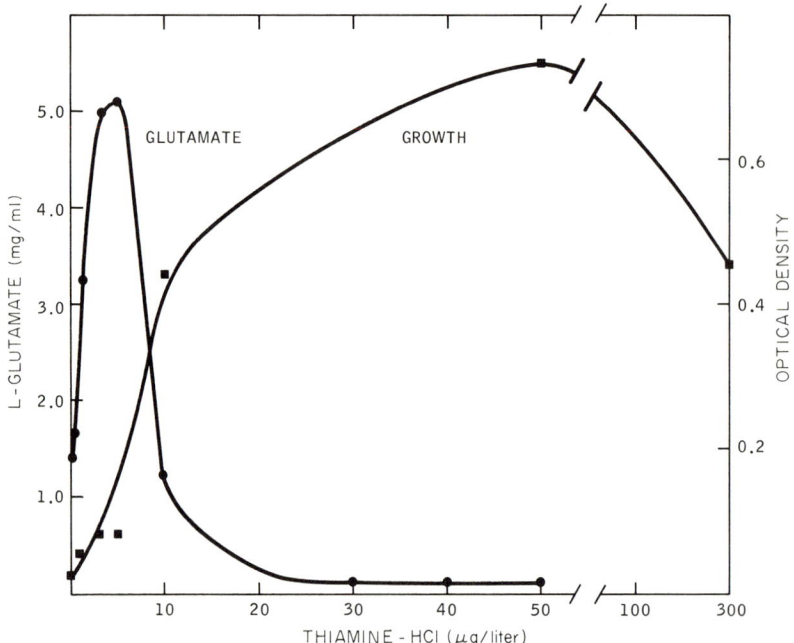

FIG. 3. Effect of thiamine concentration on growth and glutamate production by a thiamine-requiring strain of *Corynebacterium*. Redrawn from Takahashi et al. (335).

enzymes" is limited, but data on the accumulation of α-ketoglutarate from alkanes by thiamine-requiring yeast are consistent with this interpretation (see Section II,C,1).

In contrast to the preceding report, is a more recent report by Shiio and Uchio (304) indicating that glutamate production by a thiamine-requiring strain of *Corynebacterium hydrocarboclastus* was not influenced by thiamine concentration. In this study, penicillin was used to increase cellular permeability; however, when the effect of thiamine was tested, the medium lacked penicillin. Thus, if low thiamine levels favored increased glutamate production, the absence of penicillin (resulting in low permeability) may have prevented the expression of the enhanced synthesis. An alternative explanation is that this particular strain, like *C. glutamicum*, exhibits little or no α-ketoglutarate dehydrogenase activity.

The initial attack on alkane substrates is believed to be catalyzed by a mixed function oxidase which incorporates molecular oxygen into a terminal methyl group (165, 219, 221, 322). The resulting alcohol is then oxidized to a carboxy acid and degraded to acetate fragments by β-oxidation, Fig. 4 (114, 165, 368, 392). Alkane hydroxylations by protein fractions of a *Pseudomonas* sp. require ferrous or ferric ion for activity (144). Imada et al. (146) reported that glutamate accumulation from alkanes also required the presence of ferrous ion. When the terminal alcohol or fatty acid analogs of alkanes were employed as substrates, iron was no longer required. Thus, the iron required for glutamate production from alkanes appeared to be needed only for the initial hydroxylation reaction.

Recognizing that the initial oxidation of alkanes requires molecular oxygen, Imada et al. calculated the theoretical incorporation of $^{18}O_2$ into the α- and γ-carboxyl groups of glutamic acid based on the pathway shown in Fig. 4 (144, 145). They reported that the ratio of labeling between α- and γ-carboxyl groups was 44:56 compared to a theoretical 50:50 ratio. The actual magnitude of incorporation, however, was about one-half the predicted value. This does not disprove the hypothetical pathway since the discrepancy may be due to isotopic exchange reactions during cultivation and glutamate isolation.

3. *Lysine and Diaminopimelic Acid*

Lysine yields approaching 30 mg/ml have been obtained from an auxotroph of *Corynebacterium glutamicum* that utilizes glucose (48). This mutant requires homoserine or a mixture of methionine and threonine. The optimum concentration of these metabolites relaxes the regulatory control on aspartokinase, an enzyme required for lysine

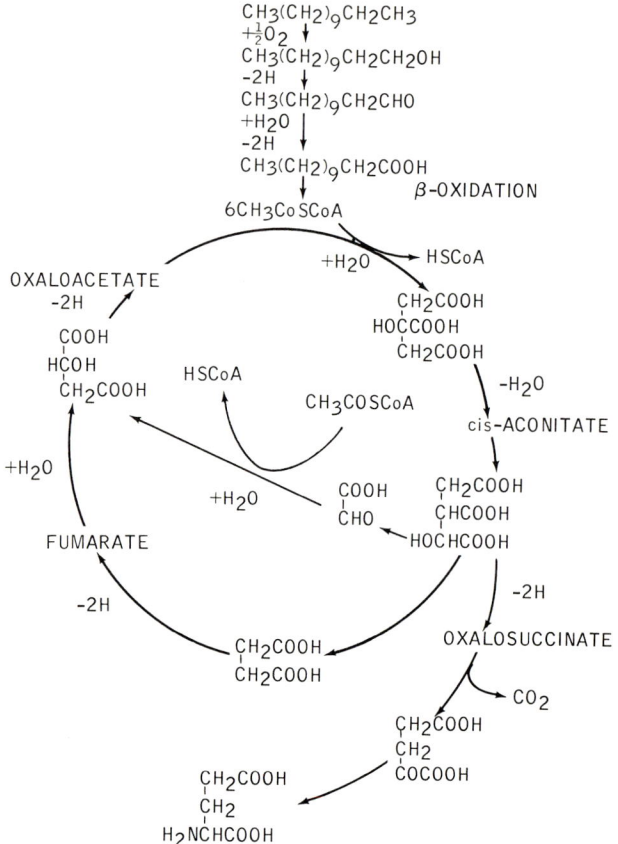

FIG. 4. Postulated pathway of glutamate synthesis from dodecane. Redrawn from Imada *et al.* (145).

synthesis. Similar mutants of *Arthrobacter paraffineus* and other hydrocarbon-utilizing organisms have been isolated (181, 192). These strains, as anticipated, accumulate lysine from a paraffin fraction, but maximum yields were only 10 mg/ml (Table V).

Diaminopimelic acid (DAP), the immediate precursor of lysine, is accumulated by lysine auxotrophs of *A. paraffineus* (199). These organisms lacked the enzyme DAP decarboxylase and were, therefore, incapable of converting DAP to lysine. The lysine requirement also relaxed the regulatory control on dihydropicolinic acid synthetase, an enzyme required for DAP synthesis. Commercially, DAP is valued as a specialty chemical and as a lysine precursor.

In the first commercial process for producing lysine a two-step fermentation was used in which DAP was accumulated by a lysine

TABLE V
AMINO ACIDS PRODUCED FROM ALKANES

Product	Substrate	Organism	Yield	Remarks	References
Lysine	10% v/v C_{12}–C_{18} Paraffins	*Brevibacterium ketoglutamicum* ATCC 15587	10 mg/ml	Homoserine required for growth.	181, 192
Lysine	10% v/v C_{12}–C_{18} Paraffins	*Arthrobacter paraffineus* ATCC 15591	10 mg/ml	Homoserine required for growth.	181, 192
Lysine and homoserine	5% v/v C_{12}–C_{14} Paraffins	*Corynebacterium hydrocarboclastus* ATCC 21267	3.0 mg/ml lysine 6.2 mg/ml homoserine	Threonine required for growth.	212
Threonine	5% v/v C_{11}–C_{18} Paraffins	*Arthrobacter paraffineus* ATCC 19558	1.3 mg/ml	Diaminopimelic acid required for growth.	211
Threonine	5% v/v C_{11}–C_{18} Paraffins	*Arthrobacter paraffineus* ATCC 19558	5.1 mg/ml	Isoleucine required for growth.	193
Threonine and valine	10% v/v C_{11}–C_{18} Paraffins	*Arthrobacter paraffineus* ATCC 21220	7.5 mg/ml	Isoleucine or α-ketoglutarate required for growth. Similar yield obtained from auxotrophs of *C. hydrocarboclastus* and *B. ketoglutamicum*.	206
Valine	5% v/v C_{11}–C_{18} Paraffins	*Arthrobacter paraffineus* ATCC 19559	8.8 mg/ml	Isoleucine, threonine, or homoserine required for growth.	191
Valine	5% v/v Tridecane	*Corynebacterium hydrocarboclastus* ATCC 19561	7.5 mg/ml	Isoleucine required for growth.	191
Phenylalanine	13% v/v Pentadecane	*Corynebacterium* sp. KY 7146	10 mg/ml	Tyrosine required for growth.	370
Phenylalanine	5% v/v C_{12}–C_{14}	*Corynebacterium hydrocarboclastus* ATCC 21226	4.5 mg/ml	Tyrosine required for growth.	197

(*Continued*)

TABLE V (Continued)

Substrate	Organism	Yield	Remarks	References
Threonine	10% v/v C_{11}–C_{18} Paraffins *Arthrobacter paraffineus* ATCC 21317	18.9 mg/ml	Methionine required for growth.	218
Isoleucine	1% wt/v C_{11}–C_{14} Paraffins *Micrococcus paraffinolyticus* ATCC 15582	1.55 mg/ml	α-Aminobutyric acid added to medium. Several other organims were also used.	207
Tyrosine	5% v/v C_{12}–C_{14} Paraffins *Arthrobacter paraffineus* ATCC 21377	3.2 mg/ml	Phenylalanine required for growth.	207
Tyrosine	5% v/v C_{12}–C_{14} Paraffins *Brevibacterium ketoglutamicum*	3.1 mg/ml	Phenylalanine required for growth.	207
Tryptophan	10% v/v C_{11}–C_{18} Paraffins *Brevibacterium ketoglutamicum* ATC 15587	1.5 mg/ml	Anthranilic acid added to medium.	202
Tryptophan	10% v/v C_{11}–C_{18} Paraffins *Candida tropicalis* ATCC 15114	2.0 mg/ml	Anthranilic acid added to medium.	202
Tryptophan	10% v/v Kerosene *Arthrobacter paraffineus* ATCC 21333	0.4 mg/ml	Histidine required for growth.	204
Serine	5% v/v C_{12}–C_{14} *Arthrobacter paraffineus* ATCC 21218	3.3 mg/ml	Isoleucine and methionine required for growth	198
Diaminopimelic acid	5% v/v C_{12}–C_{14} *Arthrobacter paraffineus* ATCC 21087	10.1 mg/ml	Lysine required for growth.	199

Product	Substrate	Organism	Yield	Remarks	Ref.
Diaminopimelic acid	5% v/v Kerosene	*Brevibacterium ketoglutamicum* ATCC 21089	2.1 mg/ml	Lysine required for growth.	199
Homoserine	10% v/v Octadecane	*Corynebacterium* sp. ATCC 7142	12 mg/ml	Threonine required for growth.	48, 267
Methionine and lysine	1.5% wt/v Decane	*Mycobacterium tuberculosis* ATCC 15073	3.1 mg/ml Lysine 0.6 mg/ml Methionine	—	249
Ornithine	10% v/v Tetradecane	*Corynebacterium hydrocarboclastus* R-7	9 mg/ml	Arginine required for growth.	382
Ornithine	5% wt/v C_{12}–C_{14} Paraffins	*Arthrobacter paraffineus* ATCC 21901	7.2 mg/ml	Arginine required for growth.	209
Ornithine	5% wt/v C_{12}–C_{14} Paraffins	*Brevibacterium ketoglutamicum* ATCC 21092	8.6 mg/ml	Arginine required for growth.	209
Citrulline	5% C_{12}–C_{14} Paraffins	*Corynebacterium hydrocarboclastus* ATCC 21242	7.1 mg/ml	Arginine required for growth.	210
Citrulline	5% C_{12}–C_{14} Paraffins	*Arthrobacter paraffineus* 2411-u-41	10.2 mg/ml	Arginine required for growth.	210

auxotroph, then decarboxylated by a second organism that contained the enzyme DAP decarboxylase (34). Using paraffin substrates for DAP accumulation, more than 10 mg/ml of this product was accumulated by *Arthrobacter paraffineus*. Lysine auxotrophs of *Brevibacterium ketoglutamicum* and *A. roseoparaffineus* also accumulated DAP from alkanes, although yields were somewhat lower (Table V). Coupling other mutational blocks (e.g., for homoserine or threonine and methionine) with the requirement for lysine should result in much higher DAP yields.

4. Homoserine

A threonine-requiring mutant of *Corynebacterium hydrocarboclastus* accumulated as much as 6.2 mg/ml of homoserine from a paraffin fraction (212). Lesser quantities of lysine (3.0 mg/ml) also were formed. Homoserine production should increase if a DAP requirement is also induced in *C. hydrocarboclastus* since this would prevent the diversion of carbon from homoserine to lysine and DAP synthesis. However, in another study, it was found that a DAP requirement was not needed to prevent lysine and DAP accumulation by a threonine-requiring mutant of *Corynebacterium* strain K47142 (267). This organism produced 12 mg/ml of homoserine from octadecane and only small quantities of isoleucine and valine. A difference between strain KY7142 and *C. hydrocarboclastus* may be in the susceptibility of dihydropicolinic acid synthetase to feedback inhibition by lysine. This control mechanism which is normally present in *Escherichia coli* prevents the oversynthesis of lysine; whereas its absence, as in *C. glutamicum* (249) and presumably in *C. hydrocarboclastus*, facilitates lysine accumulation.

5. Threonine and Valine

Threonine is difficult to accumulate in substantial quantities because of numerous interrelated metabolic controls which microorganisms have evolved to prevent oversynthesis. Until recently, the maximum quantity of threonine accumulated from sugars was about 4 mg/ml. To obtain these yields, it was necessary to employ a double auxotroph of *Escherichia coli* with requirements for diaminopimelic acid (DAP) and methionine. More recently, analog-resistant mutants were isolated that produced 13 mg/ml of threonine (301).

Various auxotrophs of *Arthrobacter paraffineus* produced threonine from alkane substrates (Table V). A DAP-requiring strain, ATCC 21322, accumulated 1.3 mg/ml of threonine from a C_{11}–C_{18} paraffin mixture (211). When sorbitol was used in place of the paraffins, only about 0.15 mg/ml of product accumulated. In another study, an isoleucine-requiring strain (ATCC 19558) of *A. paraffineus* produced 5.1 mg/ml threonine from a paraffin mixture (193). Similar yields were also obtained when an auxotroph of *Corynebacterium hydrocarboclastus* was employed (193).

Higher threonine yields have been reported using mutants of various paraffin-utilizing organisms which require isoleucine and either α-ketoglutarate or α-aminobutyrate (206). These mutants accumulated large quantities of valine in addition to threonine (Table V). Production was highest with *A. paraffineus* which yielded 12.6 mg/ml of valine and 8.0 mg/ml of threonine from tetradecane. Similar yields were obtained from a C_{11}–C_{18} paraffin mixture but use of a kerosene substrate resulted in a significantly reduced accumulation. Threonine yields were increased to 10 mg/ml by adding 5 mg/ml of various organic acids, primarily TCA (tricarboxylic acid) cycle intermediates, to the medium 48 hours after inoculation. Valine production remained at about 10 mg/ml in the presence of the additives. Although another process has been described specifically for valine production from alkanes (191), the yields were somewhat lower than those reported above.

A methionine-requiring mutant of *A. paraffineus* (ATCC 21317) accumulated large quantities of L-threonine from a C_{11}–C_{18} paraffin mixture (218). In a defined salts medium containing 3 mg/liter of thiamine and 200 mg/liter of L-methionine, 18.9 mg/ml of L-threonine was accumulated in 4 days. This yield, which exceeds the maximum obtained in nonhydrocarbon fermentations, may be sufficient to warrant commercial production.

6. *Phenylalanine*

Phenylalanine is accumulated from alkanes by a tyrosine-requiring mutant of *Corynebacterium* KY4309 (370). Production was favored by the addition of phenylpyruvic and shikimic acids and 500 μg/ml of tyrosine. The latter addition probably influences a regulatory control; the former additives function as product precursors. Proper pH control was essential for maximum yields. At pH 7, 2.7 mg/ml of tyrosine and 1.9 mg/ml of phenylpyruvic acid were produced. When the pH was lowered to 6.0, phenylpyruvic acid no longer accumulated and phenylalanine yields reached 8.6 mg/ml. Under optimum conditions phenylalanine accumulated to 10 mg/ml (Fig. 5). Tyrosine-requiring mutants of *Arthrobacter paraffineus* and *Corynebacterium hydrocarboclastus* have also been found to produce phenylalanine from

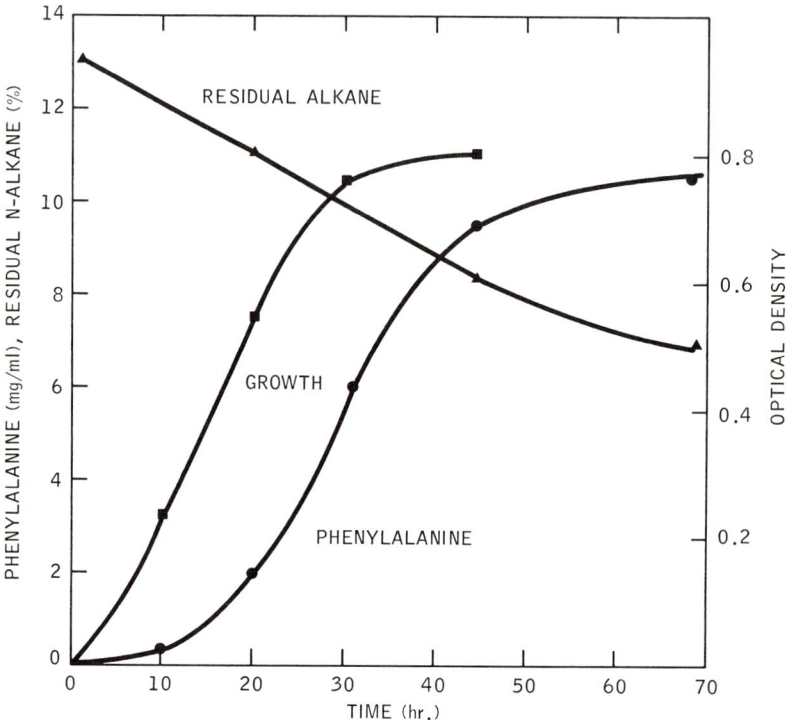

FIG. 5. Time course of phenylalanine production from alkanes. Redrawn from Tokoro et al. (370).

alkanes (Table V). However, maximum product yield in these fermentations was only 4.5 mg/ml.

7. Isoleucine and Serine

Several species of bacteria, actinomycetes, and yeast produced isoleucine from alkanes (207). In these fermentations α-aminobutyric acid was added as a precursor of the product. In its absence isoleucine yields were quite low. A maximum accumulation of 1.55 mg/ml of isoleucine was obtained with a strain of *Micrococcus paraffinolyticus* (207).

Arthrobacter paraffineus ileu⁻, met⁻ accumulated up to 3.3 mg/ml of L-serine from a paraffin mixture (198). Lesser amounts of serine (2.5 mg/ml) were formed by an isoleucine-DAP auxotroph of this organism. Isoleucine auxotrophs of other organisms were also found to produce about 2.0 mg/ml of serine. Sorbitol could replace alkanes as a carbon source although serine yields were somewhat lower on the non-hydrocarbon substrate.

8. Tyrosine and Tryptophan

Tyrosine is accumulated by phenylalanine-requiring mutants of *Arthrobacter paraffineus* (208). Similar mutants of *Corynebacterium hydrocarboclastus* and *Brevibacterium ketoglutamicum* also produced tyrosine (208). The phenylalanine requirement was satisfied with either 500 μg/ml of phenylalanine or 0.5% N-Z amine. The latter, however, gave rise to greater product yields. Tyrosine accumulation from a paraffin substrate was about 3.2 mg/ml whereas, only 1.2 mg/ml of tyrosine was produced from kerosene.

Two processes have been described for the production of tryptophan from hydrocarbons (202, 204). In one, a histidine auxotroph of *Arthrobacter paraffineus*, cultivated on a kerosene substrate, produced 0.4 mg/ml of tryptophan after 4 days incubation (204). In the second, a hydrocarbon-assimilating bacterium and a mold accumulated tryptophan in the presence of its precursor, anthranilic acid (202). Relatively large quantities of anthranilate were added, but maximum tryptophan yield was only 1.5 mg/ml.

9. Ornithine and Citrulline

Arginine-requiring mutants of *Corynebacterium hydrocarboclastus* produced ornithine from alkanes (382). Optimum arginine concentration was about 1.0 mg/ml and the most effective substrates were the C_{14} to C_{16} alkanes. The selection of the proper nitrogen source was essential for maximum product formation. The use of ammonium acetate caused the medium to become alkaline, resulting in glutamic acid and alanine accumulation. An alkaline culture broth also developed with KNO_3; however, the products accumulated were a-ketoglutarate and pyruvate. Ornithine production (and growth) was promoted by ammonium phosphate. Under optimal condition, with the latter nitrogen source, 9 mg/ml of ornithine was obtained in 80 hours corresponding to about 32.5 weight percent conversion of the 1% tetradecane substrate (382).

Brevibacterium ketoglutamicum auxotrophs that required arginine accumulated up to 8.6 mg/ml of ornithine from a C_{12}–C_{14} paraffin mixture (209). Similar auxotrophs of *A. paraffineus* also produced ornithine (7.2 mg/ml) from paraffins (209). In the latter process, arginine was supplied by N-Z amine and NH_4NO_3 was used as a nitrogen source. Ammonia was added during the fermentation to maintain the pH between 6 and 8.

The yields of the alkane-dependent processes were lower than those reported for ornithine production from glucose. Auxotrophs of *Corynebacterium glutamicum* accumulated more than 26 mg/ml of ornithine from glucose after 72 hours incubation (214). Similar auxotrophs and production media were employed in this fermentation as in the alkane process. The reasons for the improved yields from glucose are not immediately apparent.

The arginine requirement of the above auxotrophs could also be satisfied by citrulline. The metabolic block, therefore, was between ornithine and citrulline. Another group of arginine-requiring mutants incapable of utilizing citrulline also have been isolated. These auxotrophs accumulated citrulline and, thus, were blocked somewhere between citrulline and arginine (Table V). Such arginine-requiring mutants of *Arthrobacter paraffineus, Brevibacterium ketoglutamicum*, and *Corynebacterium hydrocarboclastus* produced up to 10.2 mg/ml of citrulline from a C_{12}–C_{14} paraffin mixture (210). These fermentations also used NH_4 as a nitrogen source and arginine was supplied through the addition of N-Z amine or yeast extract. Citrulline yields, as noted previously for ornithine yields, were somewhat higher from nonhydrocarbon substrates (216).

C. Organic Acids

1. α-Ketoglutaric Acid

Interest in the production of α-ketoglutarate (α-KG) stems from its potential use in chemical synthesis and its ready conversion to glutamic acid. Tsugawa *et al.* (379) screened several hundred hydrocarbon utilizers for their ability to accumulate α-KG and found 14 strains capable of accumulating 1–2 mg/ml. These strains were all classified as yeasts and a representative species was identified as *Candida lipolytica*. In another study, 141 yeast strains were examined and only those identified as *C. lipolytica* produced high levels of α-KG (380). Subsequently, a variety of genera were found to accumulate α-KG from alkanes (Table VI).

Maximum α-KG production by the *Candida* species required the addition of 0.02% CSL (379). This concentration of CSL was lower than that needed for maximum growth. Thiamine was the only growth factor required by *C. lipolytica*, and α-KG production in a defined medium containing 3 µg/liter of thiamine was comparable to that in a medium containing CSL (378). The effect of thiamine on α-KG degradation was studied by comparing the degradation of α-KG by cells grown in thiamine-limited and thiamine-sufficient media (378). Thiamine-sufficient cells degraded α-KG more rapidly than did thiamine-limited cells. Also, the addition of thiamine to cells from a thiamine-limited culture greatly increased the rate of α-KG degradation. Succinic acid was detected as the degradation product of α-KG.

TABLE VI
α-Ketoglutarate Production from Alkanes

Substrate	Organism	Yield	Remarks	References
8% v/v C_{12}–C_{17} Paraffins	Candida lipolytica AJ 5004	56 mg/ml	Medium contained low levels of thiamine or cornsteep liquor (CSL).	378–380
8.7% v/v Alkanes	Corynebacterium KY 4439	85.5% Conversion	Medium contained low levels of thiamine or cornsteep liquor.	354
10% v/v C_{12}–C_{14} Paraffins	Arthrobacter paraffineus ATCC 15591	60 mg/ml	Mineral salts medium containing 0.01% CSL was employed.	189
6.7% v/v Kerosene	Arthrobacter paraffineus ATCC 15591	18 mg/ml	Mineral salts medium containing 0.01% CSL was employed.	189
5% v/v Dodecane	Corynebacterium hydrocarboclastus ATCC 15592	11 mg/ml	Brevibacterium ketoglutamicum ATCC 15588 and Micrococcus paraffinolyticus ATCC 15589 produced 8.0 and 9.0 mg/ml respectively, α-KG from same substrate.	352
10% v/v C_{12}–C_{14} Paraffins	Arthrobacter simplex ATCC 15799	18 mg/ml	pH regulated by $CaCO_3$. Glutamate (1 mg/ml) also accumulated.	190
8% v/v Tetradecane	Micrococcus paraffinolyticus ATCC 15589	42 mg/ml	Medium contained salts plus 0.01% CSL.	352
8% v/v C_{11}–C_{18}	Arthrobacter hydrocarboglutamicus ATCC 15583	29 mg/ml	Medium contained salts plus 0.01% CSL.	352
8% v/v C_{11}–C_{18} Paraffins	Brevibacterium ketoglutamicum ATCC 15587	30 mg/ml	Glutamic acid (10 mg/ml) also accumulated.	352
5% v/v Tetradecane	Corynebacterium hydrocarboclastus S10B1	16 mg/ml	Alanine and glutamate also accumulated.	147
1% v/v Decane	Mycobacterium lacticolum	—	Production favored by limiting nitrogen concentrations.	306
2% v/v Tetradecane	Mycobacterium mucosum	0.5 mg/ml	Production favored by nitrogen limitation. Pyruvate (0.2 mg/ml) also accumulated.	178

Another effect of thiamine limitation was observed during growth of *C. lipolytica* on glucose. In the presence of a limiting thiamine concentration (3 μg/liter), pyruvate as well as α-ketoglutarate accumulated (378). Since pyruvate is also oxidized by a thiamine-requiring enzyme, it seems clear that thiamine functions by regulating the activity of enzymes which require thiamine as a cofactor.

Examination of various heavy metal ion requirements revealed that ferrous ion stimulated both growth and α-KG production (379). Zinc and ferrous ion together exhibited a synergistic effect so that by supplying an optimal concentration of each, 47 mg/ml of α-KG accumulated. As with glutamate production, the importance of iron is probably in the initial hydroxylation reaction of alkane oxidation.

Maximum α-KG production by *Candida lipolytica* occurred with alkanes from C_{17} to C_{19} (378), although production from a C_{12}–C_{17} paraffin mixture was comparable to that from nonadecane or heptadecane. The conversion of paraffin to α-KG increased from 58% to 83% as the concentration of paraffins was decreased from 8% to 1%. Tanaka *et al.* (354) using a *Corynebacterium* (KY 4439) reported a similar relationship between paraffin concentration and α-KG yield. In the latter study α-KG production was stimulated by suboptimal thiamine concentrations and by ferrous ion and various trace metal ions. Under optimal conditions strain KY 4439 converted 85.5% of the paraffins which comprised 8.7 volume percent of the medium into α-KG.

Soviet investigators reported the production of small amounts of α-KG and pyruvate from alkanes by a *Mycobacterium* species (178, 306). Pyruvate and α-KG accumulations were favored by nitrogen concentrations lower than those required for maximal growth. Addition of nitrogen to cells growing in a nitrogen-deficient medium resulted in a disappearance of accumulated keto acids (178). Presumably, in the presence of excess nitrogen, the keto acids are aminated. In contrast, another report by Soviet workers studying keto acid production by *Candida* yeasts (72) indicated that significant quantities of keto acids accumulated in a hexadecane-salts medium only if nitrogen was present at a higher concentration than that required for maximum growth. A similar relationship also characterized α-KG production by *C. lipolytica* (379).

The formation of α-KG by an *Arthrobacter* sp. can be regulated by various organic alcohols (190). Alcohols such as mannitol, sorbitol, glycerol, butanol, and octanol lowered the production of α-KG and increased the accumulation of glutamic acid. Use of ammonia to control the pH of the fermentation also favored a shift in product accumulation from α-KG to glutamate. It was postulated that organic alcohols inhibit glutamate oxidation and that compounds such as mannitol and sorbitol indirectly provide amino radicals since many *Arthrobacter* species are capable of producing amino acids from sugar alcohols (190).

An example of the effect of ammonia and organic alcohols on the course of the fermentation was illustrated with *Arthrobacter simplex* ATCC 15799 (190). In a medium where calcium carbonate was used to control pH, 18 mg/ml of α-KG and 1 mg/ml of glutamate accumulated. In a similar medium pH was controlled by ammonia and 20 mg/ml of α-KG and 20 mg/ml of glutamate were produced. The addition of 2% mannitol together with pH control by ammonia resulted in the accumulation of 39 mg/ml of glutamate and only 8.0 mg/ml of α-KG.

2. Citric Acid

Presently, citric acid is produced commercially by large-scale surface or submerged fungal fermentation of crude carbohydrate carbon sources. An important factor for high yields is the level of trace metal ions, a variable that is difficult to control because of the undefined nature of the medium. The complex medium also complicates product harvest, which accounts for a significant portion of process cost. In an attempt to improve both of these aspects of the fermentation, hydrocarbons have been studied as potential substrates.

Surprisingly little information has appeared on citric acid production from hydrocarbons even though yields can be quite high (Table VII). To date, information on this fermentation is found mainly in foreign patents and foreign language journals. Much more information may begin to appear shortly, since citric acid possesses certain ecological advantages over phosphates and NTA as sequesterants in detergents. Potential simplification of product harvest afforded by hydrocarbon substrates may make a detergent-grade citric acid commercially feasible.

Citric acid production from hydrocarbons has been reported for bacteria, yeasts, and molds (Table VII). *Arthrobacter paraffineus* accumulated up to 28 mg/ml of citric acid from a C_{12}–C_{14} paraffin fraction (184). Yields were increased by the addition of a surfactant, and the medium contained a small quantity of CSL or thiamine. As with most citrate fermentations, pH was controlled with calcium carbonate. Several species of *Penicillium* and *Aspergillus* also produced citric acid from alkanes (195). *Aspergillus elegans* ATCC 20108 accumulated only 1 mg/ml of citrate, but *Penicillium restrictum*

TABLE VII
CITRATE AND ISOCITRATE PRODUCTION FROM ALKANES
A. Citrate Production

Substrate	Organism	Yield	Remarks	References
6% v/v Hexadecane	*Candida lipolytica* No. 228	34 mg/ml	Employed salts medium containing 0.1% CSL.	330
4% v/v Kerosene	*Candida lipolytica* No. 228	2 mg/ml	Employed salts medium containing 0.1% CSL.	330
6.7% v/v C_{12}–C_{14} Paraffins	*Arthrobacter paraffineus* ATCC 15591	28 mg/ml	Employed salts medium containing 0.2% urea and 0.01% CSL and a surfactant.	184
20% v/v Kerosene	*Penicillium restrictum* ATCC 13155	68 mg/ml	Carefully defined salts medium containing CSL was utilized.	195
10% v/v Decane	*Candida lipolytica* IFO-1437	22 mg/ml	Employed salts medium containing 0.1% yeast extract.	349
5% v/v Tetradecane	*Candida tropicalis* IFO-0587	16.6 mg/ml	pH controlled with NaOH.	349
5% C_{16}–C_{20}	*Candida parapsilosis* IFO-0708	18.1 mg/ml	Medium contained 1% calcium acetate.	349

Substrate	Organism		Remarks	References
5.66% wt/v Hexadecane	Candida lipolytica ATCC 8661	56.8 mg/ml	—	36
50.0% wt/v Hexadecane	Candida lipolytica ATCC 8661	110.4 mg/ml[a]	—	36
4% v/v Octadecane	Corynebacterium sp. IFO-12730	44.2 mg/ml	—	341
5% wt/v C_{10}–C_{20} Paraffins	Candida lipolytica ATCC 16617	39.2 mg/ml	Thiamine or biotin added to medium.	10
8% wt/v C_{13}–C_{15} Paraffins	Candida lipolytica IFO-1464	112 mg/ml	Used mutant unable to degrade citric acid.	340

B. Isocitrate Production

Substrate	Organism	Yield	Remarks	References
10% C_{12}–C_{15} Paraffin mixture	Candida zeylanoides ATCC 15585	28 mg/ml	Similar amount of citric acid produced simultaneously.	215
6% wt/v Hexadecane	Candida brumptii IFO-0744	30 mg/ml	—	337

[a]Yield based on aqueous volume.

(ATCC 13155) produced as much as 68 mg/ml. In the latter process kerosene was used as a substrate and pH was maintained at 5 by a mixture of calcium hydroxide and sodium hydroxide. This is one of the few fermentations in which high product yields were obtained from a kerosene substrate.

Candida species, particularly *Candida lipolytica,* are able to accumulate large amounts of citric acid. Tabuchi et al. (330) described a strain of *Candida lipolytica* (No. 228) which accumulated citric acid from a wide variety of substrates, including glucose, fatty acids, and hydrocarbons. Although highest yields were obtained from glucose, yields of 34 mg/ml were obtained from hexadecane. Citric acid was also produced from C_{10}, C_{12}, and C_{14} alkanes and from kerosene, although yields on the latter substrate were low.

In another report (10) *Candida lipolytica* AJ 4541 (ATCC 16617) accumulated 38.6 mg/ml of citric acid, representing 77.2% conversion of a $C_{10}-C_{20}$ paraffin mixture. Product accumulation was improved by the addition of small quantities of thiamine, biotin, or CSL. These additives did not appear to function in a regulatory manner as they did in the glutamic acid fermentations. That is, unlike glutamate synthesis, citric acid production was enhanced by growth factor concentrations which permitted maximum growth. In another study (349) species of *Candida* including *C. tropicalis, C. intermedia, C. parapsilosis,* and *C. guilliermondii* were found to accumulate more than 10 mg/ml of citric acid.

Corynebacteria can also produce citric acid from paraffins (341). An unidentified species, No. 803 (IFO 12730), accumulated 44.2 mg/ml of citric acid from octadecane after just 2 days incubation. Several other strains accumulated similar quantities of citrate from various paraffin mixtures, and conversions as high as 115% by weight were reported.

The Chas. Pfizer Co. (36) was granted a patent on citric acid production using a strain of *Candida lipolytica* (ATCC 8661). This organism produced the acid in a defined salts medium although yields were higher when CSL was added. The C_{12} to C_{18} alkanes were the most effective substrates and substantial quantities of product were also produced from long-chain alkenes. 1-Alkanols did not support citrate production. The percent conversion of alkane to citrate varied with the hydrocarbon concentration. Maximum conversion was about 94 weight percent which corresponded to 56.8 mg/ml of citric acid. When 50% hexadecane was incorporated into the production medium, only about 11% was converted to citric acid, but the product concentration in the aqueous portion of the medium reached 110.4 mg/ml.

Higher citric concentrations and substrate conversion efficiencies were reported when mutants of *C. lipolytica* were employed (340). These mutants were selected for their inability to utilize citric acid as a sole carbon source. One such mutant, which also could not metabolize L-lactic acid, was cultivated in a defined salts medium containing a C_{12}–C_{18} paraffin mixture. After just 3 days incubation, 112 mg/ml of citrate accumulated representing 138% conversion of the substrate. Isocitrate, the most abundant by-product of citrate production, accumulated to only 1.3 mg/ml.

These high citrate yields and short fermentation times should make the paraffin-based processes strong competitors of the conventional molasses fermentations. Heat evolution and oxygen consumption may be negative factors hindering widespread adoption of paraffin substrates. However, the magnitude of these factors will not be as great as demonstrated for glutamate production (see Section II,H) because of the much higher yields and substrate conversion efficiencies associated with citrate production.

3. *Isocitric Acid*

Isocitric acid was detected originally as an unknown organic acid accumulating in yeast culture that produced citric acid (4). It was isolated and identified as the lactone of threo-D_s-isocitric acid (5, 9). Subsequently, many yeasts were found to produce isocitric and citric acid concomitantly from glucose or hexadecane substrates (5). One isolate, *Candida brumptii* No. 325, produced only isocitric acid. This organism was studied under a variety of conditions and compared with other citrate-producing organisms. Strain 325 did not accumulate citric under any of the conditions tested but the ratio of citrate to isocitrate produced by other *Candida* species varied. A thiamine-requiring strain of *C. lipolytica* produced large amounts of citric and isocitric acids in a medium containing high thiamine concentrations. In thiamine-deficient media a large amount of α-ketoglutarate accumulated with smaller quantities of pyruvate, citrate, and isocitrate.

Candida zelanoides (ATCC 15585) produced almost equal quantities of citric and isocitric acids from paraffin substrates (215). As much as 35 mg/ml of each was formed after 4 days incubation. The addition of 2% methanol to the medium shifted the citric–isocitrate ratio so that 85 mg/ml of citric and only 10 mg/ml of isocitric were produced.

Takeda Chemical Industries has patented a process for producing isocitric using *C. brumptii* (337). Yields as high as 30 mg/ml were obtained from hexadecane after 4 days incubation. Many other *Can-*

dida species were cited as isocitrate producers in addition to members of the genera *Pichia, Hansenula,* and *Debaryomyces* (337). The last three genera, however, utilized glucose as a substrate.

4. Fumaric and Malic Acids

In a recent study, it was found that a *Candida* species produced large quantities of fumaric acid from paraffins (398). This species differed in several characteristics from previously described organisms; thus, the new designation *Candida hydrocarbofumarica* was suggested. Fumaric acid production by this strain was equally effective from the C_{12}–C_{15} alkanes, but lower yields were observed with other paraffins. By optimizing the concentration of heavy metal ions, nitrogen source, and other medium constituents fumaric acid yields approached 40 mg/ml corresponding to conversion efficiencies of 65% based on paraffin added.

The fumaric acid produced by *C. hydrocarbofumarica* could be converted to malic acid by reinoculating the culture broth with a yeast that exhibited high fumarase activity. Thus, after 5 days incubation, during which fumaric acid accumulated, the culture was reinoculated with *Candida utilis* and incubated for an additional 5 days. During the second incubation period the fumarate was converted to malate; and the latter accumulated to 35 mg/ml (90).

5. Dipicolinic Acid

Dipicolinic acid is produced from various hydrocarbons by a *Penicillium* sp. (123). The product was isolated from boiled whole culture broth after 48 days of stationary incubation. The maximum quantity of dipicolinic acid isolated was 1.6 gm from 200 ml of culture.

D. NUCLEIC ACID-RELATED PRODUCTS

The production of nucleic acid-related products has received much attention because of the flavor-enhancing characteristics of the purine ribonucleotides-5′ monophosphates. Guanylic acid (GMP) is the most potent flavor enhancer followed by inosinic (IMP) and xanthylic (XMP) acids. Auxotrophic mutants of various organisms produce up to 15 mg/ml of these nucleotides under optimal environmental conditions from carbohydrate carbon sources. More often, however, microorganisms hydrolyze the phosphate from the nucleotide and excrete the resultant nucleoside. The successful nucleotide fermentations usually employ organisms lacking phosphohydrolases. Some of the phosphohydrolases can also be inhibited by low concentrations of heavy metal ions or the presence of phosphate. High cellular permeability, also required for nucleotide excretion, can be achieved by controlling the level of manganous ion.

Nucleoside production from hydrocarbons was first described by Iguchi, Watanabe, and Takeda (137). They isolated an adenine auxotroph of *Corynebacterium petrophilium* SB 4082 that accumulated inosine and smaller amounts of hypoxanthine from hexadecane (138, 139). Inosine formation varied with the adenine concentration of the medium. Adenine concentrations less than 100 µg/ml restricted inosine production and concentrations greater than 100 µg/ml were inhibitory (Fig. 6). Inhibition by excess adenine could be prevented by the addition of guanine. This suggests that the enzyme inosinicase, which catalyzes the formation of the IMP essential for GMP and AMP synthesis, was subject to repression by excess adenine (135).

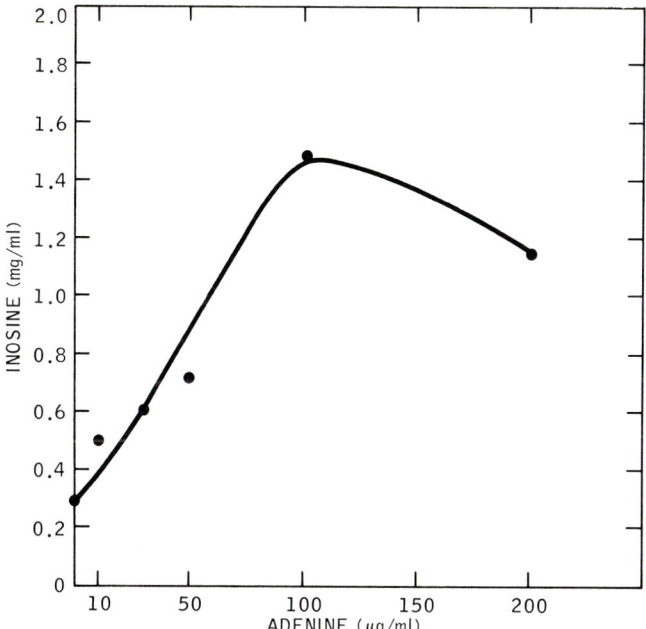

FIG. 6. Effect of adenine concentration on inosine production by *Corynebacterium petrophilium*. Redrawn from Iguchi *et al.* (139).

An interesting relationship was found between inosine production and the presence of glucose and/or hydrocarbon in *Corynebacterium petrophilum* cultures. Inosine was not produced by *C. petrophilum* from glucose, and the addition of glucose to hexadecane medium inhibited inosine production. The adenine requirement of glucose-grown cells was met only by adenine, but hydrocarbon-grown cells utilized adenosine and adenylic acid in addition to adenine. Also, larger quantities of ultraviolet-absorbing material were released from

TABLE VIII
NUCLEIC ACID-RELATED PRODUCTS PRODUCED FROM ALKANES

Product	Substrate	Organism	Yield	Remarks	References
Inosine	5% v/v Hexadecane	*Corynebacterium petrophilum* SB 4082 u-1542	1.72 mg/liter	Organism is an adenine auxotroph. Smaller amounts of hypoxanthine also accumulated.	135, 137, 138, 140
5'-Inosinic acid	5% v/v Kerosene	*Arthrobacter simplex* ATCC 15799	1.9 mg/ml	Hypoxanthine added to medium.	185
5'-Guanylic	5% v/v Kerosene	*Arthrobacter simplex* ATCC 15799	9.9 mg/ml	Guanine added to medium.	185
5'-Adenylic acid	5% v/v Kerosene	*Arthrobacter simplex* ATCC 15799	1.3 mg/ml	Adenine added to medium.	185
5'-Adenylic acid	5% v/v Kerosene	*Candida utilis* ATCC 16321	1.2 mg/ml	Adenine and yeast extract added to medium.	185
5'-Inosinic acid	5% v/v Kerosene	*Aspergillus oryzae* ATCC 16450	0.25 mg/ml	Adenine added to medium. Organism required histidine and methionine for growth.	185
Inosine and 5'-inosinic acid	30% wt/v Gas oil	*Corynebacterium simplex* ATCC 19565	0.24 mg/ml Inosine 0.10 mg/ml 5'-Inosinic acid	Adenine required for growth.	338

EXTRACELLULAR ACCUMULATION OF METABOLIC PRODUCTS

Product	Substrate	Organism	Amount	Note	Ref.
Inosine	10% wt/v C_{14}–C_{21} Paraffins	Corynebacterium simplex ATCC 19565	1.8 mg/ml	Adenine required for growth.	338
Inosine	5% wt/v C_{12}–C_{14}	Brevibacterium ketoglutamicum ATCC 21169	1.5 mg/ml	Adenine required for growth.	213
Inosine	10% wt/v C_{12}–C_{14} Paraffins	Arthrobacter paraffineus ATCC 21161	2.8 mg/ml	Adenine required for growth.	213
Orotic acid and orotidine	10% v/v Paraffin mixture	Arthrobacter paraffineus KY 7122	6 mg/ml Orotic 3.5 mg/ml Orotidine	Uracil required for growth.	163
Orotic acid and orotidine	5% v/v Kerosene	Brevibacterium ketoglutamicum ATCC 21187	1.9 mg/ml Orotic 1.2 mg/ml Orotidine	Uracil required for growth.	197
Orotic acid and orotidine	5% v/v C_{12}–C_{14} Paraffins	Corynebacterium sp. ATCC 21188	4.98 mg/ml Orotic 3.7 mg/ml Orotidine	Uracil required for growth.	197

boiled glucose cells than from similarly treated hydrocarbon cells. All these observations indicated that hydrocarbon-grown cells were more permeable than cells grown on glucose. Thus, it was postulated that the reduced permeability of glucose-grown cells was responsible for the lack of inosine production in the presence of this substrate (135).

Inosine and inosinic acid were also accumulated by an adenine-requiring mutant of *Corynebacterium simplex* (338). Crude carbon sources, such as gas oil, were used as substrates although somewhat higher product yields were obtained from a more refined paraffin fraction (Table VIII). Paraffin fractions also were used as substrates for inosine production by adenine-auxotrophs of *Brevibacterium ketoglutamicum* and *Arthrobacter paraffineus* (213). The production of 2.8 mg/ml of inosine by the latter organism represents the highest reported accumulation of this nucleoside from hydrocarbons. The highest reported nucleotide accumulation from hydrocarbons, 1.9 mg/ml of inosinic acid, was observed in the culture broth of *Arthrobacter simplex* (185). In the latter study kerosene was supplied as the substrate and 0.5 mg/ml of hypoxanthine was added to the fermentation 24 hours after inoculation. The addition of this base served to direct the synthesis of the nucleotide. As a result, when hypoxanthine was replaced by guanine or adenine, guanylic or adenylic acids accumulated. This relationship suggests that the nucleotide may be synthesized extracellularly from the base in a manner analogous to "salvage" synthesis (250) by nucleotide-producers utilizing carbohydrates.

Two other nucleic acid-related compounds, orotic acid and orotidine, can be produced by fermentation of hydrocarbon substrates (163, 197). These products accumulated simultaneously in the culture broth of uracil-requiring auxotrophs of *Arthrobacter paraffineus* (Table VIII). Product accumulation varied with the initial uracil concentration in a manner similar to the relationship of adenine concentration to inosine production (compare Figs. 6 and 7) (163). The addition of certain amino acids stimulated product accumulation and yields of 6.0 mg/ml of orotic acid and 3.5 mg/ml of orotidine were reported. When sorbitol was substituted for the paraffins, orotidine accumulation reached 7.5 mg/ml and orotic acid exceeded 8 mg/ml. By cultivating *A. paraffineus* in fermentors, instead of in a shake flask, orotic acid accumulated to 20 mg/ml.

Uracil auxotrophs of *Arthrobacter hydrocarboglutamicus* and a *Corynebacterium* sp. also excreted orotic acid and orotidine (197). These products were produced from either paraffins or carbohydrate

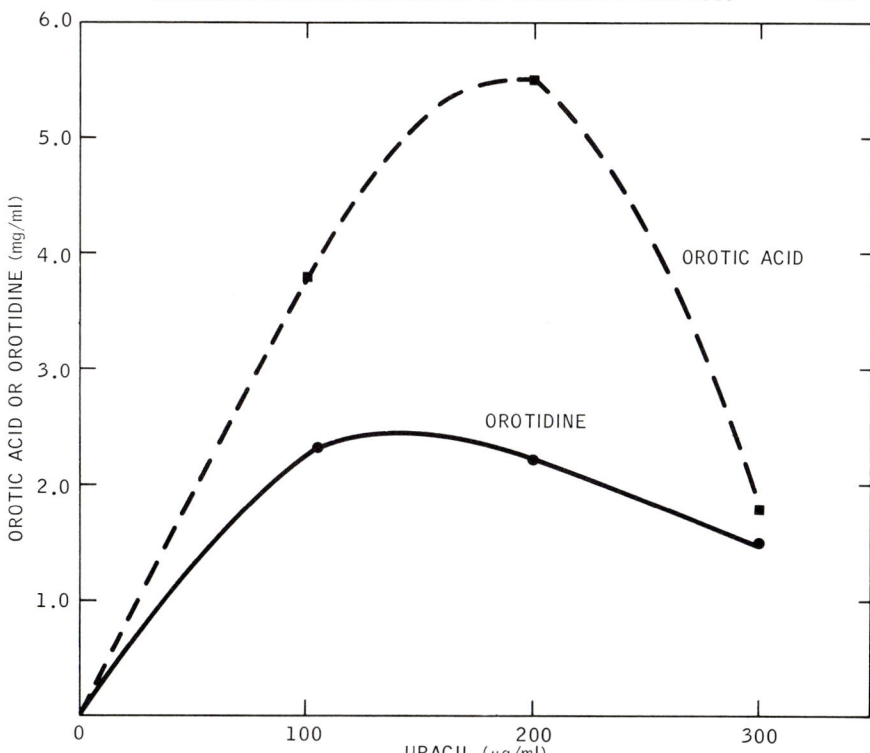

FIG. 7. Effect of uracil concentration on orotic acid and orotidine accumulation from paraffins by *Arthrobacter paraffineus*. Redrawn from Kawamoto et al. (163).

substrates and, as with *A. paraffineus*, yields were somewhat higher from the latter carbon sources.

E. VITAMINS AND PIGMENTS

Vitamins are produced by a wide variety of molds, yeasts, and bacteria; but extracellular accumulation rarely exceeds a few micrograms per liter. The notable exceptions are riboflavin, β-carotene, and vitamin B_{12} which are produced from carbohydrates in sufficient quantity to make industrial production feasible (108). These vitamins as well as biotin and vitamin B_6 also accumulate in the culture broths of various microorganisms growing on hydrocarbons (Table IX).

1. Biotin

Early studies on biotin production from carbohydrates revealed that addition of pimelic or azelaic acid greatly increased biotin accumulation (70, 358, 396, 397). Subsequent investigations implicated pimelic as an intermediate in biotin synthesis (69, 71). These obser-

TABLE IX
VITAMINS AND PIGMENTS PRODUCED FROM ALKANES

Product	Substrate	Organism	Yield	Remarks	References
Biotin (desthiobiotin)	2% v/v Kerosene	Pseudomonas strain 5-2	1 μg/ml	Yields were increased by the addition of pimelic or azelaic acid.	376
Biotin	6% v/v Undecane	Pseudomonas strain 5-2	1.5 μg/ml	Undecane was the best substrate for biotin production while heptadecane supported the highest cell yields.	377
Riboflavin (vitamin B_2)	7.5% v/v Paraffin fraction	Pichia guilliermondii	117 μg/ml	Dodecane was the best alkane for B_2 production.	253
Riboflavin	2.58% wt/v	Pichia miso	51 μg/ml		237
Riboflavin	2.0% v/v Hexadecane	Mycobacterium brevicale	1 μg/ml	Flavins are released into the medium during prolonged incubation.	240
Riboflavin	1.0% v/v C_{10}–C_{24} Paraffin fraction	Candida scatti	9 μg/ml	Riboflavin was obtained by continuous cultivation.	61, 386
Vitamin B_{12}	1.5% v/v Hexadecane	Corynebacterium simplex ATCC 6946	0.6 μg/ml	Production stimulated by $CoSO_4$.	87

Coproporphyrin III	—	*Mycobacterium smegmatis* IFO-3080	0.06 μg/ml	Low aeration favored production.	345
Porphyrins	—	*Corynebacterium simplex*		Porphyrin was bound to a peptide and contained cobalt atoms.	352
Carotenoids	2.5% v/v Mineral oil	*Mycobacterium lacticola*	4.28 mg/ml of mineral oil	β-Carotene was among the carotenoids identified.	104
Carotenoids	2.0% v/v Paraffin fraction	*Mycobacterium smegmatis*	7.10 μg/ml	Main component of carotenoids was 4-beta-γ-carotene.	344, 347
Pyocyanine	1.0% v/v Kerosene	*Pseudomonas aeruginosa* ATS-14	140 μg/ml	—	226
Vitamin B₁₂	1.0% v/v Decane	*Pseudomonas aeruginosa*	0.045 μg/ml	—	245
Vitamin B₆	1.0% v/v Hexadecane	*Candida albicans*	0.4 μg/ml	—	343, 348
Gibberellin	1.1% v/v Crude oil	*Gibberella fujikuroi*	—	—	396
1-Phenazinecarboxylic acid	2% v/v Kerosene	*Pseudomonas* sp.	350 mg/ml	—	115, 116

vations prompted Tsuboi et al. (376, 377) to screen for biotin-producing organisms in hydrocarbon media since paraffin degradation can proceed via dicarboxylic acid intermediates. From more than 600 isolates tested, 35 were found to accumulate more than 0.1 µg of biotin/ml. One of these, *Pseudomonas* sp. (strain 5-2), was selected for closer study. The addition of pimelic or azelaic acid, and increased aeration, favored biotin accumulation by strain 5-2. This isolate grew faster on glucose than on kerosene; but biotin production was much greater from the latter substrate. When alkanes from C_6 to C_{20} were tested as substrates, it was found that heptadecane supported the highest cell population while biotin accumulation was greatest from undecane. Generally, odd-chain alkanes (C_9, C_{11}, C_{13}) were the best substrates for biotin production. It seems likely that the effectiveness of odd-chain alkanes is associated with their role as immediate precursors of azelaic and pimelic acids. Both dicarboxylic acids have been identified in the culture broth during biotin production from undecane.

2. Riboflavin (Vitamin B_2)

Riboflavin is a major vitamin requirement of poultry and swine. It is produced commercially by fermentation of carbohydrates although chemical synthesis is now the predominant method of production.

There are relatively few reports on B_2 biosynthesis from hydrocarbons (Table IX). Those that have appeared have proceeded along one of two lines: (a) production of riboflavin as a by-product or adjunct of single cell protein and (b) development of a process wherein riboflavin is the sole product. With the latter objective, Nishio and Kamikubo (253, 254) isolated 63 microorganisms from various soils and found 13 capable of producing B_2 in a hydrocarbon medium. The most effective isolate was identified as *Pichia guilliermondii*. Dodecane was the best carbon source and production was stimulated by Tween 20 and 60. Under optimal conditions, the amount of B_2 excreted was 117 mg/liter.

Tadao et al. (331) described a similar process in which *Pichia miso* accumulated 52 mg/liter of B_2 from hexadecane in medium containing Tween. Other species of *Pichia* were also cited and various hydrocarbons (including kerosene) were mentioned as potential substrates.

Soviet investigators have described riboflavin production from hydrocarbons by a *Mycobacterium* sp. (240). They reported that flavin production increased as growth declined and that older cultures released flavins to the medium. The quantity of flavins produced per unit cell mass was about constant during the first 12 days of the fermentation and increased sharply during the remainder of the process. Total flavin content of the culture continued to increase through 19 days of incubation. Approximately 60% of the flavins were found in the culture supernatant with the remainder associated with the cells. The maximum yield of riboflavin, 2000 µg/liter, was significantly less than that which can be accumulated from carbohydrate substrates (108).

Dikanskaya *et al.* (61, 286) determined the riboflavin content of 77 yeast strains belonging to the genera *Candida* and *Torulopsis*. They reported that hydrocarbon-grown yeast contained several times as much riboflavin as yeast grown on carbohydrates. In addition, there was about 10 times as much riboflavin in the supernatant of continuously grown cultures as in extended batch culture. Also, the riboflavin content of continuously grown cells was higher than that of batch-grown cells. The authors postulated that increased riboflavin production from hydrocarbon was due to a greater requirement for flavin enzymes during hydrocarbon growth. Nevertheless, the maximum production of riboflavin was far below the yields reported for the best processes from carbohydrate substrates.

3. *Vitamin B_{12}*

Vitamin B_{12} was first produced commercially as a by-product of various *Streptomyces* antibiotic fermentations. Presently, the most widely used processes employ *Propionibacterium* species that produce B_{12} as a primary product. The fermentation requires about 7 days and part of the time it is anaerobic and part it is aerobic. In one process (313) the organism is grown anaerobically for 3 days during which time large quantities of cobinamide (presumably cobinamide-5′-deoxyadenosine) are formed. This precursor is then converted to 5,6-dimethyl-α-benzimidazolylcobamide-5′-dioxyadenosine during a subsequent 4-day aerobic incubation. Most of the B_{12} coenzyme is associated with the cells and is not found in the culture supernatant.

Fujii *et al.* (87) reported that most of the B_{12} produced by various *Mycobacterium, Nocardia,* and *Corynebacterium* species was retained in the cells cultivated on hydrocarbon media but was released to the culture supernatant by cells grown on carbohydrates. These organisms also produced more B_{12} in hexadecane or octadecane media than in glucose salts or a complex carbohydrate medium. Further studies with *Corynebacterium simplex* ATCC 6946 in hydrocarbon media revealed that the addition of $CoSO_4$ and nonionic detergents increased growth and vitamin production. The main component of the B_{12} vitamins of *C. simplex* was isolated and identified as 5,6-dimethylbenzimidazolylcobamide coenzyme (88). Under optimal conditions about 600 µg/liter of this vitamin was produced.

In contrast to the above studies Morikawa and Kamikuho (245) found that the B_{12} produced by various *Pseudomonas* species from hydrocarbons was predominantly in the culture supernatant. Vitamin B_{12} production by *Pseudomonas aeruginosa* like that by *Corynebacterium simplex* was increased by yeast extract, $CoSO_4$, and a nonionic detergent. Also, the addition of the B_{12} precursor dimethylbenzimidiazole stimulated vitamin production. The highest yields in this study were about 40 µg/liter. In comparison, *Propionibacterium shermanii* can accumulate up to 23 mg/liter of B_{12} from carbohydrates (313).

4. Vitamin B_6

Vitamin B_6 and related compounds have been found in the culture supernatants of several hydrocarbon-grown yeast (343). *Candida albicans* accumulated 300–400 µg/liter of B_6 in a synthetic medium containing hexadecane as a sole carbon source (343). Pyridoxal was the primary product, accounting for 50–80% of the total B_6 formed. The remainder consisted of pyridoxamine, pyridoxine, pyridoxamine phosphate, and pyridoxal phosphate. Vitamin B_6 also was produced from glucose by *C. albicans* in about four times the yield obtained from hydrocarbons. Most of the B_6 produced from glucose was excreted into the medium before the maximum cell population was obtained. However, the B_6 of hydrocarbon-grown cells was released only at the later stages of the fermentation. The same B_6 compounds were produced from glucose as from hydrocarbons but they differed in their time of appearance and their distribution between cells and supernatants (348).

5. Porphyrins

In the course of the above studies on B_{12} production Tanaka *et al.* (345) noted that *Mycobacterium smegmatis* excreted large amounts of a red pigment under conditions of reduced aeration. The pigment, extracted from the culture filtrate, was identified as free coproporphyrin III (347). Since the biosynthetic route of porphyrin synthesis is similar to B_{12} synthesis, a study was made of the factors affecting porphyrin production. Maximum porphyrin synthesis occurred during exponential growth. Both porphyrin and B_{12} production were greater in hydrocarbon than in glucose medium. However, unlike B_{12} and porphyrin production from carbohydrates, porphyrin synthesis from hydrocarbons required minimal aeration. Porphyrin excretion reached about 60 μg/liter after 7 days incubation.

Tanaka et al. (352) also reported that reduced aeration favored porphyrin production by a strain of C. simplex. This pigment, however, differed from coproporphyrin III since it was bound to a peptide and contained cobalt atoms.

6. Carotenoids

Carotenoids, specifically β-carotene, serve as precursors to vitamin A and are used as pigments in various foods. Large quantities can be produced by fermentation of carbohydrates using a 50–50 mixture of plus (+) and minus (−) strains of *Blakeslea trispora* [NRRL 2456 (+) and NRRL 2457 (−)] (252). The production medium usually contains lipids, β-ionone, or related compounds and a nonionic detergent such as Tween or Triton X. A major advance in the process was the finding that the addition of 5% kerosene to the medium would almost double carotene yields (39). Ciegler suggested that the beneficial effect of hydrocarbons and lipids derived from their function as an extracting solvent and storage reservoirs for β-carotene (38). This behavior also suggested that hydrocarbons might be ideal substrates for carotene production.

Carotenoid production in medium containing hydrocarbons as the sole carbon source was first reported by Haas and Bushnell (104). A culture of *Mycobacterium laticola* growing on refined mineral oil produced a yellow-orange pigment which colored the excess oil. Three fractions were separated chromatographically and one, β-carotene, performed similarly to vitamin A in a rat assay. *Mycobacterium laticola* accumulated about 2.14 mg of carotenoid pigments (as β-carotene) in 500 ml of oil within 2 weeks. Other mycobacteria, M. phlei, M. leprae, M. smegmatis did not produce carotenoids from hydrocarbons.

Tanaka et al. (344, 347), however, reported that their strain of *Mycobacterium smegmatis* accumulated significant quantities of carotenoids from hydrocarbons. The main component of these pigments was 4-keto-γ-carotene. The following were also found: phytofluene; ζ-carotene; 4-keto-γ-carotene (*cis*-isomer); neurosporene; monomethyoxy-4-keto-γ-carotene; monohydroxy-4-keto-γ-carotene; 1′, 2′-dihydroxy-2′-hydroxy-3′,4′-dehydro-4-keto-γ-carotene (*cis*-isomer).

The *cis*-isomers may have arisen from nonenzymatic interactions with light and oxygen. Since about 80% of the carotenoids were extracted into the hydrocarbon layer during cultivation, the access of the pigments to light and oxygen may have been increased.

A number of variables were examined to maximize carotenoid formation by *Mycobacterium smegmatis* (347). Total carotenoid production was greatest from hexadecane, although cells grown on undecane had a higher carotenoid content. The addition of Tween 80 and amino acids such as glutamic acid, histidine, and serine stimulated carotenoid production. Maximum carotenoid accumulation, under optimum conditions, was about 770 µg/liter (as β-carotene).

Nocardia sp. also produced carotenoids from hexadecane and octadecane but the pigment was not extracted by residual substrate and remained associated with the cells (53). With propane or butane carotenoid yields were higher and light was found to influence pigment accumulation. An effect of light on carotenoid synthesis also was found with various *Mycobacteria*, *Rhodotorula*, and *Sporabolomyces* grown on hydrocarbons (278).

Tanaka *et al.* (346) reported that the carotenoids of hydrocarbon-grown cells were extracted more easily than those of glucose-grown cells. They postulated that carotenoids near the cell membrane might facilitate penetration of hydrocarbons into the cells. That is, while hydrocarbon droplets extract carotenoids from the cells, the carotenoids of the cell may "extract" hydrocarbon from the medium.

7. *Pyocyanine*

Pyocyanine is another pigment which can be accumulated in relatively large quantities by microorganisms that metabolize hydrocarbons (226). *Pseudomonas aeruginosa* ATS-14, when grown on a kerosene-mineral salts medium, accumulated up to 145 µg/ml of this pigment in 10 days. Most pigment production occurred during the stationary growth phase. Optimum temperature was about 35°C, and hexadecane and heptadecane were the best substrates for the pyocyanine production.

8. *1-Phenazinecarboxylic Acid*

Higashihara and Sato (115, 116) isolated a yellow pigment produced by a pseudomonad cultivated on kerosene. The pigment, which accumulated in crystalline form on agar plates, was identified as 1-phenazinecarboxylic acid from its infrared and ultraviolet spectra and from melting point and elemental analysis determinations. Pigment yields in liquid culture were 320–350 mg/liter when kerosene was used as a substrate. However, when hexadecane or tetradecane were used pigment production was less than 150 mg/liter. The pigment was also produced from glucose but accumulated to only 20–40 mg/liter.

Several potential uses have been reported for 1-phenazinecarboxylic acid (115). It inhibits the growth of some plant pathogens and *Mycobacterium tuberculosis*. It also controls the growth of algae and toxic plants.

9. *Gibberellin*

Gibberellin and/or gibberellic acid have been identified in culture broths of *Gibberelli fujikuroi* grown on crude oil (122). A 30-day incubation period was required during which about 50 mg/liter of product was formed.

F. SUGARS AND POLYSACCHARIDES

The extracellular polysaccharides produced by bacteria presumably function as capsular material. They enable microorganisms to retain moisture in a dry environment and provide protection against bacteriophage infection. The capsule or slime is often stripped from a cell under conditions of vigorous agitation and accumulates extracellularly.

Raymond and Davis (281) reported that a *Nocardia* sp. converted 25% of an octadecane substrate into an extracellular high molecular weight polysaccharide (Table X). This polymer, when used at lower temperatures as a water loss preventative in drilling mud formulas, compared favorably with conventional additives, e.g., quebracho tannins (53).

Pseudomonas methanica produces slime material during growth on methane (68). Slime production was greater during ethane or propane oxidation by this organism even though these substrates would not support growth (221). An unidentified pseudomonad accumulated up to 6 mg/ml of a glucose polysaccharide in a mineral salts medium containing a C_6–C_{16} paraffin fraction. Species of *Corynebacterium* and *Brevibacterium* also were found to accumulate large quantities of polysaccharides from a paraffin fraction (161). The latter polymers were identified as acid heteropolysaccharides. Those produced by the *Brevibacterium* contained lauric, lactic, levulinic, and succinic acids, with glucose and mannose as the sugar components. The polymer from the *Corynebacterium* was similar except that it contained galactose and lacked succinic acid.

Suzuchi *et al.* (328) examined 20 strains of hydrocarbon-utilizing organisms and found that all, to varying degrees, accumulated the disaccharide trehalose. As mentioned earlier (Section II,A,3) penicillin facilitated accumulation by preventing incorporation of trehalose into a glycolipid component of the cells. Trehalose production was also enhanced by the addition of other antibiotics (187, 188).

TABLE X
SUGARS AND POLYSACCHARIDES PRODUCED FROM ALKANES

Product	Substrate	Organism	Yield	Remarks	References
Glucose-polysaccharide	3% v/v C_6–C_{16} Paraffins	*Pseudomonas* sp.	6 mg/ml	—	160
Sugars (glucose, trehalose)	13.3% v/v C_{10}–C_{18} Paraffins	*Arthrobacter paraffineus* ATCC 15591	15 mg/ml	Penicillin added during fermentation. Without penicillin only 6 mg/ml sugar produced.	187, 188
Sugar	Kerosene	*Corynebacterium hydrocarboclastus* ATCC 15592	6 mg/ml	Penicillin added during fermentation.	328
Acid heteropolysaccharides	10% v/v C_{12}–C_{17} Paraffins	*Corynebacterium* sp. No. 1645	≈ 10 mg/ml	—	161
Polysaccharide	0.4% Octadecane	*Nocardia* sp.	25% of utilized substrate	—	281

TABLE XI
OTHER PRODUCTS PRODUCED FROM ALKANES

Product	Substrate	Organism	Yield	Remarks	References
Dipicolinic acid	5% v/v Dodecane	Penicillium sp.	8.0 mg/ml	—	122
Fumaric acid	8% v/v C_{10}–C_{15} Paraffins	Candida hydrocarbofumarica	39 mg/ml	Fumaric acid was converted to malate by reinoculation with another yeast species.	398
Lipase	2.5% v/v Kerosene	Brettanomyces lambicus	0.95 units/ml	Tween 20 added to medium.	334
Arthronic acid $C_{32}H_{64}O_3$	20% v/v C_{11}–C_{15} Paraffin mixture	Arthrobacter paraffineus ATCC 15591	1.1 mg/ml	—	214
Corynomycolenic acid $C_{32}H_{62}O_3$	20% v/v C_{11}–C_{15} Paraffin mixture	Corynebacterium hydrocarboclastus ATCC 15592	0.25 mg/ml	Products produced simultaneously	214
Corynomycolic $C_{34}H_{64}O_3$	—		0.65 mg/ml	—	
n-Docosanoic $C_{22}H_{44}O_2$	10% v/v C_{11}–C_{15} Paraffin mixture	Mycobacterium smegmatis ATCC 362	0.60 mg/ml	Products produced simultaneously.	214
n-Ditriacontanoic $C_{34}H_{64}O_2$	—		0.85 mg/ml	—	
L-Homoseryl-L-lysine	Paraffins	Corynebacterium hydrocarboclastus	3.4 mg/ml	Products produced simultaneously.	216

G. OTHER PRODUCTS

1. Enzymes

There are very few reports describing extracellular enzyme production by microorganisms metabolizing hydrocarbons (Table XI). It is unclear whether this is attributable to some aspect of hydrocarbon metabolism or simply due to a paucity of studies directed toward isolating the appropriate organism. The latter seems more likely on the basis of the results of Takahashi *et al.* (334) who tested 26 strains of hydrocarbon-utilizing organisms and found 5 which produced an extracellular lipase. One of these cultures was identified as the yeast *Brettanomyces lambicus*. This organism utilized kerosene as a substrate and after 2 days of incubation the culture supernatant contained 0.95 units/ml of lipase activity.

2. Higher Fatty Acids

A recent patent describes the production of high molecular weight fatty acids by hydrocarbon-grown microorganisms (214). *Arthrobacter paraffineus*, in a paraffin-based medium accumulated about 1 mg/ml of an acid with an empirical formula of $C_{32}H_{64}O_3$. The acid, designated arthronic acid, was branched at the ω-position and contained a hydroxyl group at the β-position. A mixture of acids was isolated from a *Corynebacterium* culture (Table II). Palmitic and oleic acids were identified along with acids with empirical formulae $C_{32}H_{62}O_3$ and $C_{32}H_{64}O_3$. This fermentation and others for high molecular weight acids required the addition of penicillin. The latter apparently inhibited glycolipid synthesis with the resultant release of the lipid moiety to the culture medium. In the absence of penicillin, glycolipids consisting of glucose or trehalose and the above acids were isolated (217).

3. Dipeptide

Corynebacterium hydrocarboclastus produced the dipeptide L-homoseryl-L-lysine from a paraffin substrate (216). This report is notable in that 3.4 mg/ml of the product accumulated. The mechanism of accumulation has not been determined.

H. CELL YIELDS, OXYGEN REQUIREMENT, AND HEAT EVOLUTION

When the numerous processes for the production of hydrocarbon products are surveyed, it becomes apparent that several of these should be competitive with their analogous nonhydrocarbon fermentations in terms of maximum product concentration and weight percent conversion of substrate to product. This is particularly evident in the yields reported for glutamic acid, α-ketoglutarate, threonine, and citric acid. The question arises, therefore, as to why more extensive use of paraffin substrates is not encountered in commercial processes. Part of the answer resides in the increased heat evolution and oxygen consumption of cells metabolizing alkanes. Both of these factors are important economic considerations which can contribute greatly to process cost. The increased oxygen consumption is particularly significant since fermentor power requirements are dependent on the rate of oxygen transfer, which ultimately determines the productivity (grams of product per liter per hour) of the fermentor. An exception occurs with media employing pentane or heptane as substrates. For these fermentations, there is evidence that the alkane transfer, not oxygen transfer, may be the rate-limiting step (32, 107, 384).

Various studies on single cell protein production revealed that about two and one-half times more oxygen is required and two to three times as much heat is evolved in producing equivalent amounts of cells from alkanes as from carbohydrates. Guenther (102) calculated that a yield coefficient from alkanes of 1.5 gm of cells/gm of alkane would be necessary for the heat evolution and oxygen consumption during growth to approach that usually encountered with carbohydrates. Yield coefficients on alkanes are generally about 1.0 gm of cells/gm of alkanes and those from a carbohydrate such as glucose 0.5 gm of cells/gm of substrate (49). Mayberry *et al.* (238) have measured cell yields on a variety of substrates and found that these yields were all about equal when expressed as a function of the state of reduction of the substrate. For each "available electron" (electrons available for transfer to oxygen) in an organic substrate, about 3.14 gm of cells were produced. This constant also was applicable to data in numerous earlier reports of cellular yields (269). The constant did not hold for hydrocarbon and primary alcohol substrates. This is apparent in Fig. 8 where the yield coefficients predicted by 3.14 gm of cells per available electron are plotted as a function of the number of available electrons per mole of substrate. As the available electrons per mole increases by reduction of the terminal functional group, large increases in yield coefficient are predicted. However, when the available electrons per mole of substrate increase by the addition of methylene groups, the predicted increases per available electron are not as great. The latter simply reflects the increase in substrate molecular weight as available electrons are added as CH_2 groups; whereas, the molecular weight decreases when available electrons are added by reduction of oxygen-containing functional groups.

When predicted and experimentally measured yields are compared,

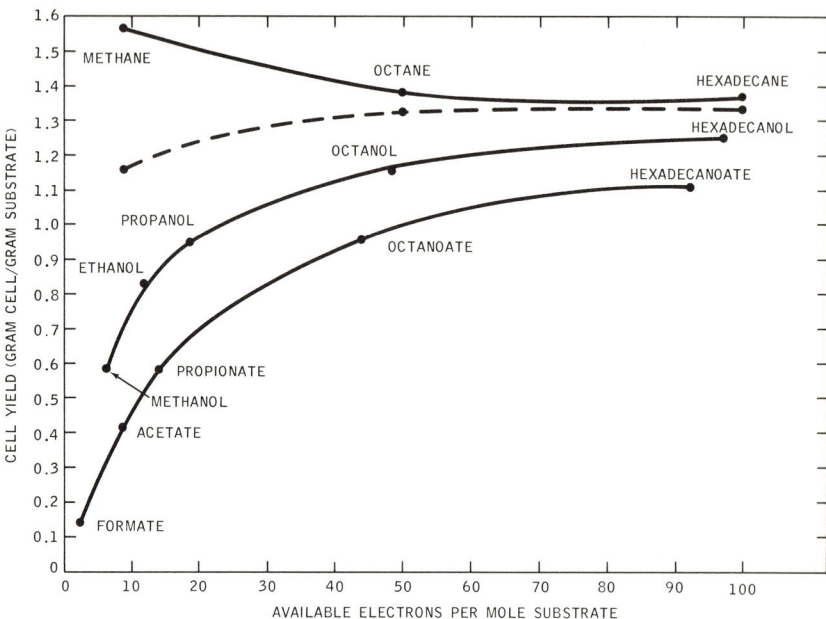

FIG. 8. Influence of the degree of substrate reduction on the yield of cells per gram of substrate assuming 3.14 grams of cells per available electron. Dashed line indicates predicted cell yields if initial alkane oxidation is not energy yielding.

it appears that the more reduced the substrate the greater the deviation from predicted yield. This is best exemplified by methane, the most reduced form of carbon. Part of the discrepancy between predicted and observed yields may be attributable to the initial alkane hydroxylation reaction. This reaction is catalyzed by a mixed function oxidase which requires a reducing agent such as NADH. Conceivably, the oxidation is not a net energy-yielding step. With the latter assumption, the data for the upper curve of Fig. 8 was replotted (dotted line). The yields predicted by the new curve are in better agreement with measured yields, although still somewhat higher. More importantly, however, the revised curve indicates that yield coefficients from methane should be lower than those from higher alkanes. This agrees with most experimental results.

Irrespective of the reasons for inefficient utilization of alkanes, a necessary consequence is the greater heat evolution and oxygen requirement. These disadvantages might be circumvented in fermentations for hydrocarbon transformation products. If cell mass can be produced on a nonhydrocarbon substrate, the hydrocarbon can be conserved solely for the transformation step. Since the latter is only a partial oxidation product of the hydrocarbon, the heat evolution and oxygen requirement associated with it should be small.

The production of biosynthetic intermediates most likely could not be conducted in this manner. However, it has not been established whether the inefficient utilization of alkanes for cell mass production also applies to product production phases of hydrocarbon fermentation. This question can be answered by a comparative examination of product yields from glucose and alkanes. Glutamic acid was selected as the basis for this comparison since it can be produced in large quantities from either substrate.

The potential of alkanes as substrates is apparent when one calculates theoretical product yields assuming all substrate carbon is retained in the product. For glucose,

$$C_6H_{12}O_6 + 2.1\ O_2 + NH_3 \rightarrow 1.2\ C_5H_9NO_4 + 5.4\ H_2O$$

a maximum molar yield of 120%, corresponding to 98% by weight conversion, is possible. With an alkane (e.g., hexadecane),

$$C_{16}H_{34} + 8.45\ O_2 + NH_3 \rightarrow 3.2\ C_5H_9NO_4 + 4.1\ H_2O$$

a molar yield of 320% (208% by weight) is a theoretical maximum. Thus, on a weight basis about twice as much product potentially can be produced from alkanes.

Typical yields of glutamate from glucose are 60% molar conversion (49% by weight) (119). The highest reported yield from alkanes (C_{11}–C_{18} fraction) is 75 weight percent (182). Assuming the paraffin fraction is equivalent to pure C_{15} alkane, the molar yield is 108%. In addition to making glutamate, both substrates are used for cell production. For the glucose fermentation described above about 10 mg/ml of cell mass was produced (119). Cell yield data are not available for the alkane to glutamate fermentation, but, the assumption was made that the ratio of cells to product was the same in both fermentations. In addition, it was assumed that the growth yields were 0.5 and 0.9 on glucose and paraffin, respectively. Also, the same cell composition ($C_{3.92}H_{6.5}O_{1.94}$) is assumed for both fermentations, and the one employed is that cited by Darlington (49). With these assumptions and data, the following balances can be written: For glucose,

$$C_6H_{12}O_6 + NH_3 + 3.04\ O_2 \rightarrow 0.18\ C_{3.92}H_{6.5}O_{1.94} + 0.60\ C_5H_9O_4N \\ + 2.30\ CO_2 + 4.74\ H_2O$$

For alkanes,

$$C_{15}H_{34} + NH_3 + 16.9\ O_2 \rightarrow 0.35\ C_{3.92}H_{6.5}O_{1.94} + 1.08\ C_5H_9O_4N$$
$$+ 8.23\ CO_2 + 12.50\ H_2O$$

From these balances, the oxygen requirement of the alkane fermentation is 3.39 gm of O_2/gm of glutamic acid and for the glucose substrate 1.10 gm of O_2/gm of glutamic acid.

The heat liberated in the above fermentation is the difference in the heat of combustion of the substrate and the heat of combustion of the products (cells and glutamate). A value of 5.43 kcal per gram was used for the heat of combustion of cells (277). The heats of combustion of the other components of the system were obtained from the appropriate tables (356). When the heat evolution is calculated, it is found that the heat liberated with the alkane substrate is 10.35 kcal/gm of glutamic acid and with glucose only 2.78 kcal/gm of glutamic acid. Thus, about three and one-half times as much heat must be removed to produce glutamate from alkanes and about three times more oxygen is required.

The heat removal problem could be circumvented if product yields (percent conversion) were greatly increased beyond their present values. The problem would also be minimized by employing thermophilic microorganisms.

The oxygen requirement would decrease if product yields from alkanes were increased. In the absence of higher yields, it would be desirable to facilitate oxygen transfer so as to reduce the cost of this part of the fermentation. This might be accomplished by novel fermentor designs or by coupling hydrocarbon fermentations with a crude oil dewaxing process. In dewaxing crude oils a large excess of unused hydrocarbon is present in the medium. Mimura et al. (241) have shown that the oxygen transfer rate increases with increasing concentration of hydrocarbons in the culture medium. In addition, Yoshida et al. (403) reported that increasing concentrations of compounds with positive spreading coefficients (e.g., toluene and oleic acid) improves oxygen transfer to water. The production of surfactants during growth would increase the spreading coefficients of paraffin so that they also, when present in excess, may facilitate oxygen transfer. Thus, the problems generated by inefficient hydrocarbon utilization, while formidable, may not be insurmountable.

III. Products from Cyclic Hydrocarbons

In contrast to the numerous reports of microbial product formation from aliphatic hydrocarbons and nonhydrocarbon aromatic compounds, relatively few instances concerning the industrial application of aromatic and cycloparaffinic hydrocarbon fermentations have been reported, even though knowledge of microbial growth and dissimilation of these classes of compounds have been known since 1915 (30). Industrial exploitation of aromatic and cycloparaffinic hydrocarbons had lagged primarily because of the difficulty in isolation of a variety of types of microbes which were able to grow on the relatively toxic cyclic compounds, slow growth rates of those strains isolated, low levels of product formation, and unavailability of sufficient quantities of inexpensive and relatively pure substrates.

During the last decade techniques were developed that allowed microbial cultivation on aromatic and cycloparaffinic hydrocarbon substrates and, at the same time, permitted product formation to reach significant levels. In addition, during the past two decades the capacity for formation and the commercial availability of these compounds increased greatly. In 1969 benzene, toluene, ethylbenzene, styrene, cyclohexane, and cumene were among the 50 largest-volume chemicals (13). Thus, the potential of cyclic hydrocarbons as substrates for microbial growth, with the resultant formation of products with detergent and medicinal applications and for use as food additives, sequestrants, dye intermediates, perfume bases, biocides, organic polymer intermediates, and other specialty chemicals recently has been recognized. The industrial exploitation of aromatic and cycloparaffinic hydrocarbons is currently in its infancy; however, the large number of products which have been seen in a relatively short period, as well as the potential applications for new products in the future, will assure continued expansion of research in this area.

A. Biosynthetic Products

Reports of biosynthetic products (as opposed to transformation products) from cyclic hydrocarbons are relatively few and, therefore, will be mentioned here only briefly. Douros and Raymond (66) reported the accumulation of low levels of the amino acids lysine and methionine from a *Streptomyces* species growing at the expense of either toluene or tetrahydronaphthalene. The production of L-glutamic acid by *Nocardia globerula* was accomplished using alkylbenzenes and polycyclic aromatic hydrocarbons as substrates for growth (62). Yields, however, were less than 2.5 mg/ml. In another study, aromatic amino acids were observed in supernatant fluids of *Pseudomonas fluorescens* cultures following growth on *n*-butylbenzene (46). Stimulation of the production of particular amino acids by *Pseudomonas* species grown on various aromatic and cycloparaffin hydrocarbons has been achieved by selecting mutants resistant to specific growth inhibitory amino acid analogs (243).

Riboflavin production from benzene and naphthalene by yeast species belonging to the genus *Pichia* has also been claimed (237). Enzymes capable of converting aliphatic, aromatic, and alicyclic alcohols and aldehydes to their corresponding acids were isolated from both bacteria and actinomycetes which had grown on aromatic and cycloparaffinic hydrocarbons (224, 225). Recently, Mabe and Brannon (229) reported the production of 0.21 mg per liter of the antibiotic pyrrolnitrin by *Pseudomonas multivorans* following growth on toluene. Substrate toxicity was minimized by supplying the culture with washed agar cubes containing 20% toluene. Also, various aromatic and cycloparaffinic hydrocarbons have been employed as substrates for growth of the nitrogen-fixing bacterium *Azotobacter aromaticum* (45); the cellular product was used for fertilizer (47).

In all of the previous examples product yields were quite low in comparison to those obtained from substrates other than cyclic hydrocarbons. In addition, when factors such as cost of substrate, value of product, and length of fermentation are considered the commercial feasibility for the production of biosynthetic products from cyclic hydrocarbons seems remote. The major area for application of cyclic hydrocarbon fermentations will, therefore, probably be restricted to substrate-related specialty chemicals, e.g., via metabolic conversion or transformation.

Other applications for cyclic hydrocarbons have been suggested by Traxler and Bernard (372). They demonstrated that benzene and various methyl-substituted benzenes stimulated the growth rate and increased the cell mass of *Pseudomonas aeruginosa* growing on *n*-alkanes. The aromatic hydrocarbons apparently increased the permeability of the cells, allowing more efficient utilization of the alkane substrate. This effect is analogous to that seen when hydrocarbons were incorporated into a nonhydrocarbon glutamate fermentation (383). Thus, cyclic hydrocarbons may benefit both hydrocarbon and nonhydrocarbon fermentation processes, not via their metabolism, but through their influence on cellular permeability.

B. Transformation Products

1. Mononuclear Aromatic Hydrocarbons

a. Benzene

The microbial dissimilation of benzene is initiated by insertion of oxygen at two adjacent carbon atoms so as to form the ortho hydroxylated compound (74). Further degradation occurs via either ortho or meta cleavage of the benzene ring (74). The primary products resulting from microbial growth on benzene are *cis*-benzene glycol, catechol, *cis*, *cis*-muconic acid, α-hydroxymuconic semialdehyde, β-ketoadipic acid and succinic acid (Table XII). Both phenol and *trans*, *trans*-muconic acid also have been implicated as bacterial oxidation products of benzene (173, 260, 395).

The products listed above represent only those that were found to accumulate with benzene as the main carbon source. However, the reports of numerous other products that accumulate from oxygenated intermediates of the benzene pathway (94, 98, 266, 285) increase the potential value of industrial exploitation of benzene metabolism by microorganisms.

Generally, organisms utilizing benzene as a sole carbon source completely degrade the benzene nucleus and, as a result, products accumulate only as transient intermediates at low levels. Recently, Gibson *et al.* (95) demonstrated that under appropriate conditions *cis*-benzene glycol was produced in significant quantities from benzene. The accumulation of this product was used to determine the mechanism of oxygen insertion into the benzene nucleus. The organism employed was a mutant of *Pseudomonas putida* obtained after nitrosoguanidine treatment of the parent strain and selected on the basis of its ability to grow on succinate but not on benzene. Among the various benzene mutants obtained by this procedure was the 39/D strain which had lost the ability to convert *cis*-benzene glycol to catechol. Consequently, glucose-grown, benzene-induced resting cells produced up to 1.1 mg/ml of *cis*-benzene glycol in a medium containing benzene (Table XII). Application of this mutation technique to the isolation of mutants blocked at other points of the benzene pathway should prove valuable in stimulating the accumulation of additional intermediates.

Mutants such as those described above can no longer use the hydrocarbon as the sole carbon source for growth. Thus, an alternate growth substrate, such as glucose, is required in addition to the substrate for product formation. This situation, which is analogous to "co-oxidation" (to be discussed later), provides a means for growing cells on an inexpensive substrate while conserving the cosubstrate primarily for product formation.

b. Methyl-Substituted Benzenes

The presence of alkyl side groups on the benzene nucleus provides additional sites for microbial oxidation and, consequently, increases the variety of potential products which may be obtained. The literature concerned with microbial oxidation of methyl-substituted ben-

TABLE XII
MICROBIAL PRODUCTS FROM BENZENE

Substrate	Concentration (mg/ml)	Organism	Product	Yield	Remarks	Refs.
Benzene	—	*Mycobacterium rhodochrous*	Catechol; *cis,cis*-Muconic acid; β-Ketoadipic acid; Succinic acid	—	—	235
	—	*Pseudomonas aeruginosa* *Micrococcus sphaeroides* *Nocardia corallina*	*trans,trans*-Muconic acid; Phenol	—	Evidence for phenol from sequential induction experiments	173, 395
	—	Gram-positive rod	Phenol	—	—	260

	Organism	Products		Conditions	Ref.
—	Pseudomonas putida B	cis-Benzene glycol; Catechol	2-Hydroxy-muconic semialdehyde	Washed cell suspensions and cell-free extracts employed	97
3	Pseudomonas putida 39/D	cis-1,2-Dihydro-1,2-dihydroxybenzene		1.1 mg/ml; Glucose-grown, benzene-induced resting cells used	95

zenes has, in its initial phases, been concerned primarily with sequential induction experiments and is quite sketchy with regard to the detailed mechanisms of the formation of particular products. A large variety of products were obtained from these substrates, but with the possible exception of the formation of dimethyl muconic acid and dihydroxy-*p*-toluic acid from *p*-xylene, little attention has been directed toward establishing the potential for commercial production of a particular product.

Microbial growth on toluene is initiated via the oxidation of either the methyl group or the ring carbon atoms, depending upon the organism employed (41, 66). Various intermediates of toluene degradation accumulated, but again, as with benzene, only at low levels as transient intermediates (Table XIII, Section A). Gibson *et al.* (96) have demonstrated that the level of a particular intermediate can be increased by use of the proper mutant. Utilizing the techniques described in the previous section they demonstrated the accumulation of 1.5 mg/ml of (+)*cis*-2,3-dihydroxy-1-methylcyclohexa-4,6-diene (Table XIII, Section A) by *Pseudomonas putida* 39/D grown on glucose in the presence of toluene vapors. In another study diphenylmethane was used as a model substrate for DDT metabolism (80) and a *Hydrogenomonas* species formed phenylacetic acid, benzoic acid, and phenylglyoxylic acid from this substrate (Table XIII, Section C).

Recently, Japanese workers initiated systematic studies of aromatic growth substrates and were interested particularly in the potential of *p*- and *m*-xylenes as well as 1,2,4-trimethylbenzene (pseudocumene) as sole carbon sources for the formation of microbial transformation products (255, 262, 263). The major products produced from these substrates were *p*- and *m*-toluic acids and 3,4-dimethylbenzoic acid, respectively (Table XIII, Section B). In addition, 3-methyl salicyclic acid was isolated from cultures grown on *m*-xylene. An extensive search for organisms capable of utilizing *o*-xylene for the main carbon source was unsuccessful (263). Subsequent experiments (264) on the metabolic pathways of *p*- and *m*-xylene and 1,2,4-trimethylbenzene degradation indicated that other intermediates also accumulated in small amounts (Table XIII, Section B). Through optimization of aeration, agitation, and medium composition, and the utilization of $CaCO_3$ for pH control the yields of *p*-toluic acid were increased from 0.4 to 3.4 mg/ml (262). Studies concerned with further improvement of *p*-toluic acid yields or increasing yields of other intermediates have not been reported.

The co-oxidation technique provided an important method for stimulation of both the quantity and the variety of products formed from methyl and alkyl benzenes. The term co-oxidation was originally developed by Foster (84) on the basis of the work with *Pseudomonas methanica* (220, 221). In its original sense co-oxidation was defined as the process whereby "non-growth hydrocarbons are oxidized when present as co-substrates in a medium in which one or more different hydrocarbons are furnished for growth." Leadbetter and Foster (220, 221) found that *Pseudomonas methanica* was dependent on methane or methanol for growth and could not grow on ethane, propane, or butane. This organism, however, partially oxidized the latter non-growth substrates with the resultant formation of alcohols, aldehydes, acids, and ketones. The oxidation of propane and butane required actively growing cells, but ethane was oxidized by washed cell suspensions in the absence of methane.

The majority of co-oxidation studies to date have fit the definition of co-oxidation as presented by Foster. However, co-oxidation-type studies have been carried out in which neither the growth substrate nor the cosubstrate was a hydrocarbon. Steroid transformations exemplify such nonhydrocarbon co-oxidations. Another type of co-oxidation occurs with alkylbenzenes containing side chains longer than eight carbon atoms. These compounds can support growth, but the benzene ring is only partially degraded or not degraded at all. Thus, the alkyl side chain functions as the growth substrate and the ring portion as the cosubstrate.

With some types of co-oxidations it is uncertain whether the cosubstrate need be present during growth or the growth substrate be present during co-oxidation. For example, in the original work by Leadbetter and Foster (221) the co-oxidation of both propane and butane required the presence of methane, the growth substrate, while ethane oxidation could be accomplished by resting cell suspensions. Such variations from Foster's definition of co-oxidation suggest that a more inclusive definition be formulated. The important aspect of this concept which should be retained is that a co-substrate is only partially oxidized and not totally degraded into carbon for cellular synthesis.

From an industrial standpoint the co-oxidation technique offers a valuable method for obtaining new products and for increasing yields of other products. From an ecological view co-oxidation may explain the fate of compounds introduced into nature and the difficulty in isolation of pure cultures capable of growth on a particular compound as the sole carbon source. Co-oxidation has provided a valuable tool for the study of products accumulating from both aromatic and cycloparaffinic hydrocarbons. Previously, the types of organisms which were reported to grow on these substrates as the sole carbon source were restricted to gram-negative rods, primarily species of *Pseudomonas*.

TABLE XIII
Microbial Products from Methyl-Substituted Benzenes

Sec.	Substrate	Concentration (mg/ml)	Organism	Product	Yield	Remarks	Refs.
A	Toluene	—	*Pseudomonas aeruginosa*	Benzoic acid (COOH)	—	Cell-free extracts	255
		—	*Pseudomonas aeruginosa*	Benzaldehyde (CHO), Benzoyl alcohol (CH₂OH)	—	Toluene-adapted, acetone-dried cells	166
		—	*Achromobacter*, *Pseudomonas* sp., *Pseudomonas putida*	3-Methylcatechol	3.9 mg/liter	Washed cell suspensions of toluene-grown cells	41, 82, 97
		—	*Pseudomonas*	Methyl-α-hydroxy-muconic semialdehyde	—	—	83
			Achromobacter, *Pseudomonas*	Pyruvic acid (CH₃C(O)COOH)	0.27 mg/ml	—	41
		0.46	*Pseudomonas mildenbergii*	Acetic acid (CH₃COH)	0.26 mg/ml	—	257

A (contd)	toluene (contd)	—	*Pseudomonas putida* 39/D	(+)-*cis*-2,3-Dihydro-2,3-dihydroxytoluene (CH₃ structure with OH, OH, H, H)	1.5 mg/ml	Parent organism grows readily on toluene, but 39/D mutant requires glucose or yeast extract for growth	96
B	*p*-Xylene	17.2	*Pseudomonas Achromobacter Pseudomonas* S668B2	*p*-Toluic acid	3.4 mg/ml	Product formation during log phase (before 20 hours)	52, 255, 262
		17.2	*Pseudomonas aeruginosa* S668B2	*p*-Cresol, *p*-Hydroxy benzoic, 4-Methyl-catechol	—	Products accumulate in low quantities	264
	m-Xylene	17.2	*Pseudomonas aeruginosa* S668B2	3-Methylsalicylic acid	120 mg/liter	—	263

TABLE XIII (*Continued*)

Sec.	Substrate	Concentration (mg/ml)	Organism	Product	Yield	Remarks	Refs.
B (contd)	*m*-Xylene (contd)	—	*Pseudomonas aeruginosa* S668B2	*m*-Toluic acid (CH₃–C₆H₄–COOH)	1% molar conversion	—	255, 263
	1,2,4-Trimethyl benzene	18	*Pseudomonas aeruginosa* S668B2	A. 3,4-Dimethyl-benzoic acid; B. 3,4-Dimethyl-phenol	Product A: 120 mg/liter	—	263
C	Diphenyl methane	5	*Hydrogenomonas* sp.	Phenylacetic acid, Benzoic acid, Phenylglyoxylic acid	—	—	80

Growth rates were relatively slow and product accumulation generally low. With the advent of co-oxidation a large group of organisms belonging to, or related to, the genus *Nocardia* gained prominence as having the potential of forming many new microbial products from cyclic hydrocarbons.

Much of the pioneering work on co-oxidation products of aromatic compounds can be attributed to the research group at Sun Oil Company. They found that several *Nocardia* species co-oxidized *p*-xylene and formed significant quantities of either *p*-toluic acid (PTA) and dihydroxy-*p*-toluic acid (DHPT) or α,α'-dimethylmuconic acid (DMMA) (153, 282, 283), depending on the pH of the medium. Furthermore, methyl-substituted benzenes containing at least two consecutive unsubstituted ring carbon atoms adjacent to the methyl group were susceptible to similar oxidation. Thus, a wide variety of products related to those above were reported from methylbenzenes ranging from toluene to the tetramethylbenzenes (Table XIV, Section A). Metabolic studies (130) suggested two routes for the degradation of *p*-xylene by these *Nocardia* species (Fig. 9). The regulation of these pathways was thought to be mediated by the pH specificity of the methyl group oxygenase (283). Thus, at pH 8.0 the methyl group oxygenase was functional and primarily *p*-toluic acid and dihydroxy-*p*-toluic acid were formed. At pH 6.5 the methyl group oxygenase system became nonfunctional resulting in the formation of the methyl-substituted muconic acid by direct dihydroxylation and cleavage of the benzene ring. The dihydroxylation reaction appeared to require molecular oxygen since addition of a second hydroxyl group on such monohydroxylated substrates as 2-hydroxy-4-methylbenzoic acid and 3-hydroxy-4-methylbenzoic acid did not occur.

Since the products formed by the *Nocardia* isolates possessed both unique structure and functionality they were envisioned as being of value as chemical intermediates and for specialty chemicals such as sequestrants. Therefore, a closer examination of the *p*-xylene fermentation was initiated in order to increase the yields of the various products and evaluate their economic potential.

The fermentation was conducted in a 40-liter fermentor by incubating *Nocardia corallina* ATCC 19070 with *n*-hexadecane and trace amounts of *p*-xylene (282). After 24 hours of growth, portions of a 90:10 mixture of *p*-xylene and *n*-hexadecane were added periodically over 82 hours. The pH was maintained between 6.5 and 7.0 and the *p*-xylene concentration was kept within the range of 80–350 ppm. The resulting fermentation yielded 13.4 mg/ml of DMMA. In another ex-

TABLE XIV
Co-Oxidation Products from Methylbenzenes

Sec.	Substrate	Concentration (mg/ml)	Organism	Product	Yield	Remarks	Refs.
A	Toluene	2.5	*Nocardia corallina* *Nocardia salmonicolor* *Nocardia minima*	A. 2,3-Dihydroxybenzoic acid B. α-Methylmuconic acid	Product B: 0.2 mg/ml	n-Hexadecane for cell growth	127, 153, 282
		—	*Nocardia corallina* ATCC 19707	α,α′-Dimethylmuconic acid	16 mg/ml	n-Hexadecane for growth. Substrate concentration maintained between 80 and 135 ppm	126, 130, 153, 282
	p-Xylene	—	*Nocardia salmonicolor* ATCC 19149	A. 2,3-Dihydroxy-p-toluic acid B. p-Toluic acid	Product A: 5.0 mg/ml Product B: 8.3 mg/ml	n-Hexadecane for cell growth, pH 7.5–8.0	126, 130, 282, 283
		—	*Nocardia corallina* A6	A. 2,3-Dihydroxy-p-toluic acid B. p-Toluic acid	Product A: 1.3 mg/ml Product B: 24.7 mg/ml	n-Hexadecane for cell growth; IRA-45 anion-exchange resin employed to remove product	127, 153, 284

A (contd)						
p-Xylene (contd)	—	*Nocardia corallina* ATCC 19147	2,3-Dihydroxy-p-toluic acid (A.); p-Toluic (B.); 3,6-Dimethyl-catechol (C); α,α′-Dimethyl-muconic acid (D.)	Product A: 14.6 mg/ml; Product B: 0.8 mg/ml; Product C: 0.8 mg/ml; Product D: 0.6 mg/ml	n-Hexadecane for cell growth; IRA-93 anion-exchange resin employed to remove product	133, 153, 284
m-Xylene	—	*Nocardia corallina* ATCC 19070, 19071, 19148; *Nocardia salmonicolor* ATCC 19149; *Nocardia minima* ATCC 19150	2,3-Dihydroxy-m-toluic acid; α,β′-Dimethyl-muconic acid	—	n-Hexadecane for cell growth	127, 282

(Continued)

TABLE XIV (Continued)

Sec.	Substrate	Concentration (mg/ml)	Organism	Product	Yield	Remarks	Refs.
A (contd)	o-Xylene	—	Nocardia corallina ATCC 19070, 19071, 19148	2,3-Dihydroxy-o-toluic acid; α,β-Dimethyl-muconic acid	—	—	127, 282
	1,2,3-Trimethylbenzene (hemimellitene)	—	Nocardia salmonicolor ATCC 19149	2,3-Dihydroxy-5,6-dimethyl benzoic acid; α,β,β'-Trimethyl-muconic acid	—	n-Hexadecane for cell growth	127, 282
	1,2,4-Trimethylbenzene (pseudocumene)	—	Nocardia minima ATCC 19150	2,3-Dihydroxy-4,6-dimethyl-benzoic acid; 2,3-Dihydroxy-4,5-dimethyl-benzoic acid; α,α',β-Trimethyl muconic acid	—	—	127, 153, 282

A (contd)	1,2,3,4-Tetra-methylbenzene (prehnitene)	—	Nocardia corallina ATCC 19070, 19071, 19148 Nocardia salmonicolor ATCC 19149 Nocardia minima ATCC 19150	2,3-Dihydroxy-4,5,6-trimethyl-benzoic acid $\alpha,\alpha',\beta,\beta'$-Tetra-methyl muconic acid	—	n-Hexadecane for cell growth	127, 282
	m-Xylene	2.5	Nocardia corallina Mycobacterium rhodochrous	m-Toluic acid	0.7 mg/ml		16, 153
	o-Xylene	25.8	Nocardia corallina Nocardia salmonicolor Nocardia minima Mycobacterium rhodochrous	o-Toluic acid	Trace	n-Hexadecane and n-decane for cell growth	153, 283
B	Toluene	—	Mycobacterium rhodochrous	Benzoic acid	Trace	—	16

(Continued)

TABLE XIV (Continued)

Sec.	Substrate	Concentration (mg/ml)	Organism	Product	Yield	Remarks	Refs.
B (contd)	1,3,5-Trimethylbenzene (mesitylene)	20	*Mycobacterium rhodochrous*	3,5-Dimethylbenzoic acid		n-Decane for cell growth. Concentration, washed-cell suspension employed	16
	1,2,4,5-Tetramethylbenzene (Durene)			2,4,5-Trimethylbenzoic acid	Trace		
	Ditolyl			3-Carboxy-4'-methyldiphenyl			

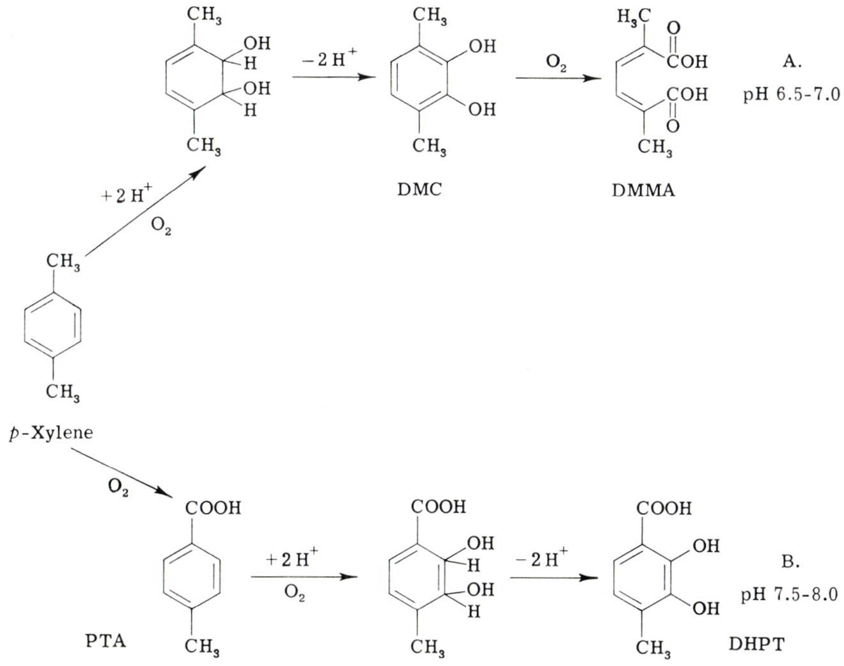

FIG. 9. Proposed pathways for p-xylene oxidation by *Nocardia*.

periment *Nocardia salmonicolor* ATCC 19149, grown under similar conditions except that the pH was maintained between 7.5 and 8.0, accumulated 5 mg/ml of DHPT and 8.3 mg/ml of PTA in 75 hours (Table XIV, Section A). At this pH no DMMA accumulated in the medium. Washed cell suspensions of *Nocardia corallina*, incubated in a buffered p-xylene:n-hexadecane solution at pH 8.0 under non-growth conditions (i.e., no nitrogen), also accumulated DHPT and PTA.

Subsequent data from shake-flask studies indicated that the product yield, as well as the variety of products accumulated, was increased by a weakly basic anion-exchange resin (IR-45, Rohm & Haas) added either as a single batch or incrementally during the fermentation (133, 284). Thus, the concomitant formation of 14.6 mg/ml of DMMA, 0.8 mg/ml of PTA, and 0.8 mg/ml of DHPT by *Nocardia corallina* ATCC 19071 occurred when IR-45 was added to shake flasks. In addition, 0.6 mg/ml of a new product, 3,6-dimethyl catechol (DMC), also accumulated in this system. Control shake flasks, without resin, yielded only 0.2 mg/ml of DMMA (133). Thus, on a shake-flask scale IR-45 led to significant yield improvements. However, shake flasks without ion-exchange resin yielded considerably less product than the 40-liter fermentor (0.62 mg/ml versus 13.4 mg/ml). It would seem important to establish whether ion-exchange resins would increase DMMA formation in larger fermentors. Hosler (129) reported a 25% increase in PTA and DHPT with ion-exchange resin addition to a 3-liter fermentor, but DMMA yields were not reported. In recent publications Hosler and Eltz (130) and Hosler (128) reported the scale-up of the *p*-xylene fermentation to a 520-liter pilot plant fermentor. Optimization of the fermentation was accomplished by use of off-gas hydrocarbon analysis, continuous controlled addition of the growth substrate (octadecane), and pH control. The off-gas analysis was used to maintain the *p*-xylene concentration between 3 and 20 mg/liter and the pH was controlled between 6.5 and 7.0. Under these conditions *Nocardia corallina* ATCC 19070 accumulated 16 mg/ml of DMMA in 50 hours. The time course of this fermentation was considerably shorter than that previously required (106–120 hours) (282) and the DMMA still appeared to be accumulating when the fermentation was terminated. Product removal by use of anion-exchange resins was not investigated. Pilot scale production of 2.5 mg/ml of DHPT and 5.5 mg/ml of PTA by *Nocardia salmonicolor* ATCC 19149 by co-oxidation of *p*-xylene was also reported. Additional studies on the production of these novel co-oxidation products were apparently terminated because of the lack of a suitable market for the products.

Mycobacterium rhodochrous also oxidized various methylbenzenes, producing benzoic or methylbenzoic acids (16) (Table XIV, Section B). The formation of 4-carboxy-4'-methyl-diphenyl from ditolyl by *M. rhodochrous* was also reported (16).

c. Alkyl Benzenes

Davis and Raymond (54) reported that *Nocardia salmonicolor* growing on *n*-alkyl benzenes containing an odd number of carbon atoms on the alkyl chain accumulated low levels of phenylacrylic and phenylpropionic acids. When *N. salmonicolor* was cultivated on *n*-dodecylbenzene small quantities of phenylacetic acid were produced (Table XV, Section A). These data were interpreted as indicating that alkylbenzenes were degraded by β-oxidation. Evidence for the subsequent degradation of phenylacrylic and phenlypropionic acids to benzoic acid and oxygenated ring intermediates was also presented. Phenylacetic acid, however, proved recalcitrant to further degradation by this strain.

Webley (391, 392) reported that two *Nocardia* strains grew on various long-chain (greater than 9 carbon atoms on the side chain) alkylbenzenes as their sole carbon and energy source. The strains were incapable of growth on n-ethyl- to n-butylbenzenes. In experiments using glucose-grown resting cells the authors reported that *Nocardia opaca* T_{16}, unlike the *Nocardia* of Davis and Raymond (54), was capable of oxidizing the phenylacetic acid which arose from growth on alkylbenzenes containing an even number of carbon atoms. As a result, o-hydroxyphenylacetic acid was detected in the medium (Table XV, Section B). Their *Nocardia* (spp-P_2) behaved similarly to Davis and Raymond's strain (54) in that phenylacetic acid was the major product from growth on 1-phenyldecane, 1-phenyldodecane, and 1-phenyloctadecane (Table XV, Section B). In addition, 3-phenyleicosane was converted to 2-phenylbutyric acid by a similar reaction (Table XV, Section B) (392).

A patent issued to Coty (46) revealed that a *Pseudomonas* species growing on n-butylbenzene as a sole carbon source accumulated low levels of phenylacrylic and phenylpropionic acid. This organism may possess a novel alkyl oxidation system since these products appear to result from a C_1 oxidative decarboxylation of the terminal methyl group (Table XV, Section C).

Initial co-oxidation studies demonstrated only low yields of transformation products from alkylbenzene cosubstrates (54, 55). These studies reported phenylacetic acid to be formed from n-ethyl-, n-butyl-, and n-dodecylbenzene cosubstrates, and either phenylacrylic or benzoic acid from n-propylbenzene (Table XVI, Section A). The shorter-chain alkylbenzene, containing C_2 to C_4 side chains, could not support growth of the *Nocardia* strains, but co-oxidation of these substrates was accomplished when the paraffin growth substrate and nitrogen were present.

More recent studies have indicated that certain bacteria can efficiently co-oxidize short-chain alkylbenzenes (63–65). Douros and Frankenfeld at Esso reported that various gram-negative rods co-oxidized alkylbenzenes, containing odd-carbon side chains, to *trans*-cinnamic acid (phenylacrylic acid) and 5-phenylvaleric acid (63–65). An organism identified as a *Cellulomonas* sp. converted n-amylbenzene to *trans*-cinnamic acid (5 mg/ml) in 100% molar yields (Table XVI, Section A). This product was also produced from other alkylbenzenes; however, the yields were lower than those obtained with n-amylbenzene. An *Alcaligenes* species formed primarily 5-phenylvaleric acid at a yield representing a 60% molar conversion of n-amylbenzene (63).

Conversion of the co-substrate to the above products required the presence of a paraffin growth substrate. Exposure to the co-substrate alone resulted in the induction of aromatic ring-splitting oxygenases and the complete degradation of the n-amylbenzene to cellular carbon and CO_2. Thus, in this case, the co-oxidation technique proved valuable in permitting the selective oxidation of the aliphatic portion of the substrate, while leaving the aromatic portion intact.

The variety of products arising from similar enzymatic reactions could be considerable. For instance, substitution of the alkylbenzene ring with additional groups would increase the variety and functionality of products. The conversion of two m-tolylparaffins to m-tolylpropionic and m-tolylacrylic acids (Table XVI, Section A) exemplifies this concept (69).

2. Polynuclear Hydrocarbons

a. Naphthalene

Reports of bacterial growth and dissimilation of naphthalene appeared as early as 1910 (101, 357, 361). Strawinski (324) first identified salicylic acid as a naphthalene degradation product of *Pseudomonas* cultures. Walker and Wiltshire (390) and Treccani *et al.* (375) confirmed these findings and, in addition, isolated and identified D-*trans*-1,2-dihydro-1,2-dihydroxynaphthalene and catechol in *Pseudomonas* and *Nocardia* cultures. Sequential induction studies also implicated coumarin, 1,2-naphthoquinone, and 1,2-dihydroxynaphthalene as intermediates of naphthalene metabolism (390). Subsequent experimentation revealed that 1,2-naphthoquinone was produced by a nonenzymatic oxidation of 1,2-dihydroxynaphthalene (247). Also, sequential induction studies did not implicate 1,2-dihydroxynaphthalene as a precursor of either 1,2-napthoquinone or the ring fission product (24). However, Fernley and Evans (77) were able to demonstrate the oxidation of 1,2-dihydroxynaphthalene by a pseudomonad when precautions were taken to minimize the nonenzymatic formation of the toxic 1,2-naphthoquinone. From naphthalene-grown cultures they isolated coumarin and detected o-hydroxy-*trans*-cinnamic acid (o-coumaric acid) and o-hydroxyphenylpropionic acid (melilotic acid) in addition to salicylic acid (Table XVII, Section A). The products, o-coumaric and melilotic acids, were not postulated as direct intermediates of the naphthalene degradation pathway since resting cultures displayed only low oxygen uptake in their presence.

Coumarin was recognized as an artifact, arising during isolation from the oxidative decarboxylation of the postulated labile ring fission product, o-carboxy-*cis*-cinnamic acid. However, subsequent sequential induction studies by Davies and Evans (51) indicated *cis*-o-hydroxybenzalpyruvate, not o-carboxy-*cis*-cinnamic acid, was the ring fission product. In addition, salicylaldehyde and α-hydroxymuconic semialdehyde were detected in the cultures. From these results a degradation pathway for naphthalene was postulated (Fig. 10) and all

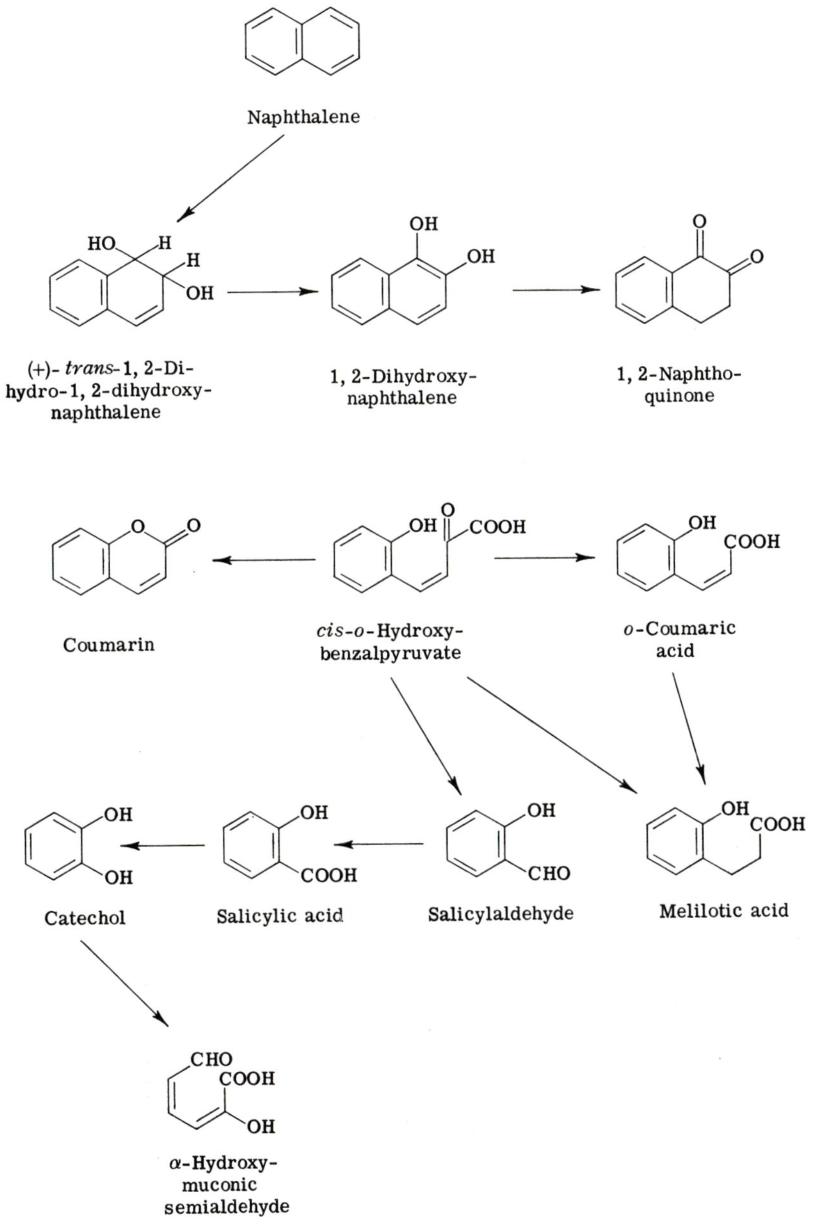

FIG. 10. Proposed naphthalene degradation pathway for *Pseudomonas* sp.

TABLE XV
Microbial Products from Alkyl Benzenes as the Sole Carbon and Energy Source

Sec.	Substrate	Concentration (mg/ml)	Organism	Product	Yield	Remarks	Refs.
A	$CH_2-(CH_2)_7CH_3$ n-Nonyl benzene	10	Nocardia salmonicolor	CH_2CH_2COOH Phenylpropionic acid; $CH=CHCOOH$ Phenylacrylic acid	0.62 mg/ml	—	
	$CH_2(CH_2)_{10}CH_3$ n-Dodecylbenzene	1.1		CH_2COOH Phenylacetic acid	0.49 mg/ml	—	54
B	$CH_2(CH_2)_{10}CH_3$ n-Dodecylbenzene	—	Nocardia opaca T_{16}	CH_2COOH, OH o-Hydroxyphenylacetic acid	—		
	$CH_2(CH_2)_{10}CH_3$ n-Dodecylbenzene	1	Nocardia opaca P_2	CH_2COOH Phenylacetic acid	Molar conversion 85%	Glucose-grown resting cells used. Product also formed from 1-phenyldecane and 1-phenyloctadecane	391, 392

B (contd)	3-Phenyleicosane CH$_3$CH$_2$CH–(CH$_2$)$_{16}$CH$_3$	—	Nocardia opaca	Phenylethyl acetic acid CH$_3$CH$_2$CHCOOH–C$_6$H$_5$	Molar conversion 56%	Glucose-grown resting cells	391, 392
C	n-Butylbenzene C$_6$H$_5$–CH$_2$CH$_2$CH$_2$CH$_3$	—	Pseudomonas sp.	β-Phenyl-acrylic acid C$_6$H$_5$–CH=CHCOOH; β-Phenyl-propionic acid C$_6$H$_5$–CH$_2$CH$_2$COOH	Trace	Aromatic amino acids also formed	46

TABLE XVI
Co-Oxidation Products from Alkyl Benzenes

Sec.	Substrate	Concentration (mg/ml)	Organism	Product	Yield	Remarks	Refs.
A	CH_2-CH_3 — Ethylbenzene	—	Nocardia salmonicolor	CH_2-COOH — Phenylacetic acid	8.4 mg/80 ml substrate		54, 55
	$CH_2CH_2CH_3$ — n-Propylbenzene	—	Nocardia salmonicolor	COOH — Benzoic acid		n-Hexadecane for cell growth	55
	$CH_2CH_2CH_3$ — n-Propylbenzene	—	Nocardia	$CH=CH-COOH$ — Phenylacrylic acid (cinnamic acid)			54
	$CH_2CH_2CH_2CH_3$ — n-Butylbenzene	—	Nocardia salmonicolor	CH_2-COOH — Phenylacetic acid	10 mg/80 ml substrate		54, 55

Substrate		Organism	Product	Yield	Notes	Ref.
A. $CH_2-(CH_2)_3-CH_3$ (phenyl), n-Amylbenzene	5.0	Pseudomonas sp. ATCC 15523; Cellulomonas sp. ATCC 15528	$CH=CH-COOH$ (phenyl), Phenylacrylic acid (cinnamic acid)	5.0 mg/ml 100% molar conversion	n-Hexadecane for cell growth. Product also formed from C_3-C_{15} n-alkyl benzenes	63, 64
$CH_2-(CH_2)_3CH_3$ (phenyl), n-Amylbenzene	8.6	Alcaligenes sp. ATCC 15525	$CH_2-(CH_2)_3-COOH$ (phenyl), 5-Phenylvaleric acid	60% molar conversion		63
$CH_2-(CH_2)_5-CH_3$ (phenyl), n-Heptylbenzene	8.6	Micrococcus cerificans S-16	$CH=CH-COOH$ (phenyl), trans-Cinnamic acid	0.3 mg/ml	n-Hexadecane for cell growth	63
$CH_2-(CH_2)_{10}-CH_3$ (phenyl), n-Dodecylbenzene	0.44	Nocardia	CH_2-COOH (phenyl), Phenylacetic acid	81 mg/liter	n-Octadecane or n-decane for cell growth	55
$CH_2-(CH_2)_n-CH_3$ with CH_3 on ring, m-Tolyl paraffin, n = 5 or 7	8.6	Micrococcus cerificans S-16	A. CH_2-CH_2COOH with CH_3 on ring, m-Tolylpropionic acid; B. $CH=CH-COOH$ with CH_3 on ring, m-Tolylacrylic acid	A: 0.25 mg/ml B: 0.14 mg/ml	n-Hexadecane for cell growth	64

TABLE XVII
Products from Polycyclic Aromatic Hydrocarbons

Sec.	Substrate	Concentration (mg/ml)	Organism	Product	Yield	Remarks	Ref.
A	Naphthalene	—	*Pseudomonas desmolytica* *Nocardia* strain R	1,2-Dihydro-1,2-dihydroxy-naphthalene; Salicylic acid; Catechol	—	—	375, 390
		10	*Pseudomonas arvilla* *Pseudomonas rathonis*	A. 1,2-Naphthoquinone; B. Salicylic acid	Product B: 0.65 mg/ml	Product A results from a nonenzymatic reaction	247
		—	*Pseudomonas* sp.	Coumarin; o-Hydroxy-*trans*-cinnamic acid; Melilotic acid; Salicylic acid	—	—	77

A (contd)	Naphthalene (contd)	—	*Pseudomonas* sp.	α-Hydroxy-muconic semialdehyde; *cis-o*-Hydroxy-benzalpyruvate; Salicylaldehyde; Salicylic acid	—		53, 55
		—	*Nocardia* NRRL 3385	A. *d-trans*-1,2-Dihydro-1,2-dihydroxy-naphthalene; B. *o*-Hydroxy-benzalpyruvate	Product A: 5 mg/ml		393
		50	*Pseudomonas desmolytica*	Gentisic acid (2,5-dihydroxybenzoic acid)	—		258
B		10	*Pseudomonas aeruginosa*	Salicylic acid	9.3 mg/ml	Initial pH 8.0. High phosphate concentration Ca^{2+}, Cu^{2+}, and Fe^{2+} ions required.	169, 325

(Continued)

Table XVII (Continued)

Sec.	Substrate	Concentration (mg/ml)	Organism	Product	Yield	Remarks	Ref.
B (contd)	Naphthalene	60	Corynebacterium renale Pseudomonas aeruginosa Pseudomonas fluorescens Pseudomonas rathonis	Salicylic acid	25.6 mg/ml	pH Control and specific growth factors resulted in increased yields	28, 99, 125,151, 288,394, 404
		35	Pseudomonas aeruginosa Pseudomonas fluorescens		30 mg/ml	Increased population density, maintenance of population viability during production and product removal via agar or ion-exchange resin allow increased yields	168, 289
		—	Pseudomonas aeruginosa		6.4 mg/ml	Production via continuous culture.	167
		—	Pseudomonas fluorescens		206 mg/ml	Production via semicontinuous dialysis culture	2
C		10	Corynebacterium n. sp. ATCC 15570		2.2 mg/ml	Glucose used as growth substrate; chemical mutation resulted in 85% yield increase	117, 118
		—	Corynebacterium		5.7 mg/ml	Hexadecane used as growth substrate. Yields increase by use of anion-exchange resins	133

1-Methyl-naphthalene (CH₃ on naphthalene)	5	*Pseudomonas aeruginosa*	2-Hydroxy-3-methylbenzoic acid	77 mg/liter		292, 374
	5	*Pseudomonas aeruginosa* N₁-a	2-Hydroxy-4-methylbenzoic acid	65 mg/liter		292
	2.5	*Nocardia corallina*	β-Napthoic acid	1.6 mg/ml		153, 284, 292
2-Methyl-naphthalene (H₃C on naphthalene)	—	*Pseudomonas desmolytica*	(+)-*trans*-7-Methyl-1,2-dihydro-1,2-dihydroxynaphthalene; 2-Naphthyl-carbinol; 4-Methyl-salicylic acid; 4-Hydroxymethyl-salicylic acid		Trace of 2-naphthoic acid also present	33, 374

(Continued)

TABLE XVII (Continued)

Sec.	Substrate	Concentration (mg/ml)	Organism	Product	Yield	Remarks	Ref.
D (contd)	1,3-Dimethyl-naphthalene	2.5		1-Methyl-3-naphthoic acid	0.95 mg/ml		
	2,3-Dimethyl-naphthalene	4	Nocardia corallina	3-Methyl-2-naphthoic acid	0.2 mg/ml	n-Hexadecane for cell growth	284
	1,6-Dimethyl-naphthalene	2.5		1-Methyl-6-naphthoic acid	0.34 mg/ml		
	2,6-Dimethyl-naphthalene	—	Nocardia corallina Streptomyces achromogenes ATCC 15077	2-Methyl-6-naphthoic acid	2.7 mg/ml	n-Hexadecane or glucose for cell growth	67, 153, 289

Substrate		Organism	Product	Yield	Remarks	Ref.
2,7-Dimethyl-naphthalene (H₃C–naphthalene–CH₃)	—	*Nocardia corallina* *Streptomyces achromogenes* ATCC 15077	2-Methyl-7-naphthoic acid (H₃C–naphthalene–COOH)	6.2 mg/ml	*n*-Hexadecane for cell growth	67, 153, 289
2,7-Diethyl-naphthalene (CH₃CH₂–naphthalene–CH₂CH₃)	—	*Streptomyces achromogenes* ATCC 15077	2-Ethyl-2-naphthyl acetic acid (CH₃CH₂–naphthalene–CH₂COOH)	—	*n*-Hexadecane for cell growth	67
2,6-Dipropyl-naphthalene (CH₃CH₂CH₂–naphthalene–CH₂CH₂CH₃)	—		6-Propyl-2-naphthoic acid (HOOC–naphthalene–CH₂CH₂CH₃)	—	*n*-Hexadecane or glucose for cell growth	
1-(1-Naphthyl)-undecane (Naphthyl–CH₂–(CH₂)₉–CH₃)	—	*Nocardia opaca* T₁₆ *Nocardia opaca* P₂	3-(1-Naphthyl)-propionic acid (Naphthyl–CH₂–CH₂–COOH); 3-(1-Naphthyl)-acrylic acid (Naphthyl–CH=CH–COOH)	—	Glucose-grown resting cells	392
Naphthalene	60	*Nocardia coeliaca* ATCC 21146 *Nocardia n.* sp. ATCC 21145 *Streptomyces* ATCC 21147	4-(2-Hydroxyphenyl)-2-ketobutyric acid	23 mg/ml	Cerulose or hexadecane for cell growth. Use of dibasic anion-exchange resin to trap product	326

(Continued)

TABLE XVII (Continued)

Sec.	Substrate	Concentration (mg/ml)	Organism	Product	Yield	Remarks	Ref.
E (contd)	1-Methyl-naphthalene	—		4-(3-Methyl-2-hydroxy-phenyl)-2-ketobutyric acid	—	n-Hexadecane for cell growth	326
	1,2-Dimethyl-naphthalene	—	*Nocardia* n. sp. ATCC 21145	4-(3,4-Dimethyl-2-hydroxy-phenyl)-2-ketobutyric acid	—		
	1,3-Dimethyl-naphthalene	—	*Streptomyces* ATCC 21147	4-(3,5-Dimethyl-2-hydroxy-phenyl)-2-ketobutyric acid	—	Cerulose for cell growth	
	2-Ethyl-naphthalene	—		4-(4-Ethyl-2-hydroxy-phenyl)-2-ketobutyric acid	—	n-Hexadecane for cell growth	

F					
Anthracene	10	*Corynebacterium* n. sp. ATCC 15570	2-Hydroxy-3-naphthoic acid (structure: naphthalene with OH and COOH)	0.25 mg/ml	Glucose used for cell growth	117
Anthracene	—	*Flavobacterium* sp. *Pseudomonas aeruginosa*	A. 2-Hydroxy-3-naphthoic acid; B. Salicylic acid; C. 1,2-Dihydro-1,2-dihydroxyanthracene	Product A: 4.4 mg/liter	—	42, 43, 75, 291
Phenanthrene	5	*Flavobacterium* sp. *Pseudomonas aeruginosa* *Corynebacterium* n. sp. ATCC 15570	1-Hydroxy-2-naphthoic acid	0.53 mg/ml	Glucose used for cell growth of *Corynebacterium*	42, 290, 291

(*Continued*)

TABLE XVII (Continued)

Sec.	Substrate	Concentration (mg/ml)	Organism	Product	Yield	Remarks	Ref.
F (contd)	Phenanthrene	—	*Flavobacterium* sp.	1-Hydroxy-2-naphthoic acid; Salicylic acid; 3,4-Dihydro-3,4-dihydroxyphenanthrene	—	—	43
	Phenanthrene	—	—	7,8-Benzocoumarin; 3,4-Dihydroxyphenanthrene	—	Cell-free extracts employed.	75

naphthalene-utilizing organisms examined were found to employ this scheme. Recent studies by Gibson (94a) with *Pseudomonas putida* 39/D indicated that the *cis-*, not the *trans*-isomer of 1,2-dihydroxynaphthalene is the initial oxidation product. A more recent study (258), that reported formation of gentisic acid from naphthalene (Table XVII, Section A), suggests that alternate degradative pathways cannot be discounted.

The commercial potential for salicylic acid production stimulated efforts to increase fermentor yields. Initially, it was established that a highly buffered medium (2–2.5% K_2HPO_4), an initial pH of 8.0, and the presence of Cu^{++}, Fe^{++}, and Ca^{++} ions were essential for maximum salicylate accumulation (169, 247, 325). Under these conditions, 9.3 mg/ml of salicylic acid were formed from 10 mg/ml of naphthalene, representing 100% molar conversion of substrate to product (Table XVII, Section B). Subsequent work indicated pH to be a critical factor in the salicylate fermentation and that in place of an excessively buffered medium, pH control enabled higher yields of salicylic acid. The addition of 1.5% $CaCO_3$, to neutralize acidity from salicylate, and 1.75% NaCl permitted yields of 20 mg/ml (394). Further salicylate accumulation was prohibited by the toxicity of both Ca^{++} and salicylic acid. Removal of excess Ca^{++} by addition of Na_2SO_4 to the culture medium resulted in accumulation of 35–40 mg/ml (394). On a production scale, urea was found more practical for pH control than $CaCO_3$ or excessive buffering (151). In a kinetic study of salicylate formation Hosler (125) found that pH control at 6.5 favored maximum production. At a higher pH, salicylate was degraded rapidly; below the optimum pH less salicylate was produced.

Optimum salicylate production also was dependent on the presence of excess naphthalene (151, 371) since upon substrate exhaustion salicylic acid was rapidly degraded. The addition of both glycine (38 μg/ml) and calcium pantothenate (1.0 μg/ml) increased yields from 14.1 to 25.6 mg/ml (288). The use of aluminum and boron compounds also has been claimed to stimulate salicylate production by low-yielding strains of *Pseudomonas* (1.0 mg/ml salicylate in 4–6 days), however, their effectiveness in other high-yield salicylic acid fermentations was not established (99, 404). Loss of cell viability during the latter stages of the fermentation caused a decline in the salicylate production rate (289). Addition of growth factors such as yeast extract and trypticase prevented premature cell death and permitted continued salicylate synthesis with increased yields.

Hill and Gordon (118) isolated mutants of a *Corynebacterium* species that produced 85% more salicylic acid than the parent strain (Table XVII, Section C). Co-oxidation, utilizing glucose as the cosubstrate, further increased yields with this organism (117). However, neither mutation nor co-oxidation has been used with the higher-yielding *Pseudomonas* fermentations. Mutants of the latter organisms, capable of naphthalene but not salicylate oxidation, and resistant to high concentrations of salicylic acid, might produce exceptional amounts of product.

Failure to achieve further significant increases in salicylate yields has been traced to product inhibition by salicylate itself (168). This kinetic analysis of salicylic acid fermentation revealed that product accumulation rather than substrate limitation governed salicylate production. To minimize the effect of product toxicity anion-exchange resins were added directly to the culture medium in order to adsorb the salicylic acid. As a result, a threefold increase in total salicylate production and yields approaching 30 mg/ml were obtained (133,168, 371). Salicylate removal by adsorption to agar also led to yield increases (289).

The application of dialysis culture for salicylate removal during growth on napthhalene was explored by Abbott and Gerhardt (2). A dialysis culture vessel inoculated with *Pseudomonas fluorescens* was incubated for 72 hours during which time salicylic acid accumulated to 7 mg/ml. Dialysis then was initiated by filling the reservoir, separated from the culture vessel by a cellophane membrane, with mineral salts solution. Incubation was continued for an additional 96 hours before the reservoir fluid was replaced. Culture pH was maintained by the addition of $CaCO_3$ to the culture vessel and naphthalene was added to insure an excess. After 15 days and four reservoir changes, 31 gm of salicylic acid had accumulated from the 150-ml culture vessel (206 mg/ml) (Table XVII, Section B). The 31 gm, however, had dialyzed into 4400 ml of reservoir fluid so that in terms of total operating volume the effective yield was on the order of 7 mg/ml. Kinetic analysis of the above dialysis fermentation revealed that product formation became progressively more dependent on maintenance metabolism with time (3). Twenty-six of the 31 gm of salicylic acid produced (84%) was attributed to the maintenance metabolism of a large nonviable, but metabolically active cell population.

The cost and volume of ion-exchange resins and the mechanical limitations of dialysis culture make the current use of these techniques for large-scale commercial production of microbial products unlikely. The toxicity of many of the products of aromatic hydrocarbon metabolism may, however, necessitate development of adequate techniques such as these for efficient product removal and for the protection of the microbial population during fermentation.

Salicylate production via continuous fermentation also has been explored (167). Initial results, employing a tower fermentor divided into five vertical chambers by perforated plates, demonstrated the continuous formation of 6.4 mg/ml of salicylic acid from 5 mg/ml naphthalene at a dilution rate of 0.06 hour. (Table XVII, Section B). This value represented a 98% conversion of substrate to product and was 71% of the maximum concentration attained by batch fermentation (9 mg/ml).

Studies concerned with yield increases of other intermediates of naphthalene degradation have been limited. Wegner (393) reported the accumulation of 5 mg/ml of 1,2-dihydro-1,2-dihydroxynapthalene and higher than usual yields of o-hydroxybenzolpyruvate by a *Nocardia* strain (Table XVII, Section A).

b. Methylnaphthalenes

From naphthalene soil enrichments Gray and Thornton (101) originally isolated a strain of *Pseudomonas desmolyticum* capable of growth on 1- and 2-methylnaphthalenes. Sequential induction studies established that microbial attack of these substrates proceeded in a manner analogous to that of naphthalene degradation (292, 374). The primary oxidation products were 2-hydroxy-3-methyl- and 2-hydroxy-4-methylbenzoic acids, respectively. Ring oxidation of methyl-substituted naphthalenes was initiated on the unsubstituted ring with the resultant accumulation of detectable amounts of the methyl-substituted, dihydroxylated ring fission precursor (33) (Table XVII, Section D). In addition to direct ring oxidation, oxidation of the methyl substituents to alcohols and carboxylic acids with the formation of naphthoic acids (292, 374) and 2-naphthalcarbinol and 4-hydroxymethylsalicylic acid (32) have been indicated by sequential induction experiments (Table XVII, Section D). Ring oxidation and cleavage of naphthalenes containing substituents of both rings has not been reported.

Co-oxidation of methyl- and ethyl-substituted naphthalenes by hexadecane-grown *Nocardia* and *Streptomyces* generally results in the oxidation of only one of the methyl or ethyl substituents to the corresponding carboxylic acid (67, 153, 283). Direct oxidation of the naphthalene ring was not accomplished by these strains. Raymond *et al.* (283) reported the co-oxidation of 2-methylnaphthalene and the 2,6-, the 2,7-, the 1,3-, the 1,6-, and the 2,3-dimethylnaphthalenes to 2-naphthoic acid and the corresponding methyl-substituted naphthoic acids (Table XVII, Section D). A stearic effect of the positioning of the methyl groups was suggested by the observation that the *Nocardia* strains could not co-oxidize either 1-methylnaphthalene or the 1,5-, the 1,8-, and the 1,4-dimethylnaphthalenes.

Recently, a unique co-oxidation of various methyl- and ethylnaphthalenes by hexadecane-grown *Nocardia* and *Streptomyces* species was reported (326). These strains were capable of direct ring oxidation and fission with the resultant accumulation of methyl- or ethyl-substituted melilotic acid derivatives (Table XVII, Section E). Examination of Fig. 10 reveals the probable mechanism for formation of these compounds. These particular organisms were either incapable of conversion of *cis-o*-hydroxybenzalpyruvate to salicylaldehyde or the co-oxidation technique permitted only the partial oxidation of the initial ring fission product to the stable melilotic acid derivatives. Products yields from this fermentation were not reported. These products find application as plant growth regulators and as intermediates for synthesis of other regulators (327).

c. *Higher Polycyclic Aromatic Hydrocarbons*

Studies of polycyclic aromatic compounds containing three or more rings have been restricted primarily to anthracene and phenanthrene. Experimentation with these compounds has centered mainly on their metabolism in order to elucidate degradative pathways. Ultimately, the determination of the dissimilatory mechanisms of the carcinogenic polynuclear aromatic hydrocarbons such as the benzpyrenes is anticipated.

Tausson (363) first reported the oxidation of anthracene and phenanthrene by soil bacteria. On the basis of the oxidation rate of suspected intermediate products he postulated the degradation of phenanthrene by cleavage of the 9—10 bond to form two molecules of *o*-hydroxybenzyl alcohol which were subsequently oxidized to salicylic acid and degraded through catechol and muconic acid (364). Rogoff and Wender (290) and Colla *et al.* (42) discounted this mode of attack and demonstrated direct end-ring oxidation with the intermediate formation of 1-hydroxy-2-naphthoic acid (Table XVII, Section F). In addition, Colla *et al.* (43) detected the formation of 3,4-dihydro-3,4-dihydroxyphenanthrene and salicylic acid from a phenanthrene-oxidizing *Flavobacterium* species.

Sequential induction studies with anthracene-oxidizing strains also indicated end-ring oxidation (42, 43, 290) with the formation of analogous reaction products (Table XVII, Section E). Evans *et al.* (75) demonstrated the presence of various additional ring fission products from enzyme extracts of a *Pseudomonas* species. An enzyme extract from phenanthrene-induced cells oxidized 3,4-dihydroxyphenanthrene to *cis*-4-(1-hydroxynaphth-2-yl)-2-oxobut-3-enoic acid, 1-hydroxy-2-naphthaldehyde, and 1,2-dihydroxynaphthalene. Also, enzyme extracts of anthracene-induced cultures oxidized 1,2-dihydroxyanthracene to cis-4-(2-hydroxynaphth-3-yl)-2-oxobut-3-enoic acid, 2-hydroxy-3-naphthaldehyde, and 2,3-dihydroxynaphthalene. These intermediates suggested the dissimilatory pathways for phenanthrene and anthracene in Fig. 11.

One co-oxidation study (117) has demonstrated the formation of 0.25 mg/ml of 2-hydroxy-3-naphthoic acid from anthracene by glucose-grown *Corynebacterium* cells (Table XVII, Section F). Microbial products from higher aromatic polycyclic hydrocarbons such as benzpyrene, benzanthracene, dibenzanthracene, pyrene, benzperylene, and perylene have not been published even though microbial growth on and degradation of these compounds is well documented (76, 234, 251, 271–273).

3. Cycloalkanes, Cyclic Monoalkenes, and Terpenes

Very few reports have appeared on the utilization of cycloalkanes as the sole carbon source for microbial growth and product formation. Imelik (148) noted the production of cyclohexanol and adipic, valeric, and formic acids in cultures of *Pseudomonas aeruginosa* growing on cyclohexane (Table XVIII, Section A), but little experimental detail was given. Colla and Treccani (44) reported that a *Flavobacterium* species produced pimelic and adipic acids while growing on decalin and, more recently, Arai and Yamada (14) isolated a strain of *Alcaligenes faecalis* that was capable of growth on ethylcyclohexane. In the latter study, 4-ethylcyclohexanol was detected as an intermediate of ethylcyclohexane degradation (Table XVIII, Section A).

Ooyama and Foster (265), in an extensive study of cycloalkane oxidation by microorganisms, isolated a Gram-negative rod capable of growth on a wide variety of normal and branched alkanes, but incapable of growth on cycloalkanes. This isolate, however, co-oxidized a series of cycloalkanes and cyclic monoalkenes while using 2-methylbutane as a growth substrate (Table XVIII, Section B). Cyclopropane was the only cosubstrate susceptible to ring fission. Ring cleavage, in this case, was thought to be related to the intrinsic instability of this highly strained ring structure. Co-oxidation of the C_5 to C_8 cycloparaffins yielded various combinations of the corresponding epoxides, alcohols, and ketones. It was suggested that this co-oxidation proceeded from the cycloalkane or alkene to the epoxide through the alcohol to the ketone, but detailed supporting evidence was not presented. The substrate specificity of this co-oxidation is interesting in that both cyclohexane and 1-cyclohexene were co-oxidized, while 1,4-cyclohexane and benzene were resistant to attack.

Alkyl cycloalkanes are also co-oxidized by microorganisms. Davis

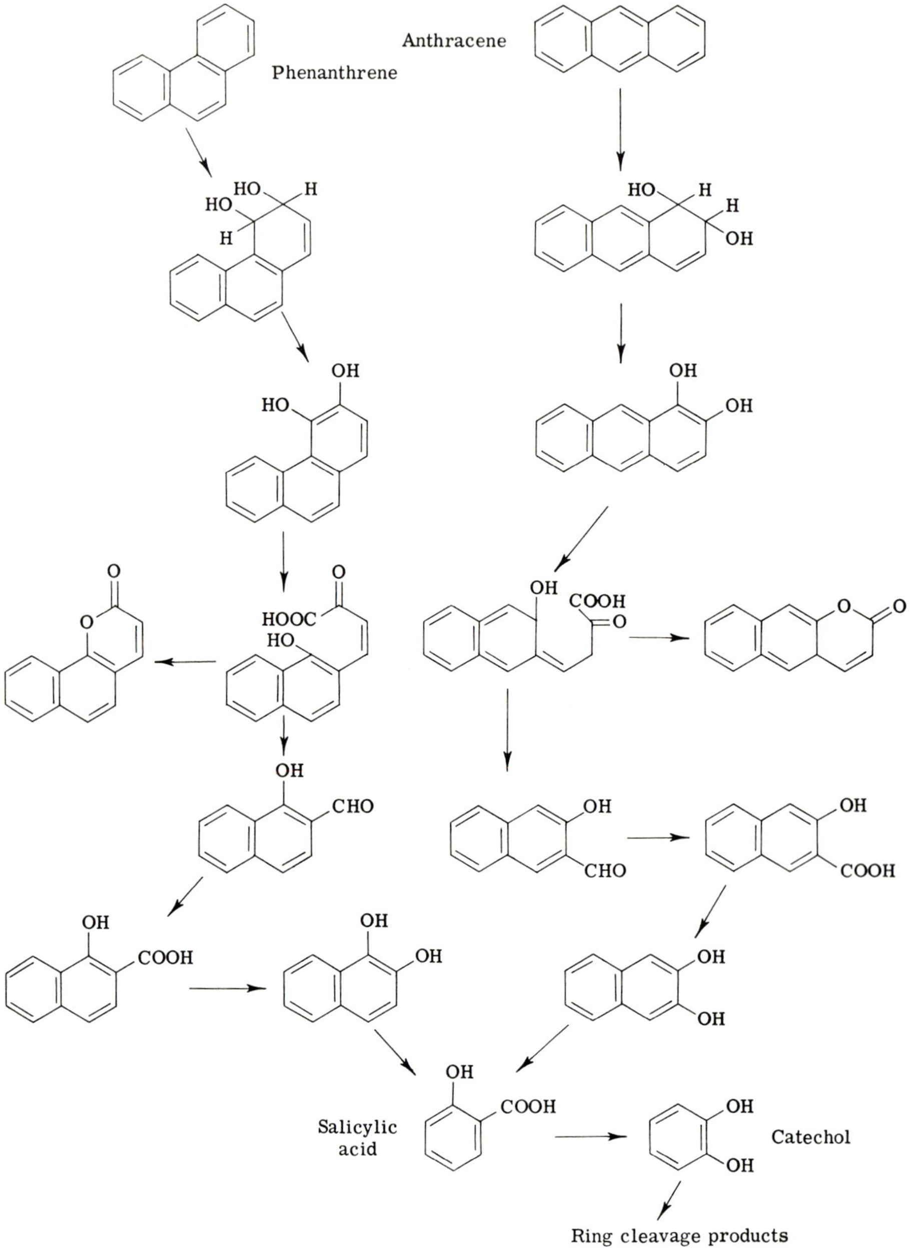

FIG. 11. Microbial dissimilation of phenanthrene and anthracene.

and Raymond (54, 55) found that a *Nocardia* species co-oxidized *n*-butylcyclohexane to cyclohexaneacetic acid (Table XVIII, Section C). Thus, the side chain and not the ring of the alkylcycloalkane was preferentially attacked. It is likely that a similar oxidation accounts for the side chain co-oxidation of alkylbenzenes.

Bicyclohexyl and related compounds represent another group of cycloalkanes that are susceptible to microbial oxidation. These substrates are co-oxidized by a large variety of fungi, bacteria, and actinomycetes. The products, which are hydroxylated derivatives of the substrate, are useful as insecticides, fungicides, protein denaturants, perfume bases, plasticizers, and dye and polymer intermediates.

Recently, microbial hydrocarbon research has been extended to the terpene area. Terpenoid compounds, chemicals whose basic structure is the isoprenoid unit (C_5), have widespread occurrence in nature and serve essential functions in each of the microbial, plant, and animal kingdoms. Carotenoids, chlorophylls, sterols, and certain essential plant oils, vitamins, and antibiotics are representative of terpenoid compounds. Research on microbial growth and oxidation of terpene hydrocarbons has centered primarily on the cyclic monoterpenes, containing two isoprenoid units.

Bhattacharyya's group in India initiated investigations into the microbial transformations of terpene hydrocarbons (22). Enrichment cultures from Czapek-Dox medium containing yeast extract, cornsteep liquor, and 0.5% α-pinene yielded a strain of *Aspergillus niger* capable of α-pinene co-oxidation. The mold was cultured in the above medium without α-pinene for 24 hours. The substrate was then added and incubation continued for an additional 8 hours. At this time the culture filtrate and mycelial mass were extracted and analysis of the combined extracts revealed three products: (1) *d-cis*-verbenol (20–25%), (2) *d*-verbenone (2–3%), and (3) *d-trans*-sobrerol (2–3%) (Table XIX, Section A). Similar products of α-pinene may arise by nonenzymatic auto-oxidation. However, the major reaction product of autooxidation is verbenone and the products are not optically pure. In subsequent studies the optimal substrate concentrations (0.6%), fermentation temperature (28.5°), and co-oxidation time (8 hours) were established. Also, four other fungal genera were isolated which were capable of α-pinene co-oxidation (274).

The transformation of carane and Δ^3-carene by *Aspergillus niger* was investigated with the goal of producing menthol or isomenthol (275). Carane was resistant to oxidation by this species, but low levels of an incompletely identified hydroxyketone were isolated from Δ^3-

TABLE XVIII
MICROBIAL PRODUCTS FROM CYCLOPARAFFINS AND CYCLIC MONOALKENES

Sec.	Substrate	Concentration (mg/ml)	Organism	Product	Yield	Remarks	Ref.
A	Cyclohexane	—	*Pseudomonas aeruginosa*	Cyclohexanol; Adipic acid; HCOOH - Formic acid; Valeric acid	—	Cyclohexane as sole carbon source	148
	Ethylcyclohexane	15	*Alcaligenes faecalis* 56B1	4-Ethylcyclohexanol	—	Ethyl cyclohexane as sole carbon source. Both *cis* and *trans* isomers formed	14
	Decalin	—	*Flavobacterium* sp.	Adipic acid; Pimelic acid	—	Decalin as sole carbon source	44
B	Cyclopropane	30% (v/v) in air	a	CH_3CH_2CHO Propionaldehyde	19 mg/liter		265
	Cyclopentane	0.3	a	Cyclopentanone	279 mg/liter	Co-oxidation 2-methyl butane as growth substrate	265

EXTRACELLULAR ACCUMULATION OF METABOLIC PRODUCTS 361

B (contd)					
Methyl-cyclopentane	1.5	a	A. 3-Methyl-cyclopentanone B. 3-Methyl-cyclopentanol	Product A: 300 mg/liter	
Cyclopentene	1.5	a	Cyclopentanone	10 mg/liter	Co-oxidation: 2-methyl butane as growth substrate 265
Cyclohexane	1.6	a	A. Cyclohexanol B. Cyclohexene epoxide	Product A: 142 mg/liter	
Methyl-cyclohexane	1.6	a	4-Methyl-cyclohexanone	250 mg/liter	
Cyclohexene	1.6	a	A. Cyclohexanone B. Cyclohexene epoxide	Product A: 114 mg/liter	
Cycloheptane	1.9	a	Cycloheptanone	37.6 mg/liter	

(Continued)

TABLE XVIII (Continued)

Sec.	Substrate	Concentration (mg/ml)	Organism	Product	Yield	Remarks	Ref.
B (contd)	Cyclooctane	2.0	a	Cyclooctanone	31.2 mg/liter	Co-oxidation: 2-methyl butane used as growth substrate	265
	Cyclooctene	2.0	a	Cyclooctanone	11 mg/liter		
C	$CH_2CH_2CH_2CH_3$ n-Butylcyclohexane	0.4	Nocardia salmonicolor	CH_2COOH Cyclohexane acetic acid	0.147 mg/ml	Co-oxidation: cell growth on n-octadecane	54, 55
	Bicyclohexyl	0.2	Phycomycetes Ascomycetes Basidiomycetes Deuteromycetes Streptomyces Corynebacterium Nocardia Mycobacterium Mycococcus Aerobacter	4,4'-Dihydroxy bicyclohexyl	60 mg/liter	Co-oxidation: growth in corn steep liquor-dextrose medium. Other oxygenated products present in lesser concentration	81
	Phenylcyclohexane	0.2	Streptomyces mediocidicus ATCC 13279 Sporotrichum sulfurescens ATCC 7159	4-Phenylcyclohexanol			

EXTRACELLULAR ACCUMULATION OF METABOLIC PRODUCTS   363

					Ref.
Phenylcyclohexane	0.2	*Rhizopus arrhizus* ATCC 11145	3-Phenylcyclohexanol	—	81
Cyclohexene	—	*Aspergillus niger* NCIM612	2-Cyclohexen-1-one, 2-Cyclohexen-1-ol, 3-Cyclohexen-1,2-diol	—	20
1-Methyl-cyclohexene	5	*Aspergillus niger* NCIM612	1-Methyl-1-cyclohexanol, 1-Methyl-1-cyclohexen-6-one, 1-Methyl-1-cyclohexen-6-ol	—	91
3-Methyl-cyclohexene	5	*Aspergillus niger* NCIM612	1-methyl-3-cyclohexen-2-one, 1-Methyl-3-cyclohexen-2-ol	—	
Tetralin	5	*Aspergillus niger* NCIM612	α-Tetralone, α-Tetralol	—	

Co-oxidation: cell growth in complex medium

[a] Small, nonmotile, Gram-negative rod (Job 5).

carene-grown cultures (Table XIX, Section A). α-Santalene, a sesquiterpene (i.e., containing three isoprenoid units), was oxidized to tere-santalic acid, tere-santalol, and a tertiary alcohol (Table XIX, Section A).

Further investigations with *Aspergillus niger* (19, 21) revealed that 2-nonene-2,3-dicarboxylic acid anhydride was formed from either camphene longifolene, β-santalone, caryophyllene, or σ-cadinene (Table XIX, Section A). It was determined, however, from tracer studies with glucose-U-^{14}C and unlabeled longifolene that the anhydride could be derived entirely from glucose without the preferential incorporation of any partial degradation product of longifolene. Nevertheless, the concentration of this product increased several thousandfold in the presence of terpenes. From a teleological viewpoint, the formation of increased anhydride levels in the presence of terpene substrates may provide a detoxification mechanism for terpenes and their degradation products.

Examination of the products formed by *Aspergillus niger* from various terpene hydrocarbons suggested that the substrate could be oxidized in three ways: (a) hydration of the double bond, (b) oxygenation at a position allylic to the double bond, and (c) oxygenation of the double bond. Cyclohexene was chosen as a model substrate for further studies on the mechanisms of these oxygenations. *Aspergillus niger* growing on cyclohexene produced three previously unreported reaction products (Table XVIII, Section D): (a) 2-cyclohexen-1-one, (b) (+)-2-cyclohexene-1-ol, and (c) (+)-3-cyclohexene-1,2-*cis*-diol. Thus, the fungal attack of cyclohexene unlike the bacterial oxidation studied by Ooyama and Foster (265) did not result in oxidation of the double bond. Based on similarities to certain chemical oxidations of cyclohexene a mechanism for the enzymatic oxidation of this substrate to the diol was presented (20).

Other model systems employing 1-methyl- and 3-methylcyclohexenes revealed both allylic (i.e., adjacent to the double bond) oxidation and hydration of the double bond (91). Thus, transformation of 1-methylcyclohexene yielded 1-methyl-1-cyclohexanol, 1-methyl-1-cyclohexene-6-one, and 1-methyl-1-cyclohexene-6-ol; 3-methylcyclohexene oxidation yielded 1-methyl-3-cyclohexene-2-ol and 1-methyl-3-cyclohexene-2-one (Table XVIII, Section D).

A unique feature of this fungal enzyme system was the hydroxylation of a double bond adjacent to a methyl substituent, as in the formation of 1-methyl-6-cyclohexanol, and the allylic oxidation to double bonds not adjacent to a methyl substituent, as with the oxidation of 3-methylcyclohexene. Another novel aspect of this system was the allylic oxidation and preferential attack of the unsaturated, and not the saturated, ring of tetralin (Table XVIII, Section D). In other studies, the fungal system was replaced by a *Pseudomonas* species capable of growth on a variety of terpenes as its sole carbon source (59). This organism oxidized limonene to seven neutral metabolites that were incapable of supporting further growth or respiration and four acidic products that were readily metabolized (Table XIX, Section B). Based on sequential induction studies and cell-free enzyme reactions (60) three pathways for limonene oxidation were postulated (Fig. 12). These involved (a) allylic oxidation (pathway 1), (b) oxidation of the 1—2 double bond (pathway 2), and (c) oxidation of the methyl group to perillic acid followed by hydroxylation of the double bond and ring cleavage (pathway 3). Most of the limonene was metabolized through the third pathway yielding carbon and energy for cell growth. Co-oxidation of limonene by glucose-grown cells indicated that the pathway-3 compounds, perillic acid and perillyl alcohol, were the major products (Table XIX, Section B).

Another soil pseudomonad, the PL strain, grew well on α- and β-pinenes, limonene, 1-p-menthene, and p-cymene, as sole carbon sources (308). The pinenes, in contrast to limonene and p-cymene, were toxic to the cells, and, therefore, had to be added incrementally during the fermentation. Fermentation of α-pinene yielded a wide variety of neutral and acidic products (Table XIX, Section B) (308). The oxidation of β-pinene yielded similar products except 4-hydroxyphellandric acid was formed instead of 4-hydroxy-dihydrophellandric acid. Perillic acid and 4,9-dihydroxy-1-p-menthen-7-oic acid were the major metabolites of both α- and β-pinene and limonene oxidation by the *Pseudomonas* strain. In other studies six to seven different metabolic pathways were implicated in the metabolism of α- and β-pinenes (307). The pathways involved either initial oxidation of the 7-methyl group or initial protonation and cleavage of the cyclobutane ring to form a p-menthene derivative.

Microbial growth on p-cymene as a sole carbon source has been reported for several *Pseudomonas* species by workers outside of Bhattacharyya's group (124, 222, 224). Horiguchi and Yamada (124) found that cumic acid accumulated to 19 mg/ml in media containing yeast extract and optimal concentration of Ca^{++}, Mn^{++}, and Fe^{++} ions (Table XIX, Section C). Davis and Raymond (54, 55) demonstrated the production of lesser quantities of cumic acid by hexadecane-grown *Nocardia* cells co-oxidizing p-cymene (Table XIX, Section C). *Pseudomonas* (PL strain) (230), however, accumulated a variety of neutral and acidic products from p-cymene (Table XIX, Section C) represented by

TABLE XIX
MICROBIAL PRODUCTS FROM TERPENE HYDROCARBONS[a]

Sec.	Substrate	Concentration (mg/ml)	Organism	Product	Yield	Remarks	Refs.
A	α-Pinene	5	Aspergillus niger NC1M612	d-Verbenone; d-cis-Verbenol; d-trans-Sorbrerol	1 mg/ml Total products	Growth in Czapek-Dox medium with cornsteep liquor and yeast extract	22, 274
	Δ³-Carene	5	Aspergillis niger NC1M612	(structures) or (structures)	—	Growth medium as for α-pinene. Product not fully identified	275
	α-Santalene	1.7	Aspergillis niger NC1M612	Teresantalic acid; Teresantalol; Intermediate oxidation product	0.25 mg/ml Total products.		
	Camphene	—	Aspergillis niger NC1M612		—		21

A (contd)	Longifolene	—	*Aspergillus niger* NC 1M612		—		21
	β-Santalene	—	*Aspergillus niger* NC 1M612	2-Nonene-2,3 dicarboxylic acid anhydride $CH_3-(CH_2)_5$ ... H_3C	—	Growth medium as for α-pinene	21
	(iso)-Caryophyllene	—	*Aspergillus niger* NC 1M612		—		19
	δ-Cadinene	—	*Aspergillus niger* NC 1M612		—		19

(*Continued*)

TABLE XIX (*Continued*)

Sec.	Substrate	Concentration (mg/ml)	Organism	Product			Yield	Remarks	Refs.
B	Limonene	8.4	*Pseudomonas* sp. strain (L)	Carvone	Dihydrocarvone	Carveol	—	Limonene as sole carbon source	59, 60
				8-*p*-Menthene-1,2-*cis*-diol	8-*p*-Menthen-1-ol-2-one	8-*p*-Menthene-1,2-*trans*-diol			
				1-*p*-Menthene-6,9-diol	Perillic acid	β-Isopropenyl pimelic acid			

B (contd)	Limonene	4.2	*Pseudomonas* sp. strain (L)	2-Hydroxy-8-*p*-menthen-7-oic acid; 4,9-Dihydroxy-1-*p*-menthen-7-oic acid; Perillic acid; Perillyl alcohol	—	Co-oxidation: glucose-grown cells	60
	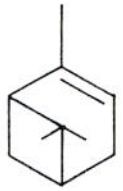 α-Pinene	8.6	*Pseudomonas* sp. strain (PL)	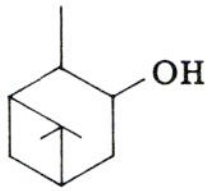 Borneol; 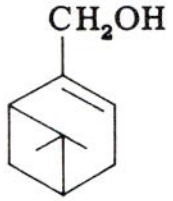 Myrtenol; 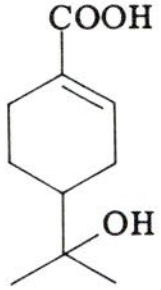 Oleuropeic acid		α-Pinene as sole carbon source	308

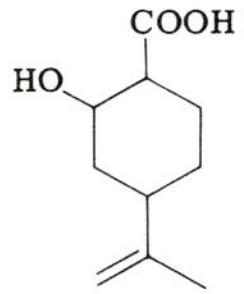

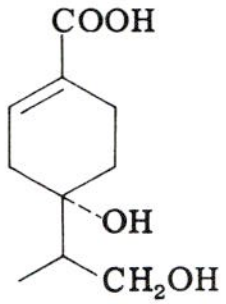

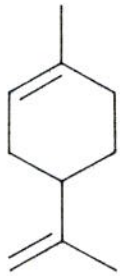

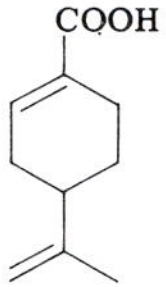

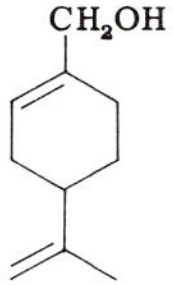

(*Continued*)

369

TABLE XIX (*Continued*)

Sec.	Substrate	Concentration (mg/ml)	Organism	Product	Yield	Remarks	Refs.
B (contd)	α-Pinene (contd)	8.6	*Pseudomonas* sp. strain (PL)	β-Isopropyl-pimelic acid; Myrtenic acid; Phellandric acid; Perillic acid; 4-Hydroxy-dihydro-phellandric acid; 4,9-Dihydroxy-1-*p*-menthen-7-oic acid	—	α-Pinene as sole carbon source	308
	β-Pinene	8.6	*Pseudomonas* sp. strain (PL)	Same products as with α-pinene except 4-hydroxyphellandric acid is formed instead of the dihydro compound. 4-Hydroxy-phellandric acid	—	β-Pinene as sole carbon source	308

c	p-Cymene	8.6	*Pseudomonas* sp. strain (PL)	Cumyl alcohol; Cumic acid; 3-Hydroxycumic acid; 2,3-Dihydroxycumic acid; β-Isopropylpyruvic acid; α-p-Tolyl-n-propanol; α-p-Tolylpropionic acid	—	p-Cymene as sole carbon source	230
	p-Cymene	43	*Pseudomonas* sp. *Pseudomonas desmolytica*	Cumic acid	19 mg/ml	p-Cymene as sole carbon source. Yeast extract stimulates yields	124, 222, 224

(*Continued*)

TABLE XIX (Continued)

Sec.	Substrate	Concentration (mg/ml)	Organism	Product	Yield	Remarks	Refs.
C (contd)	p-Cymene	3	Nocardia salmonicolor	Cumic acid (COOH-C6H4-iPr)	0.26 mg/ml	Co-oxidation: n-hexadecane for cell growth.	55, 59
D	1-p-Menthene	5.1	Pseudomonas sp. strain (PL)	β-Isopropyl-pimelic acid; Methylisopropyl-ketone	—	Concentrated cell suspension used to inoculate medium. p-Menthene as sole carbon source	134
	1-p-Menthene	—	Cladosporium sp. T₁	trans-p-Menthene-1,2-diol	—	d-Limonene as sole carbon source	246

D (contd)	d-Limonene	—	Cladosporium sp. T₇	trans-Limonene-1,2-diol	1.5 mg/ml	d-Limonene as sole carbon source. Minor quantities of cis-diol also present	246
	3-Menthene	—	Cladosporium sp. T₈	trans-p-Menthene-3,4-diol	—	3-Menthene as sole carbon source	
	d-Limonene	—	Cladosporium sp. T₁₂	d-α-Terpineol	1.0 mg/ml	d-Limonene as sole carbon source	128

[a] In Table XIX, methyl or methylene groups on ring structures are not drawn, but their presence is indicated by lines or parallel lines extending from the ring into space.

373

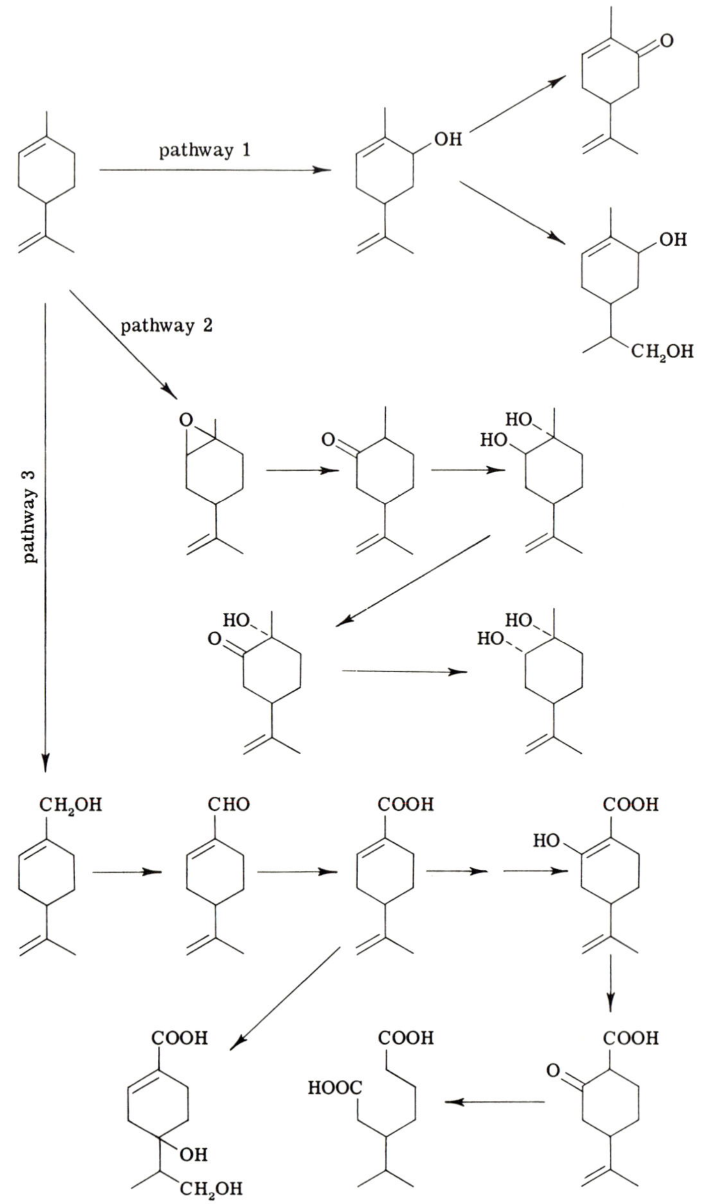

FIG. 12. Limonene oxidation by a soil *Pseudomonas* (strain L).

cumyl alcohol and α-*p*-tolyl-*n*-propanol as the major neutral products and cumic acid and 2,3-dihydroxycumic acid as the main acidic products. In addition, the co-oxidation of *p*-cymene by glucose-grown cells yielded dihydroxycumic acid. Sequential induction studies demonstrated two pathways for p-cymene metabolism (231). The main energy-yielding pathway proceeded from p-cymene → cumic acid → 3-hydroxycumic acid → 2,3-dihydroxycumic acid → β-isopropylpyruvate and to CO_2 and H_2O. The second pathway yielded p-tolylpropionic acid as the end product.

Recent studies by Bhattacharyya's group (134) have shown that microbial growth on 1-p-menthene results in the accumulation of β-isopropylpimelic acid and methylisopropylketone (Table XIX, Section D). Hill *et al.* (179, 246) reported the isolation of various *Cladosporium* strains capable of growth on several terpenes. Production of cyclic vicinal glycols was observed in cultures grown on 1-menthene, 3-menthene, and d-limonene (Table XIX, Section D). The *trans*-limonene 1,2-diol accumulated to 1.5 mg/ml. Another *Cladosporium* species converted d-limonene into d-α-terpineol at a yield of 1.0 mg/ml (128) (Table XIX, Section D). Several other bacterial and fungal cultures were found subsequently to accumulate this product.

Examination of the terpene hydrocarbon research may be indicative of the variety and scope of material which can be obtained when detailed investigations of microbial conversions of other specific chemical groups are undertaken. The application of knowledge from one group of compounds, such as the terpenes, can often be translated to other classes, such as the cycloalkenes, with the resultant discovery of totally new products. Nutritional mutants should then provide a means for increasing yields of the desired intermediates. The varied reaction mechanisms, vast array of products, and product uniqueness and functionality will continue to stimulate interest in terpene metabolism and will continue to provide a fruitful area for both academic and industrial research.

IV. Concluding Remarks

It is apparent that many of the metabolic products derived from nonhydrocarbon substrates can also be obtained from hydrocarbons. In addition, many products can be produced from hydrocarbons, via one- or two-step transformation reactions, that cannot be accumulated from nonhydrocarbon substrates. Several classes of compounds, however, rarely are found among hydrocarbon metabolities. These include the solvents (e.g., acetone and butanol), carbohydrate transformation products (e.g., gluconic acid), and antibiotics. Solvents and reduced organic acids are products of anaerobic metabolism. Thus, their absence is not surprising since microbial growth on hydrocarbons is generally considered as obligately aerobic. The absence of antibiotic production during hydrocarbon metabolism is not as readily understood. Pharmaceutical companies have undoubtedly screened for antibiotic producers that utilize hydrocarbons or have made attempts to substitute hydrocarbons in existing processes. The results of these studies have not been disclosed (see Section III, A for an exception) suggesting that they were not successful, although there is no particular aspect of hydrocarbon metabolism that would preclude antibiotic synthesis.

A variety of special techniques have been employed to increase the production of hydrocarbon metabolites. Most of these have been utilized successfully in nonhydrocarbon fermentations, although several are particularly amenable to hydrocarbon substrates. Of the latter, co-oxidation is perhaps the most promising. It offers a wide variety of potentially useful products and the possibility of 100% molar yields. Co-oxidation by itself is not sufficient to insure a high-yield fermentation. Toxicity and inhibition by product and/or substrate must also be minimized. Inhibitory co-oxidation products can often be trapped by ion-exchange resins enabling greater product yields, however, the resins may also remove important medium constituents. Dialysis or ultrafiltration provides an alternative method for removal of inhibitory metabolites. The latter are particularly attractive in that they permit removal of water soluble products from the fermentation without loss of the water insoluble hydrocarbon substrate or metabolically active cells.

Long-chain liquid paraffins can often be incorporated into fermentation media in large excess without detrimental effects, but substrate toxicity is usually encountered with the short-chain paraffins and cyclic hydrocarbons. This toxicity can be circumvented by maintaining the substrate concentrations at sufficiently low levels. In practice this may be difficult to do since substrate is utilized rapidly and only small excesses are toxic. Volatile substrates may be added as a vapor by sparging with air into the fermentor. Vapor addition would also facilitate dispersion and mass transfer of the hydrocarbon to the culture. Another approach is to embed the substrate in a matrix so that it is released slowly. Mabe and Brannon (229) embedded various aromatic hydrocarbons in agar cubes and improved pyrrolnitrin production by a pseudomonad. Less expensive materials such as higher-molecular-weight paraffins might replace agar. Alternatively, hydrocarbons might be added by diffusion through hydrophobic membranes in a manner analogous to solute diffusion through dialysis membranes. Clearly, methods are needed to minimize both substrate and product toxicity, and future progress in the area of hydrocarbon fermentation will reflect how effectively this is accomplished.

One of the least exploited yet potentially valuable techniques for accumulating hydrocarbon transformation products is the use of mutant organisms. Various auxotrophs have been employed successfully for amino acid production from paraffins, but mutants blocked at specific points in alkane degradation pathways rarely have been isolated. In some instances mutation has been used, but without selective enrichment for specific genotypes. Mutants blocked at a specific point in a metabolic pathway used exclusively for hydrocarbon dissimilation should accumulate an intermediate that precedes the block. Knowledge of the pathway by which the substrate is degraded can lead to the design of rational selective enrichment techniques for the desired mutant. These mutants would be particularly useful since they would maintain the ability to grow on a wide variety of nonhydrocarbons. This type of fermentation, therefore, is quite similar to co-oxidation except that mutation would greatly expand the number and variety of products which could be accumulated.

Permeability-altering agents have found wide use in hydrocarbon-product fermentations. They are usually essential for maximum yields and for adequate emulsification of the substrate. In several fermentations the hydrocarbon itself appeared to increase cellular permeability in addition to serving as substrate. The interaction between hydrocarbon and microorganisms may result in the extraction of lipids from the cell membrane thereby increasing permeability. This could also account for the toxicity of the cyclic and short-chain paraffins since they generally are good lipid solvents. Greater permeability may be an added benefit in many hydrocarbon fermentations. For this reason it is important to establish an optimal substrate concentration. The concentration that favors product excretion likely will be higher than that required for best growth. But by establishing these optima and employing them at the appropriate times during the course of a fermentation higher productivities should be realized.

Several developments can be anticipated from future research in hydrocarbon fermentation. The isolation of new hydrocarbon-metabolising organisms will lead to new pathways of degradation which may contain new and commercially useful intermediates. As an example, anaerobic hydrocarbon oxidation has been reported but the pathway of degradation has not been firmly established. When this is done intermediates may be found increasing the number of potential products for new fermentations. Novel screening techniques will also play an important role in uncovering products and improving yields. The procedure of Klein *et al.* (171) illustrates one approach (Section II,A,5). They reasoned that an organism that grew poorly on a hydrocarbon might only partially oxidize the substrate. Ketones were subsequently found to accumulate in a culture of an organism isolated by this "poor-growth" criterion.

Enzymatic cell-free hydrocarbon oxidation is another area which should increase in importance in coming years. As the mechanism of action of oxygenases is revealed attempts will be made to commercially exploit this important class of enzymes. Improved enzyme stability will be an essential part of this goal.

Terpenes and other cyclic hydrocarbons comprise a large number of substrates from which a vast array of products can be derived. Although few of these products accumulate in substantial quantities, further studies should reveal ways to achieve high yields. There are numerous cyclic compounds which have not been examined as substrates for microbial growth and product production. These provide a reservoir for future research and sustained interest in cyclic hydrocarbon fermentation.

Acknowledgments

The authors are grateful to Dr. A. I. Laskin for his helpful discussions, criticisms, and encouragement and to Dr. A. L. Demain for reviewing the manuscript.

References

1. Abbott, B. J., and Casida, L. (1968). *J. Bacteriol.* **96**, 925–930.
2. Abbott, B. J., and Gerhardt, P. (1970). *Biotechnol. Bioeng.* **12**, 577–589.
3. Abbott, B. J., and Gerhardt, P. (1970). *Biotechnol. Bioeng.* **12**, 591–601.
4. Abe, M., and Tabuchi, T. (1968). *Agr. Biol. Chem.* **32**, 392–393.
5. Abe, M., Tabuchi, T., and Tanaka, M. (1970). *J. Agr. Chem. Soc. Jap.* **44**, 493–498.
6. Aiba, S., Humphrey, A. E., and Millis, N. (1965). "Biochemical Engineering." Univ. of Tokyo Press, Tokyo.
7. Aida, T., and Yamaguchi, K. (1969). *Agr. Biol. Chem.* **33**, 1244–1250.
8. Ajinomoto Co. (1965). British Patent 996,544.
9. Ajinomoto Co. (1968). British Patent 1,106,956.
10. Ajinomoto Co. (1969). French Patent 2,003,199.
11. Ali Khan, M., Hall, A., and Robinson, D. (1963). *Nature (London)* **198**, 289.
12. Ali Khan, M., Hall, A., and Robinson, D. (1964). *Antonie van Leeuwenhoek, J. Microbiol. Serol.* **30**, 417–427.
13. Anonymous. (1970). *Chem. Eng. News* **48** (May 25), 18–20.
14. Arai, Y., and Yamada, K. (1969). *Agr. Biol. Chem.* **33**, 63–68.
15. Arima, K., and Shigeo, H. (1970). Japan Patent 7,024,392.
16. Atkinson, J. H., and Newth, F. H. (1967). "Symposium on Microbiology," pp. 1–10. Bartholomew Press, Dorking, England.
17. Azoulay, E., and Senez, J. C. (1960). *Ann. Inst. Pasteur, Paris* **98**, 868–879.
18. Beerstecher, E. (1954). "Petroleum Microbiology." Elsevier, Amsterdam.
19. Bhattacharyya, P. K., and Dhavalikar, R. S. (1965). *Indian J. Biochem.* **2**, 73–76.

20. Bhattacharyya, P. K., and Ganapathy, K. (1965). *Indian J. Biochem.* **2**, 137–145.
21. Bhattacharyya, P. K., Prema, B. R., Dhavalikar, R. S., and Ramachandran, B. V. (1963). *Indian J. Biochem.* **1**, 171–176.
22. Bhattacharyya, P. K., Prema, B. R., Kulbarni, B. D., and Pradhan, S. K. (1960). *Nature (London)* **187**, 689–690.
23. Birch-Hirschfeld, L. (1932). *Zentralbl. Bakteriol., Parasitenk., Infektionskr. Hyg., Abt. 2* **86**, 114.
24. Bird, C. W., and Molton, P. (1967). *Biochem. J.* **104**, 987–990.
25. Bird, C. W., and Molton, P. (1969). *Biochem. J.* **114**, 881–884.
26. Blakebrough, N. (1967). "Biochemical and Biological Engineering Science," Vol. 1. Academic Press, New York.
27. Blau, L. W. (1942). U.S. Patent 2,269,899.
28. Brillaud, A. R. (1965). U.S. Patent 3,183,169.
28a. Brillaud, A. R. Canadian Patent 741,298.
29. Bruyn, J. (1954). *Proc., Kon. Ned. Akad. Wetensch., Ser. C* **57**, 41–45.
30. Buddin, W. (1915). *J. Agr. Sci.* **6**, 417–451.
31. Bushnell, L. D., and Haas, H. F. (1941). *J. Bacteriol.* **41**, 653–673.
32. Caglar, M. A., Thompson, A. R., and Houston, C. W. (1970). *Abstr. Pap., 160th Nat. Meet., Amer. Chem. Soc.*
33. Canonica, L., Fiecchi, A., and Treccani, V. (1957). *Rend. Inst. Lomb. Sci.* **91**, 119–129.
34. Casida, L. E. (1956). U.S. Patent 2,771,396.
35. Champagnat, A., Vernet, C., Laine, B., and Filosa, J. (1963). *Nature (London)* **197**, 13–14.
36. Chas. Pfizer Co. (1970). British Patent 1,203,006.
37. Chouteau, J., Azoulay, E., and Senez, J. C. (1962). *Bull. Soc. Chim. Biol.* **44**, 671–677.
38. Ciegler, A., Arnold, M., and Anderson, R. (1959). *Appl. Microbiol.* **7**, 98–101.
39. Ciegler, A., Hall, H., and Nelson, G. (1962). U.S. Patent 3,025,221.
40. Ciegler, A., Nelson, G., and Hall, H. (1962). *Appl. Microbiol.* **10**, 132–136.
41. Claus, D., and Walker, N. (1964). *J. Gen. Microbiol.* **36**, 107–122.
42. Colla, C., Biaggi, C., and Treccani, V. (1957). *Atti Accad. Naz. Lincei* **23**, 66–69.
43. Colla, C., Fiecchi, A., and Treccani, V. (1959). *Ann. Microbiol. Enzimol.* **9**, 87–91.
44. Colla, C., and Treccani, V. (1960). *Ann. Microbiol. Enzimol.* **10**, 77–81.
45. Coty, V. F. (1967). *Biotechnol. Bioeng.* **9**, 25–32.
46. Coty, V. F. (1967). U.S. Patent 3,326,770.
47. Coty, V. F. (1968). U.S. Patent 3,387,964.
48. Daoust, D., and Stoudt, T. (1966). *Develop. Ind. Microbiol.* **7**, 22–34.
49. Darlington, W. A. (1964). *Biotechnol. Bioeng.* **6**, 241–242.
50. Davies, J. I., and Evans, W. C. (1962). *Biochem. J.* **85**, 21p–22p.
51. Davies, J. I., and Evans, W. C. (1964). *Biochem. J.* **91**, 251–261.
52. Davis, J. B. (1956). *Bacteriol. Rev.* **20**, 261–264.
53. Davis, J. B. (1967). "Petroleum Microbiology." Elsevier, Amsterdam.
54. Davis, J. B., and Raymond, R. L. (1961). *Appl. Microbiol.* **9**, 383–388.
55. Davis, J. B., and Raymond, R. L. (1962). U.S. Patent 3,057,784; British Patent 1,036,084.
56. Davis, J. B., and Updegraff, D. M. (1954). *Bacteriol. Rev.* **18**, 215–238.
57. Davis, R. S., Hossler, F. E., and Stone, R. W. (1968). *Can. J. Microbiol.* **14**, 1005–1009.
58. Demain, A. L., and Birnbaum, J. (1968). *Curr. Top. Microbiol. Immunol.* **46**, 1–25.

59. Dhavalikar, R. S., and Bhattacharyya, P. K. (1966). *Indian J. Biochem.* **3**, 144–157.
60. Dhavalikar, R. S., Rangachari, P. N., and Bhattacharyya, P. K. (1966). *Indian J. Biochem.* **3**, 158–164.
61. Dikanskaya, E., and Balabanova, A. (1966). *Int. Congr. Microbiol., Abstr. Pap. 9th, 1966* p. 258.
62. Douros, J. D., Brillaud, A. R., and Eltz, R. W. (1965). U.S. Patent 3,201,323.
63. Douros, J. D., and Frankenfeld, J. W. (1967). U.S. Patent 3,301,766.
64. Douros, J. D., and Frankenfeld, J. W. (1968). *Appl. Microbiol.* **16**, 320–325.
65. Douros, J. D., and Frankenfeld, J. W. (1968). *Appl. Microbiol.* **16**, 532–533.
66. Douros, J. D., and Raymond, R. L. (1965). U.S. Patent 3,219,543.
67. Douros, J. D., and Raymond, R. L. (1967). U.S. Patent 3,340,155.
68. Dworkin, M., and Foster, J. W. (1956). *J. Bacteriol.* **72**, 646–659.
69. Eisenberg, M. (1962). *Biochem. Biophys. Res. Commun.* **8**, 437–441.
70. Eisenberg, M. (1963). *J. Bacteriol.* **86**, 673–680.
71. Elsford, H., and Wright, L. (1962). *Fed. Proc., Fed. Amer. Soc. Exp. Biol.* **21**, 467.
72. Enmakova, I. T., Rosenfeld, S. M., Novakovskaya, N. S., Neklyndova, L. V., and Disler, E. N. (1969). *Prikl. Biokhim. Mikrobiol.* **5**, 252–255.
73. Esso Research and Engineering Company. (1968). British Patent 1,130,180.
74. Evans, W. C. (1963). *J. Gen. Microbiol.* **32**, 177–184.
75. Evans, W. C., Fernley, H. N., and Griffiths, E. (1965). *Biochem. J.* **95**, 819–831.
76. Fedoseeva, G. E., Khesina, A. J., Poglazova, M. N., Shabad, L. M., and Meisel, M. N. (1968). *Dokl. Akad. Nauk SSSR* **183**, 208–211.
77. Fernley, H. N., and Evans, W. C. (1958). *Nature (London)* **182**, 383–375.
78. Finnerty, W. R., Hawtrey, E., and Kallio, R. E. (1962). *Z. Allg. Mikrobiol.* **2**, 263–266.
79. Finnerty, W. R., Hawtrey, E., and Kallio, R. E. (1962). *Z. Allg. Mikrobiol.* **2**, 169–177.
80. Focht, D. D., and Alexander, M. (1970). *Appl. Microbiol.* **20**, 608–611.
81. Fonkin, G. S., Herr, M. E., and Murray, H. C. (1966). U.S. Patent 3,281,330.
82. Forro, J. R. (1966). Ph.D. Thesis, Pennsylvania State University, University Park, Pennsylvania.
83. Forro, J. R., and Stone, R. W. (1965). *Bacteriol. Proc.* p. 90.
84. Foster, J. W. (1962). In "Oxygenases" (O. Hayaishi, ed.), pp. 241–271. Academic Press, New York.
85. Foster, J. W. (1963). Canadian Patent 664,587.
86. Fredericks, K. (1967). *Antonie van Leeuwenhoek, J. Microbiol Serol.* **33**, 41–48.
87. Fujii, K., Shimizu, S., and Fukui, S. (1966). *J. Ferment Technol.* **44**, 185–191.
88. Fukui, S., Shimizu, S., and Fujii, K. (1967). *J. Ferment. Technol.* **45**, 530–540.
89. Fukui, S., and Tanaka, A. (1968). *Hakko Kyokaishi* **26**, 209–218.
90. Furukawa, T., Nakahara, T., and Yamada, K. (1970). *Agr. Biol. Chem.* **34**, 1833–1838.
91. Ganapathy, K., Khanchandani, K. S., and Bhattacharyya, P. K. (1966). *Indian J. Biochem.* **3**, 66–70.
92. Gerhardt, P., and Bartlett, M. C. (1959). *Advan. Appl. Microbiol.* **1**, 215–260.
93. Gholson, R. K., Baptist, J. N., and Coon, M. J. (1963). *Biochemistry* **2**, 1150–1180.
94. Gibson, D. T. (1968). *Science* **161**, 1093–1097.
94a. Gibson, D. T. (1971). Personal communication.
95. Gibson, D. T., Cardini, G. E., Maseles, F. C., and Kallio, R. E. (1970). *Biochemistry* **9**, 1631–1635.
96. Gibson, D. T., Hensley, M., Yoshioka, H., and Mabry, T. J. (1970). *Biochemistry* **9**, 1626–1630.

97. Gibson, D. T., Koch, J. R., and Kallio, R. E. (1968). *Biochemistry* **7**, 2653–2662.
98. Gibson, D. T., Wood, J. M., Chapman, P. J., and Dagley, S. (1967). *Biotechnol. Bioeng.* **9**, 33–44.
99. Goren, M. B., Zajic, J. E., and Dunlap, W. J. (1966). U.S. Patent 3,272,716.
100. Gorin, P. A. J., Spencer, J. F. T., and Tulloch, A. P. (1961). *Can. J. Chem.* **39**, 846–855.
101. Gray, P. H., and Thornton, H. G. (1928). *Zentralbl. Bakteriol., Paresitenk., Infektionskr. Hyg., Abt. 2* **73**, 74–96.
102. Guenther, K. R. (1965). *Biotechnol. Bioeng.* **7**, 445–446.
103. Haas, H. F., and Bushnell, L. D. (1941). *Trans. Kans. Acad. Sci.* **44**, 39–45.
104. Haas, H. F., and Bushnell, L D. (1944). *J. Bacteriol.* **48**, 219–231.
105. Haas, H. F., Bushnell, L. D., and Peterson, W. J. (1942). *Science* **95**, 631–632.
106. Haas, H. F. Yantzi, M. F., and Bushnell, L. D. (1941). *Trans. Kans. Acad. Sci.* **44**, 39.
107. Hamer, G. (1968). *J. Ferment. Technol.* **46**, 452–460.
108. Hanson, A. (1967). In "Microbial Technology" (H. Peppler, ed.), p. 222. Reinhold, New York.
109. Harris, R., and Strawinski, R. (1954). U.S. Patent 2,697,061.
110. Hedrick, H. G., Reynolds, R. J., and Crum, M. G. (1968). *Develop. Ind. Microbiol.* **9**, 415–425.
111. Hedrick, H. G., Reynolds, R. J., and Crum, M. G. (1969). *Develop. Ind. Microbiol.* **10**, 228–233.
112. Heinz, E., Tulloch, A. P., and Spencer, J. F. T. (1969). *J. Biol. Chem.* **244**, 882–888.
113. Henning, F. A., Sayasaka, S. S., and Klein, D. A. (1970). *Abstr. Pap., 160th Nat. Meet., Amer. Chem. Soc.*
114. Heringa, J. W., Huybregtse, R., and van der Linden, A. C. (1961). *Antonie van Leeuwenhoek, J. Microbiol. Serol.* **27**, 51–58.
115. Higashihara, T., and Sato, A. (1969). *Agr. Biol. Chem.* **33**, 1802–1804.
116. Higashihara, T., and Sato, A. (1970). *J. Ferment. Technol.* **48**, 73–78.
117. Hill, I. D. (1965). U.S. Patent 3,318,781.
118. Hill, I. D., and Gordon, A. (1967). *Biotechnol. Bioeng.* **9**, 91–105.
119. Hirose, Y., Sonoda, H., Kinoshita, K., and Okada, H. (1967). *Agr. Biol. Chem.* **31**, 1210–1216.
120. Hitzman, D. (1961). *Develop. Ind. Microbiol.* **2**, 33–42.
121. Hitzman, D. (1967). U.S. Patent 3,354,047.
122. Hitzman, D., and Mills, A. (1963). U.S. Patent 3,084,106.
123. Hodson, P., and Darlington, W. (1967). U.S. Patent 3,334,021.
124. Horiguchi, S., and Yamada, K. (1968). *Agr. Biol. Chem.* **32**, 555–560.
125. Hosler, P. (1963). *Biotechnol. Bioeng.* **5**, 243–251.
126. Hosler, P. (1969). French Patent 1,552,885.
127. Hosler, P. (1969). French Patent 1,577,859.
128. Hosler, P. (1969). U.S. Patent 3,458,399.
129. Hosler, P. (1970). U.S. Patent 3,551,297.
130. Hosler, P., and Eltz, R. W. (1969). In "Fermentation Advances" (D. Perlman, ed.), p. 789. Academic Press, New York.
131. Huang, H. T. (1964). *Progr. Ind. Microbiol.* **5**, 57–62.
132. Humphrey, A. E. (1967). *Biotechnol. Bioeng.* **9**, 3–24.
133. Humphrey, A. E., and Raymond, R. L. (1968). U.S. Patent 3,419,469; British Patent 1,111,310.
134. Hungund, B. L., Bhattacharyya, P. K., and Rangachari, P. N. (1970). *Indian J. Biochem.* **7**, 80–81.

135. Iguchi, T., Kodaira, R., and Takeda, I. (1967). *Agr. Biol. Chem.* **31**, 885–889.
136. Iguchi, T., Takeda, I., and Seno, S. (1965). *Agr. Biol. Chem.* **29**, 589–590.
137. Iguchi, T., Watanabe, T., and Takeda, I. (1966). *Agr. Biol. Chem.* **30**, 709.
138. Iguchi, T., Watanabe, T., and Takeda, I. (1967). *Agr. Biol. Chem.* **31**, 569–573.
139. Iguchi, T., Watanabe, T., and Takeda, I. (1967). *Agr. Biol. Chem.* **31**, 574–577.
140. Iida, M., and Iizuka, H. (1970). *Z. Allg. Mikrobiol.* **10**, 245–252.
141. Iizuka, H., Iida, M., and Fujita, S. (1969). *Z. Allg. Mikrobiol.* **9**, 223–226.
142. Iizuka, H., Iida, M., and Unami, Y. (1966). *J. Gen. Appl. Microbiol.* **12**, 131–138.
143. Iizuka, H., Lin, H., and Iida, M. (1970). *Z. Allg. Mikrobiol.* **10**, 189–196.
144. Imada, Y., Pakhashi, J., Yamada, K., Uchida, K., and Aida, K. (1966). *Agr. Biol. Chem.* **30**, 487–495.
145. Imada, Y., Pakhashi, J., Yamada, K., Uchida, K., and Aida, K. (1967). *Biotechnol. Bioeng.* **9**, 45–54.
146. Imada, Y., Pantskhava, E. C., and Yamada, K. (1967). *Agr. Biol. Chem.* **31**, 245–254.
147. Imada, Y., and Yamada, K. (1969). *Agr. Biol. Chem.* **33**, 1326–1332.
148. Imelik, B. (1948). *Co. R. Acad. Sci.* **226**, 2082–2083.
149. Ishii, R., Otsuka, S., and Shiio, I. (1967). *J. Gen. Appl. Microbiol.* **13**, 217–225.
150. Ishikura, T., and Foster, J. (1961). *Nature (London)* **192**, 892–893.
151. Ishikura, T., Nishida, H., Tanno, K., Miyachi, N., and Ozaki, A. (1968). *Agr. Biol. Chem.* **32**, 12–20.
152. Jacobs, S. E. (1931). *Appl. Biol.* **18**, 98–136.
153. Jamison, V. W., Raymond, R. L., and Hudson, J. O. (1969). *Appl. Microbiol.* **17**, 853–856.
154. Jarvis, F. G., and Johnson, M. J. (1949). *J. Amer. Chem. Soc.* **71**, 4124–4126.
155. Jones, D. F. (1968). *J. Chem. Soc., C* pp. 2809–2814.
156. Jones, D. F. (1968). *J. Chem. Soc., C* pp. 2827–2833.
157. Jones, D. F., and Howe, R. (1968). *J. Chem. Soc., C* pp. 2801–2808.
158. Jones, D. F., and Howe, R. (1968). *J. Chem. Soc., C* pp. 2816–2821.
159. Jones, D. F., and Howe, R. (1968). *J. Chem. Soc., C* pp. 2821–2827.
160. Jones, R. J., and Colasito, I. J. (1963). *Bacteriol. Proc.* p. A83.
161. Kanamaru, T., and Yamatodani, S. (1969). *Agr. Biol. Chem.* **33**, 1521–1522.
162. Kaserer, H. (1906). *Zentralbl. Bakteriol., Parasitenk., Infektionskr. Hyg., Abt 2* **15**, 573–576.
163. Kawamoto, I., Nara, T., Misawa, M., and Kinoshita, S. (1970). *Agr. Biol. Chem.* **34**, 1142–1149.
164. Kester, A. S. (1961). Ph.D. Thesis, University of Texas, Austin.
165. Kester, A. S., and Foster, J. (1963). *J. Bacteriol.* **85**, 859–869.
166. Kitagawa, M. (1956). *J. Biochem. (Tokyo)* **43**, 553–563.
167. Kitai, A., and Ozaki, A. (1969). *J. Ferment. Technol.* **47**, 527–535.
168. Kitai, A., Tone, H., Ishikura, T., and Ozaki, A. (1968). *J. Ferment. Technol.* **46**, 442–451.
169. Klausmeier, R. E., and Strawinski, R. (1957). *J. Bacteriol.* **73**, 461–464.
170. Klausmeier, R. E., and Strawinski, R. (1957). *Bacteriol. Proc.* p. 18.
170a. Klein, D. (1971). Personal communication.
171. Klein, D., Davis, J., and Casida, L. (1968). *Antonie van Leeuwenhoek, J. Microbiol. Serol.* **34**, 495–503.
172. Klein, D., and Henning, F. (1969). *Appl. Microbiol.* **17**, 676–681.
173. Kleinzeller, A., and Fencl, Z. (1952). *Chem. Listy* **46**, 300–302.
174. Klug, M., and Markovetz, A. (1967). *J. Bacteriol.* **93**, 1847–1852.

175. Klug, M., and Markovetz, A. (1968). *J. Bacteriol.* **96**, 1115–1123.
176. Klug, M., and Markovetz, A. (1969). *Biotechnol. Bioeng.* **11**, 427–440.
177. Kodama, K., and Yamada, K. (1965). *Hakko Kyokaishi* **23**, 347–351.
178. Koshelva, N. A., Nette, I. T., and Baikova, L. A. (1965). *Prikl. Biokhim. Mikrobiol.* **1**, 617–622.
179. Kraidman, G., Mukherjee, B. B., and Hill, I. D. (1969). *Bacteriol. Proc.* p. 63.
180. Krasil'nikov, N., and Koronelli, T. (1969). *Mikrobiologiya* **38**, 757–760.
181. Kyowa Hakko Kogyo. (1967). French Patent 1,499,336.
182. Kyowa Hakko Kogyo. (1967). French Patent 1,527,583.
183. Kyowa Hakko Kogyo. (1967). French Patent 1,562,301.
184. Kyowa Hakko Kogyo. (1967). Indian Patent 113,664.
185. Kyowa Hakko Kogyo. (1967). British Patent 1,092,597.
186. Kyowa Hakko Kogyo. (1967). British Patent 1,095,724.
187. Kyowa Hakko Kogyo. (1968). French Patent 1,508,912.
188. Kyowa Hakko Kogyo. (1968). French Patent 1,530,165.
189. Kyowa Hakko Kogyo. (1968). British Patent 1,102,905.
190. Kyowa Hakko Kogyo. (1968). British Patent 1,102,906.
191. Kyowa Hakko Kogyo. (1968). British Patent 1,117,154.
192. Kyowa Hakko Kogyo. (1968). British Patent 1,117,699.
193. Kyowa Hakko Kogyo. (1968). British Patent 1,122,508.
194. Kyowa Hakko Kogyo. (1968). French Patent 1,562,301.
195. Kyowa Hakko Kogyo. (1968). Indian Patent 116,662.
196. Kyowa Hakko Kogyo. (1969). German Patent 1,912,797.
197. Kyowa Hakko Kogyo. (1969). German Patent 1,915,855.
198. Kyowa Hakko Kogyo. (1969). German Patent 1,916,421.
199. Kyowa Hakko Kogyo. (1969). French Patent 1,556,992.
200. Kyowa Hakko Kogyo. (1969). French Patent 1,562,657.
201. Kyowa Hakko Kogyo. (1969). French Patent 1,565,232.
202. Kyowa Hakko Kogyo. (1969). French Patent 1,568,652.
203. Kyowa Hakko Kogyo. (1969). French Patent 1,577,264.
204. Kyowa Hakko Kogyo. (1969). French Patent 2,000,641.
205. Kyowa Hakko Kogyo. (1969). French Patent 2,001,118.
206. Kyowa Hakko Kogyo. (1969). French Patent 2,003,584.
207. Kyowa Hakko Kogyo. (1969). French Patent 2,003,809.
208. Kyowa Hakko Kogyo. (1969). French Patent 2,005,071.
209. Kyowa Hakko Kogyo. (1969). British Patent 1,140,827.
210. Kyowa Hakko Kogyo. (1970). British Patent 1,176,266.
211. Kyowa Hakko Kogyo. (1970). British Patent 1,186,959.
212. Kyowa Hakko Kogyo. (1970). British Patent 1,186,989.
213. Kyowa Hakko Kogyo. (1970). British Patent 1,191,397.
214. Kyowa Hakko Kogyo. (1970). British Patent 1,211,672.
215. Kyowa Hakko Kogyo. (1970). German Patent 2,005,848.
216. Kyowa Hakko Kogyo. (1970). German Patent 2,023,157.
217. Kyowa Hakko Kogyo. (1970). British Patent 1,211,674.
218. Kyowa Hakko Kogyo. (1970). German Patent 1,806,386.
219. Leadbetter, E., and Foster, J. (1959). *Nature (London)* **184**, 1428.
220. Leadbetter, E., and Foster, J. (1959). *Arch. Biochem. Biophys.* **82**, 491–492.
221. Leadbetter, E., and Foster, J. (1960). *Arch. Mikrobiol.* **35**, 92–104.
222. Leavitt, R. I. (1967). *J. Gen. Microbiol.* **49**, 411–420.
223. Leavitt, R. I. (1967). U.S. Patent 3,326,771.

224. Leavitt, R. I. (1967). U.S. Patent 3,326,772.
225. Leavitt, R. I. (1967). U.S. Patent 3,344,037.
226. Lee, G., and Walder, C. (1969). *Appl. Microbiol.* **17,** 520–523.
227. Linday, E., and McDonald, M. B. (1961). *J. Biochem. Microbiol. Technol. Eng.* **3,** 219–233.
228. Lukins, H. B. (1962). Ph.D. Thesis, University of Texas, Austin.
229. Mabe, J. A., and Brannon, D. R. (1970). *Bacteriol. Proc.* p. 4.
230. Madhyastha, K. M., and Bhattacharyya, P. K. (1968). *Indian J. Biochem.* **5,** 102–111.
231. Madhyastha, K. M., and Bhattacharyya, P. K. (1968). *Indian J. Biochem.* **5,** 161-167.
232. Malek, I., and Fencl, Z. (1966). "Theoretical and Methodological Basis of Continuous Culture of Microorganisms. Publ. House Czech. Acad. Sci., Prague.
233. Maliyanta, A. A. (1935). *Azerb. Neft. Khoz.* **6,** 89–93.
234. Mallet, M. L., Priou, M. L., and Leon, M. M. (1969). *C. R. Acad. Sci.* **268,** 202–205.
235. Marr, E. E., and Stone, R. W. (1960). *J. Bacteriol.* **81,** 425–430.
236. Mateles, R. I., and Tannenbaum, S. R. (1968). "Single Cell Protein." MIT Press, Cambridge, Massachusetts.
237. Matsubayashi, T., Urawa-shi, and Suzubi, Y. (1969). U.S. Patent 3,433,707.
238. Mayberry, W. R., Prochazka, G. J., and Payne, W. J. (1967). *Appl. Microbiol.* **15,** 1332–1338.
239. McKenna, E. J., and Coon, M. J. (1970). *J. Biol. Chem.* **245,** 3882–3889.
240. Mil'ko, E., and Rabotnova, I. (1969). *Mikrobiologiya* **38,** 264–269.
241. Mimura, A., Kawano, T., and Kodaira, R. (1969). *J. Ferment. Technol.* **47,** 229–236.
242. Miyoshi, M. (1895). *Jahrb, Wiss. Bot.* **28,** 269–289.
243. Mobil Oil Corporation. (1967). British Patent 1,071,935.
244. Moguelevskii, G. A. (1940). *Razved. Nedr.* **12,** 32–43.
245. Morikawa, H., and Kamikuho, T. (1969). *J. Ferment. Technol.* **47,** 470–477.
246. Mukherjee, B. B., Kraidman, G., and Hill, I. D. (1969). *Bacteriol Proc.* p. 63.
247. Murphy, J. F., and Stone, R. W. (1955). *Can. J. Microbiol.* **1,** 579–588.
248. Nakao, Y., Kikuchi, M., Suzuki, M., and Doi, M. (1970). *Agr. Biol. Chem.* **34,** 1865–1876.
249. Nakayama, K., Tanaka, H., Hagino, H , and Kinoshita, S. (1966). *Agr. Biol. Chem.* **30,** 611–616.
250. Nara, T., Misawa, M., Komuro, T., and Konoshita, S. (1969). *Agr. Biol. Chem.* **33,** 358–369.
251. Niaussat, M. M., Auger, C., and Mallet, M. L. (1970). *Co. R. Acad. Sci.* **270,** 1042–1045.
252. Ninet, L , Renant, J., and Tissieo, R. (1966). U.S. Patent 3,235,467.
253. Nishio, N., and Kamikubo, T. (1970). *J. Ferment. Technol.* **48,** 1–7.
254. Nishio, N., and Kamikubo, T. (1970). *Hakko Kogaku Zasshi* **48,** 1–7.
255. Nosaka, J., and Kusunose, M. (1968). *Agr. Biol. Chem.* **32,** 1033–1039.
256. Nosaka, J., and Kusunose, M. (1969). *Agr. Biol. Chem.* **33,** 962–965.
257. Noyes, R. (1969). "Protein Food Supplement." Noyes Development Corporation, Park Ridge, New Jersey.
258. Nutting, L. A., and Borjas, L. (1966). U.S. Patent 3,267,004.
259. Nyns, E. J., and Wiaux, A. L. (1969). *Agricultura* **17,** 3–56.
260. O'Conner, R. J., Weinrich, B. W., and Darlington, W. A. (1964). *Bacteriol. Proc.* p. 97.

261. Ogino, S., Yans, K., Tamura, G., and Arima, K. (1965). *Agr. Biol. Chem.* **29**, 1009-1015.
262. Omori, T., Horiguchi, S., and Yamada, K. (1967). *Agr. Biol. Chem,* **31**, 1337-1342.
263. Omori, T., and Yamada, K. (1969). *Agr. Biol. Chem.* **33**, 979-985.
264. Omori, T., and Yamada, K. (1970). *Agr. Biol. Chem.* **34**, 659-663.
265. Ooyama, J., and Foster, J. W. (1965). *Antonie van Leeuwenhoek, J. Microbiol. Serol.* **31**, 45-65.
266. Ornston, L. N., and Stanier, R. Y. (1966). *J. Biol. Chem.* **241**, 3776-3786.
267. Oshima, K., Tokoro, Y., Okii, M., Tanaka, K., and Kinoshita, S. (1969). *Hakko To Taisha* **20**, 1-7.
268. Otsuka, S., Ishii, R., Shiio, I., and Katsuya, N. (1964). *J. Gen. Appl. Microbiol.* **10**, 179-180.
269. Payne, W. J. (1970). *Annu. Rev. Microbiol.* **24**, 17-52.
270. Phillips, U. (1967). U.S. Patent 3,313,709.
271. Poglazova, M. N., Fedoseeva, G. E., Khesina, A. J., Meisel, M. N., and Shabad, L. M. (1966). *Dokl. Akad. Nauk SSSR* **169**, 1174-1177.
272. Poglazova, M. N., Fedoseeva, G. E., Khesina, A. J., Meisel, M. N., and Shabad, L. M. (1967). *Life Sci.* **6**, 1053-1062.
273. Poglazova, M. N., Fedoseeva, G. E., Khesina, A. J., Meisel, M. N., and Shabad, L. M. (1968). *Dokl. Akad. Nauk SSSR* **179**, 1460-1462.
274. Prema, B. R., and Bhattacharyya, P. K. (1962). *Appl. Microbiol.* **10**, 524-528.
275. Prema, B. R., and Bhattacharyya, P. K. (1962). *Appl. Microbiol.* **10**, 529-531.
276. Prevot, A. R. (1949). *Ann. Inst. Pasteur, Paris* **77**, 418-426.
277. Prochazka, G. J., Payne, W. J., and Mayberry, W. R. (1970). *J. Bacteriol.* **104**, 646-649.
278. Rabotnova, I., Nikitina, K., Bobkova, T., and Grechushkina, N. (1966). *Int. Congr. Microbiol., Abstr. Pap., 9th, 1966.* p. 258.
279. Ratledge, C. (1968). *Biotechnol. Bioeng.* **10**, 511-533.
280. Ratledge, C. (1970). *Chem. Ind. (London)* pp. 843-854.
281. Raymond, R., and Davis, J. (1960). *Appl. Microbiol.* **8**, 329-334.
282. Raymond, R. L., and Jamison, V. W. (1968). U.S. Patent 3,383,289.
283. Raymond, R. L., Jamison, V. W., and Hudson, J. O. (1967). *Appl. Microbiol.* **15**, 857-865.
284. Raymond, R. L., Jamison, V. W., and Hudson, J. O. (1969). *Appl. Microbiol.* **17**, 512-515.
285. Ribbons, D. W. (1966). *Annu. Rep. Progr. Chem.* **62**, 445-467.
286. Rodionova, G., Dikanskaya, E., Bravicheva, R. (1969). *In* "Continuous Cultivation of Microorganisms" (I. Malek, ed.), p. 561. Academic Press, New York.
287. Rogoff, M. (1961). *Advan. Appl. Microbiol.* **3**, 193-221.
288. Rogoff, M. (1967). U.S. Patent 3,331,750.
289. Rogoff, M. (1969). U.S. Patent 3,420,741.
290. Rogoff, M. H., and Wender, I. (1957). *J. Bacteriol.* **73**, 264-268.
291. Rogoff, M. H., and Wender, I. (1957). *J. Bacteriol.* **74**, 108-109.
292. Rogoff, M. H., and Wender, I. (1959). *J. Bacteriol.* **77**, 783-788.
293. Rudolfs, W. (1953). "Industrial Wastes: Their Disposal and Treatment." Reinhold, New York.
294. Ruinen, J., and Ieinema, M. H. (1964). *Antonie van Leeuwenhoek, J. Microbiol. Serol.* **30**, 377.
295. Senez, J., and Konovaltschokoff-Mazoyer, M. (1956). *Co. R. Acad. Sci.* **242**, 2873-2875.
296. Shah, J. H., Khalid, A. M., and Wahid, W. (1967). *Agr. Biol. Chem.* **31**, 1499-1504.

297. Shah, J. H., Sedi, M. H., and Sheikh, T. H. (1967). *Agr. Biol. Chem.* **31**, 645–650.
298. Sharpley, J. M. (1964). Tech. Doc. Rep. No. ASD-TDR-63-752. Wright-Patterson Air Force Base, Ohio.
299. Shell International Research. (1964). British Patent 1,052,779.
300. Shell International Research. (1965). Netherlands Patent 291,163.
301. Shiio, I., and Nakamori, S. (1970). *Agr. Biol. Chem.* **34**, 448–456.
302. Shiio, I., Otsuka, S., Ishii, R., Katsuya, N., and Iizuka, H. (1963). *J. Gen. Appl. Microbiol.* **9**, 23–30.
303. Shiio, I., Otsuka, S., and Takahashi, M. (1962). *J. Biochem. (Tokyo)* **51**, 56–62.
304. Shiio, I., and Uchio, R. (1969). *J. Gen. Appl. Microbiol.* **15**, 65–84.
305. Shimakara, K., and Yamashita, H. (1967). *J. Ferment. Technol.* **45**, 1172–1179.
306. Shpon'ko, R. I., Nikitina, N. I., and Nette, I. T. (1969). *Vestn. Mosk. Univ., Ser. Biol., Pochvoved., Geol., Geogr.* **24**, 106–109.
307. Shukla, O. P., and Bhattacharyya, P. K. (1968). *Indian J. Biochem.* **3**, 102–111.
308. Shukla, O. P., Moholay, M. N., and Bhattacharyya, P. K. (1968). *Indian J. Biochem.* **3**, 79–91.
309. Sohngen, N. L. (1906). *Centralbl. Bakteriol., Parasitenk., Infektionskr. Hyg., Abt. 2* **15**, 513–517.
310. Sohngen, N. L. (1913). *Centralbl. Bakteriol., Parasitenk., Infektionskr. Hyg., Abt. 2* **37**, 595–609.
311. Sohngen, N. L., and Fol, J. G. (1914). *Centralbl. Bakteriol., Parasitenk., Infektionskr. Hyg., Abt. 1: Orig.* **40**, 87–98.
312. Somerson, N., and Phillips, T. (1961). Belgian Patent 593,807.
313. Speedie, J., and Hull, G. (1960). U.S. Patent 2,951,017.
314. Spencer, J. F. T., Tulloch, A. P., and Gorin, P. A. J. (1962). *Biotechnol. Bioeng.* **4**, 271–279.
315. Spencer, J. F. T., Tulloch, A. P., and Gorin, P. A. J. (1965). U.S. Patent 3,205,150.
316. Stanier, R. Y. (1950). *Bacteriol. Rev.* **14**, 179–195.
317. Steel, R. (1958). "Biochemical Engineering." Heywood Press, London.
318. Stevens, N., Frankenfeld, J. W., and Douros, J. (1969). U.S. Patent 3,409,506.
319. Stevenson, D., Finnerty, W. R., and Kallio, R. E. (1962). *Biochem. Biophys. Res. Commun.* **9**, 426–429.
320. Stewart, J., Finnerty, W. R., Kallio, R. E., and Stevenson, D. (1960). *Science* **132**, 1254–1255.
321. Stewart, J., and Kallio, R. E. (1959). *J. Bacteriol.* **78**, 726–730.
322. Stewart, J., Kallio, R. E., Stevenson, D., Jones, A., and Schissler, D. (1959). *J. Bacteriol.* **78**, 441–448.
323. Stone, R. W., and Zobell, C. E. (1952). *Ind. Eng. Chem.* **44**, 2564–2567.
324. Strawinski, R. J. (1943). PhD. Thesis, Pennsylvania State College, State College, Pennsylvania.
325. Strawinski, R. J., and Stone, R. W. (1955). *Can. J. Microbiol.* **1**, 206–210.
326. Sun Oil Co. (1969). German Patent 1,810,026.
327. Sun Oil Co. (1970). French Patent 1,593,718.
328. Suzuki, T., Tanaka, K., and Kinoshita, S. (1969). *Agr. Biol. Chem.* **33**, 190–195.
329. Suzuki, T., Tanaka, K., Matsubara, I., and Kinoshita, S. (1969). *Agr. Biol. Chem.* **33**, 1619–1627.
330. Tabuchi, T., Tanaka, M., and Abe, M. (1969). *Nippon Nogei Kagaku Kaishi* **43**, 154–158.
331. Tadao, T., Urawa-shi, and Suzuki, Y. (1969). U.S. Patent 3,433,707.
332. Taggart, M. S. (1941). U.S. Patent 2,234,637.
333. Taggart, M. S. (1946). U.S. Patent 2,396,900.

334. Takahashi, J., Imada, Y., and Yamada, K. (1963). *Agr. Biol. Chem.* **27**, 396.
335. Takahashi, J., Kobayashi, K., Imada, Y., and Yamada, K. (1965). *Appl. Microbiol.* **13**, 1–4.
336. Takeda Chemical Industries. (1968). French Patent 1,523,599.
337. Takeda Chemical Industries. (1969). French Patent 725,417.
338. Takeda Chemical Industries. (1969). British Patent 1,155,945.
339. Takeda Chemical Industries. (1969). Belgian Patent 716,247.
340. Takeda Chemical Industries. (1970). Belgian Patent 744,416.
341. Takeda Chemical Industries. (1970). Belgian Patent 744,946.
342. Tanaka, A., Fujii, K., and Fukui, S. (1969). *J. Ferment. Technol.* **47**, 297–302.
343. Tanaka, A., and Fukui, S. (1967). *J. Ferment. Technol.* **45**, 611–616.
344. Tanaka, A., Nagasaki, T., and Fukui, S. (1968). *J. Ferment. Technol.* **46**, 477–000.
345. Tanaka, A., Nagasaki, T., and Fukui, S. (1968). *J. Ferment. Technol.* **46**, 984–991.
346. Tanaka, A., Nagasaki, T., Fukui, S., and Murakami, S. (1969). *J. Ferment. Technol.* **47**, 739–743.
347. Tanaka A., Nagasaki, T., Inagawa, M., and Fukui, S. (1968). *J. Ferment. Technol.* **46**, 468–476.
348. Tanaka, A., Ohishi, N., and Fukui, S. (1967). *J. Ferment. Technol.* **45**, 617–623.
349. Tanaka, K., Akita, S., Kimura, K., and Kinoshita, S. (1960). *J. Agr. Chem. Soc. Jap.* **34**, 600–608.
350. Tanaka, K., Iwasaki, H., and Kinoshita, S. (1960). *J. Agr. Chem. Soc. Jap.* **34**, 593–600.
351. Tanaka, K., Kimura, K., Machida-shi, and Yamaguchi, K. (1967). U.S. Patent 3,359,178.
352. Tanaka, K., Kimura, K., Machida-shi, and Yamaguchi, K. (1968). Canadian Patent 788,978.
353. Tanaka, K., Kimura, K., Machida-shi, and Yamaguchi, K. (1969). U.S. Patent 3,450,599.
354. Tanaka, K., Kimura, K., Suzuki, T., Yamaguchi, K., and Kinoshita, S. (1969). *Hakko Kogaku Zasshi* **47**, 291–296.
355. Tanaka, M., Takara, Y., Tabuchi, T., and Abe, M. (1970). *J. Agr. Chem. Soc. Jap.* **44**, 499–504.
356. Tarama, K. (1965). In "Handbook of Organic Structural Analysis" (Y. Yukawa, ed.), p. 548. Benjamin, New York.
357. Tattersfield, F. (1927). *Ann. Appl. Biol.* **15**, 57–80.
358. Tatum, E. (1945). *J. Biol. Chem.* **160**, 455–459.
359. Tausson, T. A. (1939). *Mikrobiologiya* **8**, 828–833.
360. Tausson, W. O. (1925). *Biochem. Z.* **155**, 356–368.
361. Tausson, W. O. (1927). *Planta* **4**, 214–256.
362. Tausson, W. O. (1928). *Neft. Khoz.* **14**, 220–230.
363. Tausson, W. O. (1928). *Planta* **5**, 239–278.
364. Tausson, W. O. (1928). *Neft. Khoz.* **14**, 220–230.
365. Thaysen, A. C. (1939). *J. Inst. Petrol., London* **25**, 411–415.
366. Thaysen, A. C. (1940). *Proc. Int. Congr. Microbiol., 3rd, 1939* p. 729.
367. Thijsse, G. (1964). *Biochim. Biophys. Acta* **84**, 195–197.
368. Thijsse, G., and Van der Linden, A. C. (1958). *Antonie van Leeuwenhoek, J. Microbiol. Serol.* **24**, 298–308.
369. Thijsse, G., and Van der Linden, A. C. (1963). *Antonie van Leeuwenhoek, J. Microbiol. Serol.* **29**, 89–100.
370. Tokoro, Y., Oshima, K., Okii, M., Yamaguchi, K., Tanaka, K., and Kinoshita, S. (1970). *Agr. Biol. Chem.* **34**, 1516–1521.

371. Tone, H., Kitai, A., and Ozaki, A. (1968). *Biotechnol. Bioeng.* **10**, 689–692.
372. Traxler, R. W., and Bernard, J. M. (1969). *Biotechnol. Bioeng.* **11**, 441–448.
373. Treccani, V. (1963). *Progr. Ind. Microbiol.* **4**, 3–33.
374. Treccani, V., and Fiecchi, A. (1958). *Ann. Microbiol. Enzimol.* **8**, 36–44.
375. Treccani, V., Walker, N., and Wiltshire, G. H. (1954). *J. Gen. Microbiol.* **11**, 341–348.
376. Tsuboi, T., Sekijo, C., and Shoji, O. (1966). *Agr. Biol. Chem.* **30**, 1238–1242.
377. Tsuboi, T., Sekijo, C., and Shoji, O. (1966). *Agr. Biol. Chem.* **30**, 1243–1246.
378. Tsugawa, R. (1969). *Agr. Biol. Chem.* **33**, 676–683.
379. Tsugawa, R., Nakase, T., Kobayashi, T., Yamashita, K., and Okumura, S. (1969). *Agr. Biol. Chem.* **33**, 158–168.
380. Tsugawa, R., Nakase, T., Kobayashi, T., Yamashita, K., and Okumura, S. (1969). *Agr. Biol. Chem.* **33**, 929–938.
381. Tulloch, A. P., Hill, A., and Spencer, J. F. T. (1968). *Can. J. Chem.* **46**, 3337–3351.
382. Uchio, R., Otsuka, S., and Shiio, I. (1967). *J. Gen. Appl. Microbiol.* **13**, 303–312.
383. Udagawa, K., and Kohata, M. (1968). Canadian Patent 781,446.
384. Uemura, N., Takahashi, J., and Veda, K. (1969). *J. Ferment. Technol.* **47**, 220–228.
385. Van der Linden, A. C. (1963). *Biochim. Biophys. Acta* **77**, 157–159.
386. Van der Linden, A. C., and Thijsse, G. J. E. (1965). *Advan. Enzymol.* **27**, 469–546.
387. Veldkamp, H., van den Berg, G., and Zevenhuizen, L. (1963). *Antonie van Leeuwenhoek, J. Microbiol. Serol.* **29**, 35–51.
388. Wagner, F., Zahn, W., and Buhring, U. (1967). *Angew Chem., Int. Ed. Engl.* **6**, 359–360.
389. Wagner, R. (1914). *Z. Gaerungsphysiol.* **4**, 289–319.
390. Walker, N., and Wiltshire, G. H. (1953). *J. Gen. Microbiol.* **8**, 273–276.
391. Webley, D. M. (1954). *J. Gen. Microbiol.* **11**, 420–425.
392. Webley, D. M., Duff, R. B., and Farmer, V. C. (1956). *Nature (London)* **178**, 1467–1468.
393. Wegner, G. H. (1969). *Bacteriol. Proc.* p. 62.
394. Wessley, D. F. (1967). *Pap., 154th Meet., Amer. Chem. Soc.* Q-13.
395. Wieland, T., Griss, W., and Haccius, B. (1958). *Arch. Mikrobiol.* **28**, 383–393.
396. Wright, L., and Cresson, E. (1954). *J. Amer. Chem. Soc.* **76**, 4156–4160.
397. Wright, L., Cresson, E., and Driscoll, C. (1955). *Proc. Soc. Exp. Biol. Med.* **89**, 234–236.
398. Yamada, K., Furukawa, T., and Nakahara, T. (1970). *Agr. Biol. Chem.* **34**, 670–675.
399. Yamada, K., Takahashi, J., and Kobayashi, K. (1962). *Agr. Biol. Chem.* **26**, 636.
400. Yamada, K., Takahashi, J., and Kobayashi, K. (1963). *Agr. Biol. Chem.* **27**, 773–783.
401. Yamada, K., Takahashi, J., and Kobayashi, K. (1963). *Nature (London)* **198**, 1115.
402. Yamada, K., Takahashi, J., Kobayashi, K., and Imada, Y. (1963). *Agr. Biol. Chem.* **27**, 390–395.
403. Yoshida, J., Yamane, T., and Miyamoto, Y. (1970). *Ind. Eng. Chem., Process Des. Develop.* **9**, 570–577.
404. Zajic, J. E., and Dunlap, W. J. (1966). U.S. Patent 3,274,074.
405. Zobell, C. E. (1946). *Bacteriol. Rev.* **10**, 1–49.
406. Zobell, C. E. (1950). *Advan. Enzymol.* **10**, 443–486.

AUTHOR INDEX

Numbers in parentheses are reference numbers and indicate that an author's work is referred to, although his name is not cited in the text. Numbers in italics show the page on which the complete reference is listed.

A

Abbot, M. T. J., 140, *146*
Abbott, B. J., 98, *120*, 260 (1), 271, 246 (2), 354 (3), *379*
Abe, M., 294 (330), 296 (330), 297 (4, 5), 378, *386, 387*
Abraham, E. P., 34, *45*, 123, 125, 126, 132, 133, *146, 149*
Adams, J. N., 94, *121*
Adler, L., 6, *45*
Afonso, A., 119, *120*
Agate, A. D., 75, *90*
Agre, N. S., 55, 65, *70*, 85, *90*
Ahern, J. J., 173, *179*
Aiba, S., 238, 241, *246*, 250 (6), *378*
Aida, K., 281 (144, 145), 282 (145), *382*
Aida, T., 126, 136, *147*, 264, *378*
Akita, E., 88, *90*
Akita, S., 294 (349), 296 (349), *387*
Alburn, H. E., 126, 136, *147*
Alexander, M., 107, *120*, 324 (80), 325 (80), *380*
Ali Khan, M., 255 (11, 12), *378*
Allen, J. H., 171, *180*
Alter, B. McD., 176, *180*
Ammann, A., 88, *90*
Anchel, M., 62, *69*
Andersen, A. A., 78, *90*
Anderson, P. W., 131, 132, *149*
Anderson, R., 309 (38), *379*
Andres, W. W., 126, 137, *146*
Arai, Y., 358, 360 (14), *378*
Argoudelis, A. D., 125, 126, 128, 131, 132, 139, 144, *146*, 188, 200, 220, 225, 227, *228*
Arima, K., 255 (15), 262 (15, 261), *378*, *385*
Arnold, M., 309 (38), *379*
Arnold, N., 126, *148*
Ashby, E. C., 162, *179*
Ashton, G. C., 134, *147*
Asselineau, J., 60, 62, *69*
Atkinson, J. H., 334 (16), 335 (16), 336 (16), *378*

Auger, C., 357 (251), *384*
Avery, R. J., 57, *69*
Azoulay, E., 271, *378, 379*

B

Backus, E. J., 51, 72, 82, *92*
Baikova, L. A., 291 (178), 292 (178), *383*
Balabanova, A., 304 (61), 307 (61), *380*
Balan, J., 88, *90*
Baldacci, E., 50, 54, 62, 65, *69*, *71*
Baldan, B., 65, *71*
Ball, S., 84, *90*
Ballio, A., 37, *45*
Bannister, B., 185, 186, 189, 195, 203, 207, *228, 229*
Baptist, J. N., *380*
Baráth, Z., 87, 89, *90*
Barker, H. A., 140, *147*
Baron, L. S., 163, *180*
Barr, M., 171, *181*
Barros, M., 166, *182*
Bartlett, M. C., 250 (92), *380*
Bartnicki-Garcia, S., 49, *69*
Bataille, F., 49, *69*
Batchelor, F. R., 37, *45*
Beck, D., 126, 137, *146*, 167, *180*
Becker, B., 57, 58, *69*
Becker, R. E., 161, *180*
Beerstecher, E., 250 (18), *378*
Benedict, R. G., 75, 81, *90, 91*
Bennett, R. E., 114, *121*, 135, *146*
Bentley, D. W., 128, *149*
Benveniste, R., 126, 130, 131, 132, *146, 148*
Beretta, G., 65, *72*
Bergman, I., 242, *246*
Bergy, M. E., 185, *228*
Bernard, J. M., 320, *388*
Bernheimer, A. W., 158, *181*
Berry, L. J., 164, *180*
Bessell, C. J., 84, *90*
Beswick, D. E., 177, *180*
Betina, V., 87, 88, 89, *90*
Bhat, J. V., 75, *90*

Bhattacharyya, P. K., 363 (20, 91), 359 (274, 275), 364 (19, 20, 21, 91), 366 (21, 22, 274, 275), 367 (19, 21), 368 (59, 60), 369 (60, 308), 370 (308), 371 (230), 372 (59, 134), 365 (59, 60, 230, 307, 308), 375 (231), *378, 379, 380, 381, 384, 385, 386*
Biaggi, C., 351 (42), 356 (42), *379*
Bianchi, B., 99, *122*
Binaghi, A., 88, *91*
Birch-Hirschfeld, L., 251 (23), 254 (23), *379*
Bird, C. W., 252 (24, 25), 339 (24), *380*
Birkenmeyer, R. D., 185, 186, 188, 195, 196, 199, 203, 205, 207, 213, 215, 220, 221, 222, 225, *227*, 229
Birnbaum, J., 274 (58), 278 (58), *379*
Bistis, G. N., 62, *69*
Blackwood, R. K., 123, *146*
Blakebrough, N., 250 (26), *379*
Blank, F., 57, *69*
Blank, R. H., 112, 113, *121*
Blau, L. W., *379*
Blinov, N. O., 51, *69*
Bliss, E. A., 173, 176, *180*
Bloom, B. M., 117, *120*
Bobkova, T., 310 (278), *385*
Böhni, E., 175, *180*
Bönicke, R., 64, *69*
Bohme, K. H., 117, *121*
Bohnhoff, M., 168, 175, 177, *180, 182*
Bohonos, N., 136, 137, *147, 148*
Bonanchaud, D., 128, *146*
Boone, C. J., 58, *69*
Boothroyd, B., 126, 139, *146*
Borjas, L., 345 (258), 353 (258), *384*
Borman, A., 111, *120*
Boyle, A. M., 84, *92*
Brading, D. J., 240, *246*
Bradley, S. G., 55, 56, 59, *70, 72*
Brandly, C. A., 156, *183*
Brannon, D. R., 320, 376, *384*
Braun, W., 159, *180*
Bravicheva, R., 307 (286), *385*
Brian, P. W., 124, 143, *146*
Brillaud, A. R., 275 (62), 319 (62), 346 (28), *379, 380*
Brock, T. D., 74, *90*
Brodasky, T. F., 200, *228*
Brodsky, R. F., 126, 127, *149*

Brown, G., 239, *246*
Browning, C. H., 169, *180*
Bruff, J. A., 174, *181*
Bruyn, J., 252 (29), 257 (29), *379*
Buddin, W., 319 (30), *379*
Buhring, U., 98, *122*, 271 (388), *388*
Bullen, J. J., 158, *180*
Bu'Lock, J. D., 74, *90*
Burch, M. R., 187, 190, 193, *228*
Burkholder, P. R., 79, 83, *92*
Burnell, J. M., 173, *179*
Burton, H. S., 34, *45*
Bushnell, L. D., 250 (103), 251 (105, 106), 305 (104), 309, *379, 381*

C

Caglar, M. A., 315 (32), 355 (32), *379*
Cain, R. B., 108, 110, *120, 121*
Calam, C. T., 85, *90*
Cambie, R. C., 52, *69*
Cameron, J., 155, 158, *180*
Canhan, S. C., 62, *69*
Canonica, L., 97, *122*, 347 (33), 355 (33), *379*
Čapek, A., 124, 144, *146*
Cardini, G. E., 321 (95), 323 (95), *380*
Carey, W. F., 163, 167, *180, 182*
Cartwright, N. J., 108, *120*
Casida, L. E., 57, *70*, 98, *120*, 257 (171), 260 (1), 270 (171), 271 (171), 286 (36), 377 (171), *378, 379, 382*
Cassady, J. V., 170, *180*
Chaiet, L., 86, *91*
Chain, E. B., 26, 37, *45*, 123, 126, 132, *146*
Chamberlain, R. E., 170, *182*
Chamoiseau, G., 62, *69*
Champagnat, A., 250 (35), *379*
Chance, B., 244, *246*
Chang, F. N., 113, *120*
Chapman, P. J., 321 (98), *381*
Charney, W., 119, *120*, 124, 144, *146*
Chase, F. E., 75, *90*
Chiang, C., 135, *146*
Chidester, C. G., 142, *149*, 199, *229*
Chouteau, J., 271 (37), *379*
Chun, D., 84, *90*
Ciegler, A., 309, (38, 39), *379*
Citri, N., 133, *146*
Clapp, H. W., 185, 190, 193, *228*
Claridge, C. A., 126, 135, 136, *146*

AUTHOR INDEX

Claus, D., 324 (41), 326 (41), *379*
Clutterbuck, P. W., 25, *45*
Coats, J. H., 125, 128, 131, 132, 139, 144, *146*, 225, *227, 228*
Cocito-Vandermeulen, J., 142, *149*
Colasito, I. J., 312 (160), *382*
Cole, M., 126, 135, *146*
Cole, R. J., 160, *183*
Colla, C., 351, (42, 43), 352 (43), 357, 358, 360 (44), *379*
Collier, B., 39, *45*
Coombe, R. G., 117, 119, *120, 121*
Coon, M. J., *380, 384*
Cooper, M. L., 177, *180*
Corbaz, R., 79, 83, *90*
Cordon, T. C., 78, *92*
Corke, C. T., 75, 83, *90, 91*
Coronelli, C., 88, *91*
Corso, V., 232, *246*
Coty, V. F., 319 (46), 320 (45, 47), 331 (46), 337, 341 (46), *379*
Couch, J. N., 60, *69*, 78, *90*
Craveri, A., 56, *71*
Craveri, R., 55, 56, *69, 71*
Crawford, K., 34, *45*
Crawley, G. C., 126, 138, *147*
Cresson, E., 303 (396, 397), 305 (396, 397), *388*
Crofton, J., 173, *180*
Cron, M. J., 123, *146*
Cross, T., 50, 55, 62, 65, *70*, 78, 81, 85, *90*
Cross, W. R., 171, *180*
Crum, G. F., 186, *228*
Crum, M. G., 250 (110, 111), *381*
Cruz-Camarillo, R., 160, *183*
Cummins, C. S., 57, *70*

D

Dagley, S., 321 (98), *381*
Dammin, G. J., 169, *180*
Dann, M., 112, 113, *121*
Dans, P. E., 173, *183*
Daoust, D., 281 (48), 285 (48), *379*
Darlington, W. A., 298 (123), 315 (49), 317, 321 (260), 322 (260), *379, 381, 384*
Darnell, J. E., 50, *71*
Darrah, J. A., 110, *120*
Davies, F. L., 62, *72*, 75, 83, *92*
Davies, J., 126, 130, 131, 132, *146, 150*
Davies, J. I., 338, *379*

Davies, M. C., 82, *92*
Davis, J., 250 (281), 257 (171), 269, 270 (171), 271 (171), 311, 312 (281), 377 (171), *382, 385*
Davis, J. B., 95, 96, 99, *120, 121*, 249 (53), 250 (52, 53, 56), 270, 310 (53), 311 (53), 327 (52), 336, 337 (54, 55), 340 (54), 342 (54, 55), 343 (55), 345 (53, 55), 362 (54, 55), 359 (54, 55), 372 (55), 365, *379*
Davis, R. S., *379*
Davydova, I. M., 55, *70*
Day, J. T., 234, *246*
DeBoer, C., 185, 186, *228, 229*
deGirolamo, M. G., 97, *122*
DeHaan, R. M., 197, 200, *228, 229*
Deindoerfer, F. H., 238, 240, *246*
DeKock, P. C., 97, 105, *122*
DeLey, J., 55, *70*
Demain, A. L., 125, 126, 134, 135, 136, *146, 147*, 274 (58), 278 (58), *379*
DeNavasquez, S. J., 167, *180*
Dentice di Accadia, F., 37, *45*
DeSomer, P., 163, *180*
Dettwiler, H. A., 175, 177, *182*
DeVries, W. H., 186, *228*
Dhavalikar, R. S., 364 (19, 21), 366 (21), 367 (19, 21), 368 (59, 60), 369 (60), 372 (59), 365 (59, 60), *378, 379, 380*
Dietz, A., 185, 186, *228, 229*
Dikanskaya, E., 304 (61), 307, *380, 385*
Dill, I. K., 136, *147*
Disler, E. N., 292 (72), *381*
Dodson, R. M., 112, 115, *120*
Doi, M., 277 (248), 279 (248), *384*
Doi, O., 126, 130, 131, *146, 147, 149*
Dolak, L., 213, 220, *228*
Dolezilova, L., 84, 87, *92*
Donald, M. B., 95, *121*, 258 (227), 262, *384*
Donaldson, D. M., 163, *182*
Donovick, R., 162, *182*
Dorokhova, L. A., 65, *70*
Doskocilova, D., 88, *90*
Douros, J. D., 259 (318), 269 (318), 273, 275 (62), 319 (62), 324 (66), 337 (63, 64, 65), 343 (63, 64), 348 (67), 349 (67), 355 (67), *381, 386*
Dowling, H. F., 171, 172, 173, *180, 181*
Doyle, F. P., 37, *45*
Drake, B. L., 168, *180*
Driscoll, C., 303 (397), 305 (397), *388*

Dubnau, D. A., 133, *147*
Duchamp, D. J., 196, *228*
Duff, R. B., 97, 100, 105, 107, *122*, 281 (392), 331 (392), 336 (392), 337 (392), 340 (392), 341 (392), 349 (392), *388*
Dunlap, W. J., 346 (99, 404), 353 (99, 404), *381*, *388*
Dutcher, J. D., 86, *91*
Dvonch, W., 126, 136, *147*
Dworkin, M., 311 (68), *380*

E

Ebata, M., 126, 136, *147*
Eberhardt, H., 142, *148*
Eble, T. E., 188, 200, *227*, *228*
Eckardt, K., 51, *71*, 86, *91*
Egami, F., 126, 136, *147*
Ehrlich, P., 151, *180*
Eisenberg, M., 303 (69, 70), 338 (69), *380*
Eisler, W. J., Jr., 244, *246*
Eisman, P. C., 165, *180*
Ellis, H., 162, *179*
Ellison, R., 219, *229*
El-Nakeeb, M. A., 81, *90*, 140, *147*
Elsford, H., 303 (71), 305 (71), *380*
Eltz, R. W., 101, 102, 103, *120*, 275 (62), 319 (62), 329 (130), 330 (130), 336, *380*, *381*
Engley, F. B., 159, *180*
English, A. R., 123, *146*
Enmakova, I. T., 292 (72), *380*
Eppstein, S. H., 111, *120*
Erikson, D., 96, *120*
Eschweiler, W., 242, *246*
Esplin, D. W., 163, *182*
Evans, R. H., Jr., 112, 113, *121*
Evans, W. C., 320 (74), 338, 344 (77), 351 (75), 352 (75), 356, *379*, *380*
Evreinova, T. N., 55, *70*
Eyssen, H., 163, *180*

F

Fantes, K. H., 134, 135, *147*
Farina, G., 50, 56, 59, *69*, *70*
Farmer, V. C., 97, 100, 105, 107, *122*, 281 (392), 331 (392), 336 (392), 337 (392), 340 (392), 341 (392), 349 (392), *388*
Farrar, W. E., Jr., 163, *180*
Federbush, C., 119, *120*

Fedoseeva, G. E., 357 (76, 271, 272, 273), *380*, *385*
Feingold, D., 178, *180*
Feldman, L. I., 112, 113, *121*, 126, 136, *147*
Feldman, W. H., 176, *180*
Fencl, Z., 250 (232), 321 (173), 322 (173), *382*, *384*
Fernley, H. N., 339, 344 (77), 351 (75), 352 (75), 356 (75), *380*
Ferrari, A., 245, *246*
Ferrer, E. B., 160, *183*
Fiecchi, A., 347 (33, 374), 351 (43), 352 (43), 355 (33, 374), 356 (43), *379*, *388*
Fiechter, A., 243, *246*
Filosa, J., 250 (35), *379*
Finland, M., 173, *183*
Finnerty, W. R., 96, *120*, 250 (320), 257 (320), 258 (78), 259 (319, 320), 269 (78), 270, *380*, *386*
Fisher, E., Jr., 171, *180*
FitzPatrick, F. K., 165, 167, *180*, *182*
Flaig, W., 75, 76, 84, *90*
Fleischman, A. I., 86, *91*
Fleming, A., 25, *45*
Florey, H. W., 26, *45*
Flynn, D. S., 232, *246*
Flynn, E. H., 136, *148*
Focht, D. D., 324 (80), 325 (80), *380*
Fol, J. G., 251 (311), *386*
Fonkin, G. S., 362 (81), 363 (81), *380*
Formal, S. B., 169, *180*, *181*
Forro, J. R., 99, *120*, 326 (82, 83), *380*
Foster, J., 255 (165), 257 (150), 261 (220), 262 (165), 270, 281 (165, 219, 221), 311 (221), 325 (220, 221), *382*, *383*
Foster, J. W., 99, *120*, 249 (265), 262 (85), 311 (68), 325, 357, 360 (265), 361 (265), 362 (265), 364, *380*, *385*
Fox, J. A., 188, *227*
Frankenfeld, J. W., 259 (318), 269 (318), 337 (63, 64, 65), 343 (63, 64), *380*, *386*
Fredericks, K., 257 (86), 270, *380*
Freedman, L. R., 167, *180*
Freter, R., 169, *180*, *181*
Frew, B. P., 52, *71*
Fried, J., 111, *120*
Frost, B. N., 156, 170, *180*, *182*
Füresz, S., 123, *149*
Fugono, T., 126, 129, *147*

AUTHOR INDEX

Fujii, K., 96, *120*, 304 (87), 307, 308 (88), *380*, *387*
Fujita, S., 257 (141), 271 (141), *382*
Fukagawa, Y., 126, 141, *147*, *149*
Fukui, G. M., 178, *180*
Fukui, S., 96, *120*, 304 (87), 305 (343, 344, 345, 347, 348), 307 (87), 308 (88, 343, 345, 347, 348), 309 (344, 347), 310 (346, 347), *380*, *381*
Fuquay, M. E., 170, *181*
Furgiuele, F. P., 170, *180*
Furukawa, T., 298 (90, 398), 313 (398), *380*, *388*

G

Gadebusch, H. H., 166, 177, *182*
Gaden, E. L., Jr., 232, *246*
Gale, G. O., 178, *181*
Gallo, G. G., 126, 130, *148*
Gallop, R. C., 154, *183*
Ganapathy, K., 363 (20, 91), 365 (20, 91), *379*, *380*
Gardner, A. D., 26, *45*
Garrison, D. W., 197, *228*
Gaughran, R. L., 53, *70*
Gelzer, J., 166, 177, *181*
Gerber, N. N., 51, 52, 53, 58, *70*, *71*
Gerencser, M. A., 57, *71*
Gerhardt, P., 250 (92), 346 (2), 355 (3), *378*, *380*
Gerke, J. R., 245, *246*
Gershenfeld, L., 171, *181*
Gholson, R. K., *381*
Gibson, D. T., 117, *120*, *121*, 321 (94, 98), 323 (95, 97), 324, 326 (97), 327 (96), 354, *380*, *381*
Gledhill, W. E., 57, *70*
Glenn, W. G., 161, *180*
Godfrey, J. C., 123, *146*
Godtfredsen, W. O., 126, 136, *147*, *150*
Gold, W., 178, *180*
Goldschmidt, F. K., 170, *180*
Goldstein, A., 142, *147*
Goll, P. H., 124, *147*
Gooder, H., 66, *72*
Goodfellow, M., 65, *70*
Goos, R. D., 80. *92*
Gordon, A., 346 (118), 354, *381*
Gordon, R. E., 50, 53, 57, 58, 62, *69*, *70*, 94, *120*

Gore, J. H., 234, *246*
Goren, M. B., 345 (99), 353 (99), *381*
Gorin, P. A. J., 21, 41, *45*, 259 (314), 264 (314), 265 (100, 315), *381*, *386*
Gorrill, R. H., 167, *180*
Gottlieb, D., 50, 55, *70*, 71, 81, 88, *90*, 92
Gourevitch, A., 143, *147*, 160, *181*
Grady, J. E., 185, 190, 193, *228*
Grainger, J. M., 62, *70*
Gray, J. E., 200, *229*
Gray, P. H., 338 (101), 355, *381*
Grayson, P., 232, *246*
Grechushkina, N., 310 (278), *385*
Greenberg, J., 140, *147*
Greenspan, G., 126, 136, *147*
Gregory, P. H., 78, 79, 83, *90*
Grein, A., 78, *90*
Griffith, L. J., 177, *180*
Griffiths, E., 351 (75), 352 (75), 356 (75), *380*
Griss, G., 99, *122*
Griss, W., 321 (395), 322 (395), *388*
Grivell, A. R., 160, *181*
Grosklags, J. H., 86, *90*
Grove, J. F., 140, *146*
Grover, A. A., 161, 179, *181*, *183*
Grunberg, E., 175, 178, *182*
Grundy, W. E., 84, *92*
Guenther, K. R., 315, *381*
Gunnison, J. B., 172, 174, *181*
Gutekunst, D. P., 170, *182*
Guze, L. B., 178, *181*

H

Haas, H. F., 250 (103), 251 (105, 106), 305 (104), 309, *379*, *381*
Haccius, B., 99, *122*, 321 (395), 322 (395), *388*
Hagino, H., 285 (249), 286 (249), *384*
Hale, M. E., 49, *70*
Hall, A., 255 (11, 12), *378*
Hall, H. H., 75, *90*, 309 (39), *379*
Hamada, M., 141, *149*
Hamana, K., 128, 130, 131, *149*
Hamer, G., 315 (107), *381*
Hamilton, P. B., 117, *120*
Hamilton-Miller, J. M. T., 126, 135, *147*
Hamre, D. M., 163, 175, 177, *181*, *182*
Hanč, O., 124, 144, *146*
Hanka, L. J., 187, 190, 193, *228*

Hanson, A., 303 (108), 307 (108), *381*
Hanton, W. K., 60, *70*
Harada, S., 126, 129, *147*
Harashima, K., 53, *70*
Harris, D. M., 167, *181*
Harris, H., 57, *70*
Harris, R., *382*
Harrison, D. E. F., 244, *246*
Harrison, E. F., 170, *181*
Harrison, J., 219, *229*
Harris-Smith, P. W., 154, *183*
Harwood, J. H., 126, 131, 132, *147*
Hash, J. A., 137, *148*
Hastings, J. J. H., 30, *45*
Hata, T., 86, *91*
Hausmann, W. K., 77, 85, 87, *91*
Havens, D. W., 177, *180*
Hawking, F., 151, *181*
Hawtrey, E., 96, *120*, 258 (78), 269 (78, 79), 270 (79), *380*
Hayes, J. A., 82, *92*
Hearn, W. R., 53, *72*
Heatley, N. G., 26, *45*
Heckley, R. J., 160, *181*
Hedrick, H. G., 250 (110, 111), *381*
Hegnauer, R., 48, *70*
Heinz, E., 265 (112), *381*
Hendlin, D., 156, *182*
Hengellar, C., 126, 130, *148*
Henning, F., 257 (172), 270 (172), *382*
Henning, F. A., 264, *381*
Hensley, M., 324 (96), 327 (96), *380*
Henssen, A., 65, *70*, 79, *90*
Hentges, D., 169, *181*
Herbert, T., 167, *182*
Heringa, J. W., 253, 255 (114), 281 (114), *381*
Herr, M. E., 126, 141, *147*, 362 (81), 363 (81), *380*
Herr, R. R., 185, 186, 189, 199, 203, 213, 221, 222, *228*, *229*
Herrell, W. E., 26, *45*, 151, 171, *181*, 185, *228*
Herz, J. E., 111, *120*
Herzog, H. L., 119, *120*, 124, 144, *146*
Hessburg, P. C., 171, *181*, *183*
Hesseltine, C. W., 81, *91*
Higashide, E., 126, 129, *147*
Higashihara, T., 310, 311 (115), *382*
Higgins, M. L., 57, 62, *72*
Hill, A., 266, *388*

Hill, I. D., 346 (117, 118), 351 (117), 353, 354 (117), 357 (117), 372 (246), 373 (246), 375, *381*, *383*, *384*
Hinshaw, H. C., 176, *180*
Hirose, Y., 317 (119), 374 (119), *381*
Hirsch, P., 107, *120*
Hitzman, D., 250 (120), 259 (121), 269 (121), 311 (122), 313 (122), 354 (122), 365 (120, 121), *381*
Hobby, G. L., 176, 178, *181*
Hockenhull, D. J. D., 134, 135, *147*
Hodson, P., 298 (123), *381*
Hoeksema, H., 185, 186, 189, 195, 203, 212, 222, *228*, *229*
Hörhold, C., 117, *121*
Hoerlein, H., 151, *181*
Holbert, P. E., 64, *71*
Holmlund, C. E., 112, 113, *121*, 126, 136, 137, *147*
Homma, I., 126, 141, *149*
Hooper, I. R., 123, *146*
Hopper, M. W., 161, *183*
Horan, A. C., 61, 62, 63, 64, *70*, *71*
Hori, M., 126, 141, *149*
Horiguchi, S., 324 (262), 327 (262), 371 (124), 365, *381*, *385*
Horning, E. S., 84, 86, *92*
Horton, D., 222, 224, *228*
Horvath, I., 84, 86, *90*
Horváth, J., 111, *121*
Hosler, P., 101, 102, 103, *120*, 329 (130), 330 (126, 127, 130), 332 (127), 333 (127), 334 (127), 336, 346 (125), 353, 373 (128), 375 (128), *381*
Hossler, F. E., *379*
Hou, C. T., 125, 134, *147*, *148*
Houck, C. L., 175, 177, *182*
Houston, C. W., 315 (32), 355 (32), *379*
Houtman, R. L., 194, *228*
Howarth, G. B., 222, *228*
Howe, C. W., 170, *181*
Howe, R., 126, 129, *147*, 260 (157), 265, 266 (158, 159), 267, 268, *382*
Huang, H. T., 126, 135, *147*, *381*
Huang, S. Y., 241, *246*
Hudson, J. O., 101, 103, 104, *120*, *121*, 329 (153, 283), 330 (153, 283, 284), 332 (153, 283), 333 (153, 283), 334 (153, 283, 284), 335 (284), 347 (153, 284), 348 (153, 284), 349 (153), 355 (153, 283), *382*, *385*
Hütter, R., 56, 62, *70*

AUTHOR INDEX

Hughes, D. F., 171, *182*
Hull, G., 307 (313), 308 (313), *386*
Humphrey, A. E., 238, *246*, 249 (132), 250 (6), 332 (133), 335 (133), 346 (133), *378*, *381*
Hungund, B. L., 372 (134), 375 (134), *381*
Hunt, G. A., 160, *181*
Hurst, E. W., 162, 174, *183*
Huybregtse, R., 253 (114), 255 (114), 281 (114), *381*

I

Ieinema, M. H., *385*
Ielasi, A., 161, *181*
Igarashi, S., 86, *91*
Iguchi, T., 278, 299 (135, 139), 300 (135, 137, 138), 302 (135), *382*
Iida, M., 157 (140, 141), 259 (143), 261 (142), 262 (142), 271, 300 (140), *382*
Iizuka, H., 111, *120*, 124, *147*, 257 (140, 141), 259 (143), 261 (142), 262 (142), 271, 300 (140), *382*, *386*
Ijichi, K., 88, *92*
Ikekawa, T., 88, *90*
Imada, Y., 273 (146, 402), 277 (146), 279 (335), 280 (335), 281 (144), 282, 291 (147), 313 (334), 314 (334), *382*, *387*, *388*
Imelik, B., 358, 360 (148), *382*
Inagawa, M., 305 (347), 308 (347), 309 (347), 310 (347), *387*
Inamine, E., 126, 134, 135, *146*, *147*
Ishiguro, E. E., 64, *70*
Ishii, R., 273, 278 (268), *382*, *385*
Ishikura, T., 257 (150), 346 (151, 168), 353 (151), 354 (168), *382*
Ito, M., 126, 136, *147*
Iwami, F., 88, *90*
Iwasaki, H., 278 (350), *387*
Iwasaki, S., 86, *91*

J

Jackson, B. G., 136, *148*
Jackson, J. F., 160, *181*
Jackson, R. W., 75, *90*
Jacobs, S. E., 251, *382*
Jagnow, G., 77, 79, 80, 81, 84, *90*
Jago, M., 125, 133, *149*
Jamison, V. W., 101, 103, 104, *120*, *121*, 329 (153, 282, 283), 330 (153, 282, 283, 284), 332 (153, 282, 283), 333 (153, 283), 334 (153, 283, 284), 335 (284), 336 (282), 347 (153, 284), 348 (153, 284), 349 (153), 355 (153, 283), *382*, *385*
Janjigian, J. A., 131, 132, *147*, *149*
Janssen, W. A., 178, *180*
Jarvis, B., 135, *147*
Jarvis, F. G., 265 (154), *382*
Jawetz, E., 171, 172, 173, 174, *181*
Jeffreys, E. G., 143, *147*
Jemski, J. V., 178, *181*
Jennings, M. A., 26, *45*
Jensen, H. L., 76, 81, *90*
Johnson, M. J., 99, *122*, 241, *246*, 265 (154), *382*
Johnson, S., 126, *148*
Jones, A., 258 (322), 269 (322), 270 (322), 281 (322), *386*
Jones, D. F., 126, 138, *147*, 260 (157), 265, 266 (156, 158, 159), 267, 268, *382*
Jones, J. K. N., 222, *228*
Jones, R. D., 135, *147*
Jones, R. J., 312 (160), *382*
Jourdonais, L. F., 154, *182*
Juhasz, S. E., 64, *69*

K

Kagan, F., 185, 186, 195, 196, 197, 199, 200, 203, 204, 205, 207, 213, 215, 219, 221, 222, *228*, *229*
Kaiser, D. G., 194, *228*
Kalakoutskii, L. V., 65, *70*
Kallio, R. E., 96, *120*, *121*, 250 (320), 257 (320), 258 (78, 321, 322), 259 (319, 320, 321), 269 (78, 79, 322), 270 (79, 322), 281 (322), 321 (95), 323 (95, 97), 326 (97), *380*, *381*, *386*
Kamikubo, T., 96, *121*, 304 (253), 306, *384*
Kamikuho, T., 305 (245), 308, *384*
Kamiya, K., 129, *147*
Kanamaru, T., 311 (161), 312 (161), *382*
Kaneshiro, W. M., 160, *183*
Karlson, A. G., 176, *180*
Karnes, H. A., 190, *229*
Kaserer, H., 251, *382*
Katho, T., 131, *148*
Katsuya, N., 278 (268), *385*, *386*
Katz, E., 142, *147*
Kauffman, F., 50, *70*
Kaufman, D. D., 108, *121*
Kawamoto, I., 301 (163), 302 (163), 303, *382*

Kawano, T., 318 (241), *384*
Kaye, D., 173, *181*
Keddie, R. M., 62, *70*
Keil, J. G., 123, *146*
Keith, D. D., 53, *72*
Keitt, G. W., 81, *92*
Keller, H. M., 177, *180*
Kelner, A., 83, *90*
Kemp, G. A., 178, *181*
Kempe, C. H., 176, *183*
Kent, T. H., 163, *180*
Kester, A. S., 255 (165), 262 (165), 281 (165), *382*
Khalid, A. M., 273 (296), *385*
Khanchandani, K. S., 363 (91), 364 (91), *380*
Khesina, A. J., 357 (76, 271, 272, 273), *380, 385*
Khokhlov, A. S., 51, *69*
Kikuchi, M., 277 (248), 279 (248), *384*
Kim, H. K., 161, *181*
Kimura, K., 275 (351), 277 (353), 278 (354), 291 (352, 354), 292 (354), 294 (349), 296 (349), 305 (352), 309 (352), *387*
Kinoshita, K., 317 (119), 374 (119), *381*
Kinoshita, S., 123, *147*, 261 (329), 265 (328, 329), 278 (350, 354), 283 (370), 285 (249, 267), 286 (249, 267), 287 (370), 291 (354), 292 (354), 294 (349), 296 (349), 301 (163), 302 (163, 250), 303 (163), 311 (328), 312 (328), *382, 384, 385, 386, 387*
Kirby, J. P., 77, 85, 87, *91*
Kirby, W. M. M., 173, *179*
Kiser, J. S., 178, *181*
Kishi, T., 126, 129, *147*
Kishida, Y., 135, *148*
Kitagawa, M., 326 (166), *382*
Kitai, A., 346 (167, 168), 353 (371), 354 (168, 371), 355 (167), *382, 388*
Kiyohara, H., 131, *148*
Kiyamoto, M., 130, *146*
Klausmeier, R. E., 345 (169), 353 (169), *382*
Kleemann, T., 96, *122*
Klein, D., 142, *147*, 257 (171, 172), 270, 271, 377, *382*
Klein, D. A., 264 (113), *381*
Kleinzeller, A., 321 (173), 322 (173), *382*
Kluepfel, D., 116, *121*

Klug, M., 253 (175, 176), 257 (174, 175), *382, 383*
Kobayashi, K., 273 (399, 402), 279 (335), 280 (335), *387, 388*
Kobayashi, T., 290 (379, 380), 291 (379, 380), 292 (379), *388*
Koch, J. R., 323 (97), 326 (97), *381*
Kodaira, R., 299 (135), 300 (135), 302 (135), 318 (241), *382, 384*
Kodama, K., 250 (177), *383*
Koenig, M. G., 173, *181*
Kohata, M., 320 (383), *388*
Kohn, S. R., 171, *181*
Kollár, G., 134, 135, *148*
Komuro, T., 302 (250), *384*
Kondo, E., 117, *121*
Kondo, S., 125, 126, 128, 130, 131, *147, 148, 149*
Kono, M., 127, *149*
Konopka, E. A., 166, 177, *181*
Konoshita, S., *384*
Konovaltschokoff-Mazoyer, M., 261 (295), *385*
Koronelli, T., 259 (180), 270 (180), *383*
Koshelva, N. A., 291 (178), 292 (178), *382*
Kosmachev, A. E., 79, *90*
Kotlarski, I., 161, *181*
Koyama, G., 141, *149*
Koyama, Y., 126, 136, *147*
Kozlowski, M. A., 114, *121*
Krámli, A., 111, *121*
Kraidman, G., 372 (246), 373 (246), 375 (179, 246), *383, 384*
Krasikova, N. V., 55, *72*
Krassilnikov, N. A., 50, 54, 65, *70*, 259 (180), 270 (180), *383*
Krause, R. M., 50, *70*
Kropp, H., 156, *182*
Kudo, R. R., 49, *70*
Kudrina, E. S., 85, *90*
Küster, E., 62, *70*, 75, 79, 83, *91*
Kulbarni, B. D., 359 (22), 366 (22), *379*
Kulczar, G., 80, *92*
Kumagaya, A., 111, *122*
Kunstmann, M. P., 126, 138, *146*
Kuroda, S., 55, *72*
Kusunose, M., 324 (255), 325 (256), 326 (256), 327 (256), *384*
Kutscher, J., 242, *246*
Kutzner, H. J., 75, 76, 84, *90*
Kvitek, K., 130, 131, 132, *146*

L

LaBrec, E. H., 169, *180*, *181*
Lacey, J., 78, *90*
Lacey, M. E., 78, 79, 83, *90*
Lago, B. D., 134, 135, *147*
Laine, B., 250 (35), *379*
Lamb, D. J., 190, 200, *229*
Lampen, J. O., 140, *147*
Lance, D. G., 222, *228*
Lancini, G. C., 126, 130, *148*
Lanéelle-Carrieu, M.-A., 60, *71*
Langley, L. F., 171, *181*
Langlykke, A. F., 126, 134, 144, *148*
Lawson, J. B., 197, *228*
Lawton, W. D., 178, *180*
Leadbetter, E., 261 (220), 270, 281 (219, 221), 311 (221), 325 (220, 221), *383*
Leask, B. G. S., 62, *70*
Leavitt, R. I., 320 (224, 225), 371 (222, 224), *383*, *384*
Leben, C., 81, *92*
Lechevalier, H. A., 47, 51, 54, 57, 58, 59, 61, 62, 63, 64, 65, 69, *70*, *71*, 74, 77, 81, 83, 85, 86, 87, *90*, *91*, *92*
Lechevalier, M. P., 47, 51, 57, 58, 59, 61, 62, 63, 64, 65, 69, *70*, *71*, 74, 77, 83, 85, *91*
Lee, G., 305 (226), 310 (226), *384*
Lee, S. S., 112, 113, 117, 119, *121*
Leigh, L. C., 158, *180*
Lein, J., 126, 135, 136, *146*, 160, *181*
Lenert, T. F., 176, *181*
Leon, M. M., 357 (234), *384*
Lepper, M. H., 173, *181*
LeQuesne, P. W., 52, *69*
Lester Smith, E., 21, *45*
Leurs, H., 6, *45*
Levy, R., 151, *182*
Lewin, R. A., 49, *71*
Lewis, A., 123, 133, 136, *149*
Lewis, C. N., 185, 186, 188, 190, 193, 198, 218, 219, *228*, *229*
Lewis, J. C., 88, *92*
Lewis, L., 166, 177, *181*
Lilly, D. M., 73, *91*
Lin, H., 259 (143), *382*
Lincks, M., 137, *148*
Linday, E. M., 95, *121*, 258 (227), 262, *384*
Lindenbein, W., 81, *91*
Lindenfelser, L. A., 75, *90*

Lingappa, Y., 75, *91*
Lingens, F., 142, *148*
Lipman, R. L., *181*
Littman, M. L., 86, *91*
Lloyd, A. B., 78, 80, *91*
Locci, R., 50, 62, 65, 69, *71*, 79, *91*
Lockwood, J. L., 75, *91*
Long, P. H., 173, *180*
Lorch, H., 126, 136, *150*
Lovell, R., 25, *45*
Lovrekovich, I., 84, 86, *90*
Luedemann, G., 83, *91*
Lukins, H. B., 254 (228), 270, *384*
Lummis, W. L., 199, *229*
Luria, S. E., 50, *71*
Luttinger, J. R., 126, 135, 136, *146*

M

Mabe, J. A., 320, 376, *384*
Mabry, T. J., 324 (96), 327 (96), *380*
McAleer, W. J., 114, *121*
McCarter, J. C., 165, *183*
McClung, N. M., 94, *121*
McConville, C., 170, *183*
McCormick, J. R. D., 126, *148*
McCoy, E., 78, 81, *92*
McDermott, W., 178, *181*
McDonald, T. J., 171, *182*
McDurmont, C. I., 65, *71*
MacGahren, W. J., 126, 138, *146*
Machida-shi, 275 (351), 277 (353), 291 (352), 305 (352), 309 (352), *387*
MacIver, A. M., 50, 55, *70*, 78, *90*
McKee, C. M., 163, 175, 177, *181*, *182*
Mackel, D. C., 171, *181*
MacKellar, F. A., 185, 186, 190, 193, *228*, *229*
McKenna, E. J., 97, *121*, *384*
MacLeod, C. M., 158, *181*
McMeel, J. W., 171, *182*
MacPhillamy, H. B., 163, *181*
Maddix, C., 244, *246*
Madhyastha, K. M., 371 (230), 374 (230), 375 (231), *384*
Maeda, K., 125, 126, 128, 130, 131, *148*, *149*
Maffii, G., 123, *149*
Magerlein, B. J., 123, *148*, 185, 186, 188, 195, 196, 197, 199, 200, 203, 204, 205,

207, 212, 213, 215, 218, 219, 220, 221, 222, 228, 229
Maggi, N., 123, 149
Magnani, T. J., 169, 181
Magyar, K., 84, 86, 90
Majumdar, M. K., 131, 144, 148
Majumdar, S. K., 131, 144, 148
Maksimova, T. S., 85, 90
Malby, P. G., 39, 45
Malek, I., 250 (232), 384
Maliyanta, A. A., 250 (233), 384
Mallet, L., 357 (251), 384
Mallet, M. L., 357 (234), 384
Mamoli, L., 111, 121
Manachini, P. L., 55, 56, 69, 71
Marcus, S., 163, 182
Markovetz, A., 253 (175, 176), 257 (174, 175), 382, 383
Marlatt, V., 114, 121
Marr, E. E., 322 (235), 384
Marshall, D. J., 115, 121
Martin, A. R., 174, 183
Martin, J. H., 137, 148
Martin, W. R., 168, 180
Maseles, F. C., 321 (95), 323 (95), 380
Mason, D. J., 125, 126, 131, 139, 144, 146, 185, 186, 187, 188, 190, 193, 218, 225, 227, 227, 228, 229
Mastner, J., 238, 246
Mateles, R. I., 250 (236), 384
Matsubara, I., 261 (329), 265, (329), 386
Matsubayashi, T., 304 (237), 320 (237), 384
Matsui, M., 123, 147
Matsuo, A., 131, 148
Mayama, M., 81, 91
Mayberry, W. R., 315, 318 (277), 384, 385
Meisel, M. N., 358 (76, 271, 272, 273), 380, 385
Meister, P. D., 111, 120
Metzler, C. M., 194, 229
Meyers, E., 142, 148
Meyers, S. P., 78, 90
Meynell, G. G., 168, 182
Middlebrook, G., 179, 182
Migashide, E., 129, 147
Mihm, J. M., 50, 62, 70, 94, 120
Mil'ko, E., 304 (240), 306 (240), 384
Miller, A. K., 153, 154, 156, 157, 159, 162, 163, 165, 174, 182, 183
Miller, A. L., 126, 131, 144, 148

Miller, C. P., 154, 168, 175, 177, 180, 182
Miller, I. M., 86, 91, 125, 136, 146
Miller, P. A., 126, 137, 148
Millis, N. F., 238, 246, 250 (6), 378
Mills, A., 311 (122), 313 (122), 354 (122), 381
Mimura, A., 318, 384
Miraglia, G. J., 166, 177, 182
Misawa, M., 301 (163), 302 (163, 250), 303 (163), 382, 384
Mitscher, L. A., 137, 148
Mitsuhashi, S., 128, 130, 131, 149
Miyachi, N., 346 (151), 354 (151), 382
Miyamoto, Y., 318 (403), 388
Miyazaki, T., 131, 148
Miyoshi, M., 250 (242), 384
Mizoguchi, T., 131, 148
Moguelevskii, G. A., 250 (244), 384
Moholay, M. N., 369 (308), 370 (308), 365 (308), 386
Molton, P., 252 (24, 25), 338 (24), 379
Moore, R. H., 126, 129, 138, 147
Mordarska, H., 60, 71
Morganroth, J., 151, 182
Mori, M., 235, 246
Morikawa, H., 96, 121, 305 (245), 308, 384
Morin, R. B., 136, 148
Morozowich, W., 190, 193, 200, 229
Mortimer, A., 84, 90
Mott, T. J., 162, 179
Muench, H., 162, 182
Muir, R. D., 112, 115, 120
Mukherjee, B. B., 372 (246), 373 (246), 376 (179, 246), 383, 384
Murakami, S., 310 (346), 387
Murao, S., 135, 148, 149
Murase, M., 141, 149
Murooka, H., 111, 122
Murphy, J. F., 338 (247), 344 (247), 353 (247), 384
Murray, H. C., 111, 120, 126, 141, 147, 362 (81), 363 (81), 380
Muschel, L. H., 163, 180, 182
Muto, T., 163, 182

N

Nadelson, J., 53, 72
Nagasaki, T., 305 (344, 345, 347), 308

(345, 347), 309 (344, 347), 310 (346, 347), 387
Nagatsu, J., 53, 70
Naito, A., 111, 120, 124, 147
Nakadate, M., 222, 224, 228
Nakahara, T., 298 (90, 398), 313 (398), 380, 388
Nakamiro, S., 286 (301), 386
Nakamura, G., 86, 91
Nakamura, M., 171, 180
Nakano, K., 123, 147
Nakao, Y., 277 (248), 279, 384
Nakase, T., 290 (379, 380), 291 (379, 380), 292 (379), 388
Nakayama, K., 285 (249), 286 (249), 384
Nakayama, Y., 86, 91
Namestnikova, V. P., 55, 72
Napier, E. J., 126, 139, 146
Nara, T., 301 (163), 302 (163, 250), 303 (163), 382, 384
Navalkar, R., 179, 183
Naylor, J. H. C., 37, 45
Neilands, J. B., 126, 136, 150
Neklyndova, L. V., 292 (72), 380
Nelson, G., 309 (39), 381
Nemec, P., 88, 89, 90
Nette, I. T., 291 (178, 306), 292 (178, 306), 383, 386
Nettleton, D. E., 123, 146
Newkirk, J. F., 125, 136, 146
Newth, F. H., 334 (16), 335 (16), 337 (16), 378
Newton, G. G. F., 34, 45
Niaussat, M. M., 357 (251), 384
Nicolaus, B. J. R., 88, 91
Nicolson, N. J., 244, 246
Nikitina, K., 310 (278), 385
Nikitina, N. I., 291 (306), 292 (306), 386
Nimi, O., 131, 144, 148
Ninet, L., 126, 129, 144, 148, 309 (252), 384
Nishida, H., 346 (151), 353 (151), 382
Nishikawa, M., 129, 147
Nishimura, H., 81, 91
Nishio, N., 304 (253), 306, 384
Nissen, T. V., 142, 148
Nobles, M. K., 52, 71
Nocard, E., 62, 71
Nomi, R., 81, 91, 131, 144, 148
Nonomura, H., 76, 83, 91

Norton, R. L., 244, 246
Nosaka, J., 324 (255), 325 (256), 326 (256), 327 (256), 384
Novak, E., 199, 229
Novakovskaya, N. S., 292 (72), 380
Novick, R. P., 133, 148
Noyes, R., 250 (257), 326 (257), 384
Nungester, W. J., 154, 182
Nutting, L. A., 345 (258), 353 (258), 384
Nyns, E. J., 249 (259), 250 (259), 384

O

Oblisami, G., 77, 91
O'Conner, R. J., 321 (260), 322 (260), 384
Oesterling, T. O., 190, 194, 197, 201, 229
Ogata, K., 86, 91
Ogino, S., 262 (261), 385
Ogura, M., 125, 130, 131, 147, 149
Ohara, Y., 76, 83, 91
Ohata, Y., 131, 148
Ohishi, N., 305 (348), 308 (348), 387
Ohki, E., 222, 229
Okada, H., 317 (119), 374 (119), 381
Okamoto, S., 125, 126, 127, 128, 131, 148, 149
Okanishi, M., 125, 126, 128, 130, 131, 148, 149
Okii, M., 283 (370), 285 (267), 286 (267), 287 (370), 385, 387
Okuda, S., 86, 91
Okumura, S., 290 (379, 380), 291 (379, 380), 292 (379), 388
Olitzki, L., 154, 182
Oliver, T. J., 51, 71
Olsen, A. J. C., 44, 45
Oltmanns, O., 142, 148
Omori, T., 324 (262, 263, 264), 325 (263), 327 (262, 263, 264), 384, 385
Ooyama, J., 249 (265), 357, 360 (265), 361 (265), 362 (265), 364, 385
Orleanskii, V. K., 85, 90
Ornston, L. N., 321 (266), 385
Orr-Ewing, J., 26, 45
Osbourne, R. R., 176, 183
Oshima, K., 283 (370), 285 (267), 286 (267), 287 (370), 385, 387
Ostrander, W. E., 177, 180
Otani, S., 126, 136, 150
Otsuka, S., 273 (149), 278, 285 (382), 289 (382), 382, 385, 386, 388

Ozaki, A., 346 (151, 167, 168), 354 (151, 371), 355 (168, 371), 356 (167), *383, 388*
Ozanne, B., 126, 130, 131, 132, *146*

P

Pagani, H., 65, *72*
Pakhashi, J., 281 (144, 145), 282 (145), *383*
Pang, H.-N., 50, *70*
Pansy, F. E., 86, *91*
Pantskhava, E. C., 273 (146), 277 (146), 281 (146), *382*
Paplanus, S., 167, *180*
Patel, N. C., *229*
Patterson, E. L., 136, *147*
Payne, W. J., 315 (238, 269), 318 (277), *384, 385*
Pearson, J. M. H., 166, *182*
Penn, W. P., 171, *181, 183*
Perlman, E., 111, *120*, 125, 126, 131, 134, 144, *147, 148, 149*
Perry, J. J., 86, *91*
Peterson, D. H., 111, *120, 121*
Peterson, W. J., 251 (105), *381*
Petrolini, B., 65, *71*
Pfeifer, V. F., 232, *246*
Phillips, D. H., 241, *246*
Phillips, G. B., 178, *181*
Phillips, T., 278 (312), *386*
Phillips, U., 276 (270), *385*
Pienta, P., 142, *147*
Piepoli, C. R., 161, *183*
Piffaretti, J. C., 128, *148*
Pine, L., 58, *69*
Piper, D. L., 151, *183*
Pisano, M. A., 86, *91*
Pitsch, B., 167, *182*
Pitton, J. S., 128, *148*
Pochon, J., 74, *91*
Poglazova, M. N., 358 (76, 271, 272, 273), *380, 385*
Pollock, M. R., 133, *146, 147, 148*
Porter, J. N., 75, 76, 77, 80, 81, 85, 87, *91*
Powers, K. G., 219, *229*
Pradhan, S, K., 364 (22), 366 (22), *379*
Pramer, D., 74, *91*, 124, 142, 143, *147, 148*
Prauser, H., 51, 57, 62, 63, 64, 72, 86, *91*
Prema, B. R., 359 (22, 274, 275), 364 (21), 366 (21, 22, 274, 275), 367 (21), *379, 385*
Preobrazhenskaya, T. P., 51, *71*
Prescott, N., 131, 132, *149*

Prevot, A. R., 250 (276), *385*
Price, K. E., 123, 135, *146, 149*
Pridham, T. G., 51, *71*, 75, 81, *90, 91*
Priou, M. L., 357 (234), *384*
Prochazka, G. J., 315 (238), 318 (277), *384, 385*
Prokop, J. F., 84, *92*
Pugh, G. W., Jr., 171, *182*

R

Rabotnova, I., 304 (240), 306 (240), 310 (278), *384, 385*
Ragan, E. A., 123, *146*
Rainbow, C., 73, *91*
Raistrick, H., 25, *45*
Rake, G., 162, 163, 175, 177, *181, 182*
Ramachandran, B. V., 364 (21), 366 (21), 367 (21), *379*
Rangachari, P. N., 368 (60), 369 (60), 372 (134), 365 (60), 375 (134), *380, 381*
Rangaswami, G., 77, *91*
Rao, K. V., 126, 140, *149*
Ratledge, C., 96, *121*, 258 (279, 280), 263, *385*
Raymond, R., 250 (281), 269 311, 312 (281), *385*
Raymond, R. L., 95, 99, 101, 103, 104, *120, 121*, 273, 319, 324 (66), 329 (153, 282, 283), 330 (153, 282, 283, 284), 332 (133, 153, 282, 283), 333 (153, 283), 334 (153, 283, 284), 335 (133, 284), 336 (282), 337 (54, 55), 340 (54), 342 (54, 55), 343 (55), 345 (55), 346 (133), 347 (153, 284), 348 (67, 153, 284), 349 (67, 153), 355 (67, 153, 283), 362 (54, 55), 359, 365, 372 (55), *379, 380, 381, 382, 385*
Reber, H., 57, 62, *72*
Reed, L. J., 162, *182*
Rees, R. J. W., 166, *182*
Rehacek, Z., 76, 84, 87, *91, 92*
Rehm, H. J., 84, *91*
Reisman, H. B., 234, *246*
Renant, J., 309 (252), *384*
Renn, R. W., 126, 140, *149*
Renz, K. J., 166, 177, *182*
Reynolds, R. J., 250 (110, 111), *381*
Ribbons, D. W., 321 (285), *385*
Richmond, M. H., 133, *148, 149*
Ring, K., 242, *246*
Ritter, R., 117, *121*

Roberts, A. N., 241, *246*
Robinson, D., 255 (11, 12), *378*
Rocha, H., 166, *182*
Rodgers, G. C., 53, *72*
Rodionova, G., 307 (286), *385*
Roeske, R. W., 136, *148*
Rogers, H. J., 158, *180*
Rogers, P. J., 39, 40, *45*
Rogoff, M., 250 (287), 346 (288, 289), 347 (292), 348 (289), 349 (289), 351 (290, 291), 353 (288, 289), 354 (289), 355 (292), 356, *385*
Rolinson, G. N., 37, *45*
Romano, A. H., 57, *71*
Rose, A. H., 73, *91*
Rosenfeld, S. M., 292 (72), *380*
Ross, J., 4, *45*
Routien, J. B., 84, *92*
Rowe, E. L., 190, 201, *229*
Rudolfs, W., 250 (293), *385*
Ruinen, J., *385*
Russell, H. E., 170, *182*
Ryan, K. J., 115, *121*
Rynearson, T. R., 50, *70*

S

Sabath, L. D., 125, 133, *149*
Saeki, H., 222, *229*
Saito, Y., 126, 136, *150*
Sakaguchi, K., 135, *149*
Sanders, A. G., 26, *45*
Sands, L., 128, *149*
Sano, Y., 86, *91*
Sartori, G., 126, 130, *148*
Sassiver, M. L., 123, 133, 136, *149*
Sato, A., 310, 311 (115), *381*
Sato, R., 126, 136, *147*
Savage, G. M., 186, *228*
Sawa, T., 126, 141, *147*, *149*
Sawyer, W. D., 161, *183*
Sax, K. J., 112, 113, *121*
Sayasaka, S. S., 264 (113), *381*
Schallock, M. R., 125, 134, *147*
Schissler, D., 258 (322), 269 (322), 270 (322), 281 (322), *386*
Schlecht, S., 242, *246*
Schlegel, H. G., 143, *149*
Schmidt, L. H., 175, 177, *182*, 219, *229*
Schmitz, H., 123, *146*
Schneidau, J. D., Jr., 94, 111, *121*

Schneider, H., 169, *180*, *181*
Schneller, G. H., 151, *183*
Schnepf, E., 79, *90*
Schnitzer, R. J., 174, 175, 178, *182*, *183*
Schreiber, R. H., 123, *146*
Schroeder, W., 185, 186, 195, 213, *228*, *229*
Schubert, K., 117, *121*
Schwab, G., 123, *146*
Schwartz, J. L., 126, 131, *149*
Schewe, E., 164, *182*
Sebek, O. K., 126, 131, 139, 144, *146*, 225, 227, *227*
Sedi, M. H., 273 (297), *385*, *386*
Sekijo, C., 304 (376, 377), 306 (376, 377), *388*
Selbie, F. R., 170, *182*
Senez, J., 261 (295), *385*
Senez, J. C., 271 (17, 37), *378*, *381*
Seno, S., 278 (136), *382*
Sensi, P., 123, 126, 130, *148*, *149*
Servin-Massieu, M., 160, *182*, *183*
Sesler, C., 175, 177, *182*
Seto, T. A., 126, 135, *147*
Sevceik, V., 74, *92*
Shabad, L. M., 358 (76, 271, 272, 273), *380*, *385*
Shaffer, M. F., 94, *121*
Shah, J. H., 273 (296), *386*
Shapiro, A. P., 166, *181*
Sharpley, J. M., 250 (298), *386*
Shaw, R. K., 134, *149*
Shaw, W. V., 126, 127, 128, *149*
Shay, A. J., 112, 113, *121*, 126, 137, *147*
Sheikh, T. H., 273 (297), *386*
Shepherd, P. G., 241, *246*
Shigeo, H., 255 (15), 262 (15), *378*
Shiio, I., 273 (149), 278 (268), 281, 285 (382), 286 (301), 289 (382), *382*, *385*, *386*, *388*
Shimakara, I., 259 (305), 269, *386*
Shimizu, M., 123, *147*
Shimizu, S., 96, *120*, 304 (87), 307 (87), 308 (88), *380*
Shirling, E. B., 50, 55, *70*, *71*, 81, *92*
Shoji, O., 304 (376, 377), 306 (376, 377), *388*
Shpon'ko, R. I., 291 (306), 292 (306), *386*
Shu, P., 137, *148*
Shukla, O. P., 369 (308), 370 (308), 365 (307, 308), *386*

Shull, G. M., 111, 117, *120*, *121*, 126, 135, 137, *146*, *147*
Shultz, J. S., 112, 113, *121*
Sih, C. J., 112, 113, 114, 115, 116, 117, 118, 119, *120*, *121*, *122*
Sikyta, B., 238, *246*
Silver, H. K., 176, *183*
Simon, R. D., 170, *182*
Simpson, R. B., 81, *91*
Singh, K., 115, *121*
Sinkula, A. A., 190, 193, *229*
Sjolander, N. O., 126, *148*
Slack, J. M., 57, *71*
Slomp, G., 185, 186, 189, *228*, *229*
Small, J., 166, *181*
Smith, A. H., 49, *71*
Smith, C. B., 173, *183*
Smith, D. A., 142, *148*
Smith, D. H., 126, 131, 132, *147*, *149*
Smith, D. W., 161, 179, *181*, *183*
Smith, G. N., 126, 136, 138, *149*
Smith, H., 154, *183*
Smith, R. E., 123, *146*
Sneath, P. H. A., 48, *72*
Snell, N., 88, *92*
Sohler, A., 57, *71*
Sohngen, N. L., 251 (310, 311), *386*
Sokal, R. R., 48, *72*
Sokolski, W. T., 142, *149*, 160, *183*
Solomons, G. L., 232, 233, 240, 241, *246*
Solotorovsky, M., 65, *71*
Somerson, N., 278 (312), *387*
Sommerfield, G. A., 126, 139, *146*
Sonoda, H., 317 (119), 365 (119), *381*
Soo-Hoo, G., 174, *183*
Sophian, L. H., 151, *183*
Speck, R. S., 174, 178, *181*, *183*
Speedie, J., 307 (313), 308 (313), *386*
Spencer, E. L., 86, *92*
Spencer, J. F. T., 21, 41, *45*, 259 (314), 264 (314), 265 (100, 112), 266, *381*, *386*, *388*
Stanier, R. Y., 48, *72*, 250 (316), 321 (266), *385*, *386*
Stanley, J. L., 154, *183*
Stapert, E. M., 160, *183*
Stapley, E. O., 86, *91*
Starkey, R. L., 80, 92, 142, 143, *148*, *150*
Steel, R., 250 (317), *386*
Steenken, W., Jr., 176, *183*

Stein, B., 117, *121*
Stern, K. F., 190, *229*
Stevens, N., 259 (318), 269 (318), *386*
Stevenson, D., 250 (320), 258 (322), 259 (319, 320), 269 (322), 270 (322), 281 (322), *386*
Stewart, J., 250 (320), 257 (320), 258 (321, 322), 259 (320, 321), 269 (322), 270, 281 (322), *386*
Stewart, J. E., 96, *121*
Stewart, R. H., 160, *183*
Stillwell, R. H., 73, *91*
Stone, R. W., 250 (323), 322 (235), 326 (83), 338 (247), 344 (247), 345 (325), 353 (247, 325), *379*, *380*, *386*
Stotzky, G., 80, *92*
Stoudt, T. H., 111, 114, *121*, 281 (48), 285 (48), *381*
Strawinski, R., 251, 338, 345 (169, 325), 353 (169, 325), *381*, *382*, *386*
Subbaith, T. V., 168, *182*
Sueoka, N., 55, *72*
Suhadolnik, R. J., 126, 140, *149*
Sukapanit, S., 161, *183*
Sukapure, R. S., 57, 62, *72*
Sun, F. F., 200, *229*
Šuput, J., 58, *72*
Surgalla, M. J., 178, *180*
Suzubi, Y., 304 (237), 320 (237), *384*
Suzuki, M., 277 (248), 279 (248), *384*
Suzuki, T., 126, 129, *147*, 261 (329), 265 (328, 329), 278 (354), 291 (354), 292 (354), 311, 312 (328), *388*, *389*
Suzuki, Y., 125, 126, 127, 128, 131, *149*, 306 (331), *386*
Swaminathan, R., 77, *91*
Swift, M. E., 86, *90*
Szanislo, P. J., 66, *72*
Szarek, W. A., 222, *228*
Szybalski, W., 142, *149*

T

Tabuchi, T., 294 (330), 296, 297 (4, 5), *378*, *386*, *387*
Tadao, T., 306, *386*
Tadra, M., 124, 144, *146*
Taggart, M. S., 250 (332), 251, 252 (333), *386*
Tai, H. H., 119, *121*

Takahashi, J., 273 (399, 402), 279 (335), 280, 313 (334), 314, 315 (384), 387, 388
Takahashi, M., 386
Takahashi, T., 123, 147
Takara, Y., 387
Takasawa, S., 125, 131, 149
Takazawa, S., 126, 130, 131, 149
Takeda, I., 278 (136), 299 (135, 139), 300 (135, 137, 138), 302 (135), 382, 384
Takeuchi, T., 126, 141, 147, 149
Tamura, G., 262 (261), 385
Tanaka, A., 305 (343, 344, 345, 347, 348), 308 (343, 345, 347, 348), 309, 310 (347), 380, 387
Tanaka, H., 285 (249), 286 (249), 384
Tanaka, K., 261 (329), 265 (328, 329), 275 (351), 277 (353), 278 (350, 354), 283 (370), 285 (267), 286 (267), 287 (370), 291 (352, 254), 292, 294 (349), 296 (349), 305 (352), 309, 311 (328), 312 (328), 385 386, 387
Tanaka, M., 294 (330), 296 (330), 297 (5), 378, 386, 387
Tanaka, N., 125, 126, 130, 131, 146, 147, 149
Tanaka, T., 53, 70
Tanguay, A. E., 159, 183
Tannenbaum, S. R., 250 (236), 384
Tanno, K., 346 (151), 354 (151), 382
Tarama, K., 318 (356), 387
Tarazka, A. J., 194, 228
Taraszka, M. J., 190, 229
Tardieux, P., 74, 91
Tattersfield, F., 338 (357), 387
Tatum, E., 303 (358), 387
Tausson, T. A., 252 (359), 387
Tausson, W. O., 251, 252 (360, 364), 338 (361), 356 (364), 387
Tawara, K., 81, 91
Taylor, H. F., 107, 121
Teillon, J., 54, 72
Teles, E. da S., 166, 182
Tendler, M. D., 79, 83, 92
Tetrault, P. A., 79, 92
Tewfik, E., 55, 72
Thaysen, A. C., 251, 387
Thiemann, J. E., 65, 66, 72
Thijsse, G., 253, 255 (369), 256 (367, 369), 281 (368), 387
Thijsse, G. J. E., 250 (386), 304 (386), 388

Thirumalachar, M. J., 62, 65, 72
Thoma, R. W., 111, 120
Thompson, A. R., 315 (32), 355 (32), 379
Thornton, H. G., 339 (101), 355, 381
Thorp, J. M., 174, 183
Timonin, M. I., 80, 92
Tipper, D., 126, 130, 131, 132, 148, 150
Tissieo, R., 309 (252), 384
Tokoro, Y., 283 (370), 285 (267), 286 (267), 287 (370), 385, 389
Tone, H., 346 (168), 354 (371), 354 (168, 371), 382, 388
Tranter, E. K., 110, 120, 121
Traxler, R. W., 320, 387
Treccani, V., 97. 98. 99, 121, 122, 250 (373), 338, 344 (375), 347 (33, 374), 351 (42, 43), 352 (43), 355 (33, 374), 356 (42, 43), 357, 360 (44), 379, 388
Treick, R. W., 187, 190, 193, 228
Tresner, H. D., 51, 72, 75, 82, 91, 92
Triyanond, C., 161, 183
Tronchet, J. M. J., 222, 224, 228
Truant, J. P., 171, 181, 183
Tsao, P. H., 81, 92
Tsaplina, I. A., 55, 70
Tsong, Y. Y., 113, 117, 119, 120, 121
Tsuboi, T., 304 (376, 377), 306, 388
Tsuchida, N., 53, 70
Tsuda, K., 86, 91
Tsugawa, R., 290 (378, 380), 291 (378, 379, 380), 292 (378, 379), 388
Tsyganov, V. A., 55, 72
Tulloch, A. P., 259 (314), 264 (314), 265 (100, 112, 315), 266, 381, 386, 388
Turfitt, G. E., 111, 118, 122
Tybring, L., 136, 147
Tyrell, E., 166, 181

U

Uchida, K., 281 (144, 145), 282 (145), 382
Uchio, R., 281, 285 (382), 289 (382), 386, 388
Udagawa, K., 320 (383), 388
Udagawa, S., 86, 91
Uemura, N., 315 (384), 388
Umbreit, W. W., 78, 92, 245, 246
Umezawa, H., 54, 72, 86, 88, 90, 92, 125, 126, 128, 130, 131, 141, 146, 147, 148, 149
Unami, Y., 261 (142), 262 (142), 382

Ungar, J., 174, *183*
Updegraff, D. M., 250 (56), *379*
Urawa-shi, 304 (237), 306 (331), 320 (237), *384*, *386*
Uridil, J. E., 79, *92*
Utahara, R., 125, 126, 128, 130, 131, *148*, *149*
Uzu, K., 123, *147*

V

Valient, M. E., 156, 170, *180*, *182*
Valyi-Nagy, T., 80, *92*
van den Berg, G., *388*
van der Hoeven, M. G., 135, *150*
van der Linden, A. C., 250 (386), 253 (114), 255 (114, 369), 256 (369, 385), 272 (385), 281 (114, 368), 304 (386), *381*, *387*, *388*
Van de Voorde, H., 163, *180*
Van Dijck, P., 163, *180*
Vanek, Z., 84, 87, *92*
Vangedal, S., 136, *147*
Veda, K., 315 (384), *388*
Vegetti, G., 50, *69*
Veldkamp, 62, 67, 72, 143, *149*, *388*
Venice, L. A., 171, *181*
Vercellone, A., 111, *121*
Vernet, C., 250 (35), *379*
Verrier, J., 126, 129, 144, *148*
Verwey, W. F., 153, 162, 163, 170, 174, *182*, *183*
Vézina, C., 115, 116, *121*
Villanueva, J. R., 110, 111, *122*
Vischer, E., 111, *122*
Vojnovich, C., 232, *246*
von Daehne, W., 126, 136, *147*, *150*
Vondracek, M., 88, *90*
von Meyenburg, K., 243, *246*

W

Wada, Y., 129, *147*
Wagner, F., 96, 98, *122*, 271, *388*
Wagner, J. G., 199, *229*
Wagner, R., 251 (389), *388*
Wahid, W., 273 (296), *387*
Wain, R. L., 107, *121*
Wakaki, S., 123, *147*
Wakashiro, T., 126, 141, *149*
Waksman, S. A., 25, *45*, 54, 72, 74, 76, 78, 80, 83, 84, 86, 87, 92, 94, *122*
Walder, C., 305 (226), 310 (266), *384*

Walker, J. B., 126, 131, 144, *148*
Walker, N., 98, *122*, 324 (41), 326 (41), 338 (375), 344 (375, 390), *379*, *388*
Walker, P. M. B., 56, *72*
Walton, R. B., 125, 126, 136, *146*, *150*
Wang, K. C., 113, 117, 118, 119, *120*, *121*, *122*
Warren, G. H., 161, *183*
Warren, H. B., Jr., 84, *92*
Warren, R. A. J., 126, 136, *150*
Warth, P. T., 173, *180*
Wasserman, H. H., 53, *72*
Watanabe, S., 111, *122*
Watanabe, T., 299 (139), 300 (135, 137, 138), *382*
Watson, D. W., 156, *183*
Watson, R. W., 245, *246*
Webley, D. M., 97, 100, 105, 107, *122*, 281 (392), 331 (391, 392), 336, 338 (392), 340 (391, 392), 341 (391, 392), 349 (392), *388*
Weddell, A. G. M., 166, *182*
Wegner, G. H., 98, *122*, 345 (393), 355, *388*
Weinrich, B. W., 321 (260), 322 (260), *386*
Weissburger, H. W. O., 135, *150*
Welsch, M., 84, *92*
Wender, I., 347 (292), 351 (290, 291), 355 (292), 356, *385*
Werner, A. S., 167, *180*
Wessley, D. F., 346 (394), 353 (394), *388*
West, J. M., 240, *246*
West, M. K., 162, 163, 170, *182*, *183*
Wettstein, A., 111, *122*
Weyland, H., 78, *92*
Whaley, J. W., 84, *92*, 136, *147*
Wheater, D. W. F., 162, *183*
Whitehead, B. K., 134, 135, *147*
Whitlock, H. W., Jr., 117, *120*, *121*
Whitmarsh, J. M., 118, *122*
Wiaux, A. L., 249 (259), 250 (259), *384*
Wiegeshaus, E. H., 161, 179, *181*, *183*
Wieland, T., 99, *122*, 321 (395), 322 (395), *388*
Wiggins, R. L., 170, *183*
Wilfert, J. N., 173, *183*
Wilhelm, J. J., 75, 76, 77, 80, 81, 85, 87, *91*
Wilkins, J. R., 186, *228*
Williams, R. P., 53, *72*
Williams, S. T., 62, 64, 65, 72, 75, 83, *91*, *92*

Williston, E. H., 176, *183*
Wilmer, D. L., 153, *182*
Wiltshire, G. H., 338 (375), 344 (375, 390), *388*
Wiltshire, G. M., 98, *122*
Witkamp, M., 143, *150*
Wodzinski, R. S., 99, *122*
Wolf, A. A., 154, *182*
Wolfe, R. S., 64, *70*
Wolinsky, E., 176, *183*
Wolochow, H., 178, *183*
Wood, J. M., 321 (98), *381*
Wood, R. W., 171, *182*
Woodruff, H. B., 84, *92*
Worcester, P., 219, *229*
Work, S., 151, *183*
Work, T. S., 151, *183*
Worrel, C. S., 126, 136, 138, *149*
Wright, D. N., 160, *183*
Wright, L., 303 (71, 396, 397), 305 (71, 396, 397), *382*, *388*
Wu, R. Y., 55, *72*

Y

Yajima, T., 88, *92*
Yamada, K., 250 (177), 273 (146, 402), 277 (146), 279 (335), 280 (335), 281 (144, 145, 146), 282 (145), 291 (147), 298 (90, 398), 313 (334, 398), 314 (334), 324 (262, 263, 264), 325 (263), 327 (262, 263, 264), 357, 360 (14), 371 (124), 365, *378*, *380*, *381*, *382*, *383*, *387*, *388*
Yamada, T., 126, 130, 131, 132, *146*, *150*

Yamaguchi, H., 86, *91*
Yamaguchi, K., 264, 275 (351), 277 (353), 278 (354), 283 (370), 287 (370), 291 (352, 354), 292 (354), 305 (352), 309 (352), *378*, *387*
Yamaguchi, T., 55, 57, *72*
Yamamoto, H., 126, 129, *147*
Yamane, T., 318 (403), *388*
Yamashita, H., 259 (305), 269, *386*
Yamashita, K., 290 (379, 380), 291 (379, 380), 292 (379), *388*
Yamashita, S., 235, *246*
Yamatodani, S., 311 (161), 312 (161), *382*
Yans, K., 262 (261), *385*
Yantzi, M. F., 251 (106), *381*
Yeager, R. L., 142, *149*
Yoshida, J., 318, *389*
Yoshioka, H., 324 (96), 327 (96), *380*
Youmans, A. S., 165, *183*
Youmans, G. P., 165, 176, *183*
Yukioka, M., 126, 136, *150*
Yurchenko, J. A., 161, *183*
Yurchenko, M. C., 161, *183*

Z

Zahn, W., 96, 98, *122*, 271 (388), *388*
Zajic, J. E., 346 (99, 404), 353 (99, 404), *381*, *388*
Zevenhuizen, L., *388*
Zobell, C. E., 250, *386*, *388*
Zoganas, H. C., 166, 177, *181*
Zwartouw, H. T., 154, *183*

SUBJECT INDEX

A

Acetone fermentation, 12, 13, 15–18
Actinomycetes
  antibiotic producing, 83–85
  classification, 62–69, 81, 82
  distribution, 73
  isolation methods, 74, 75
  occurrence in soils, 76, 77
  in waters, 77, 78
6-Aminopenicillanic acid, 37
Animal feed supplements from fermentation, 20
Antibacterial evaluations, *in vivo*
  eye test, 170–171
  general tests, 152–155
  infectious organisms, 156–161
  intestinal tract test, 167–168
  host species, 161–162
  skin test, 169–170
  TB tests, 165
  urinary tract test, 166–167
Antibiotic resistance *in vivo*, 175–178
Antibiotic transformations
  acylation, 126–130
  adenylation, 131–132
  chemical, 123
  deamination, 140–141
  demethylation, 139–140
  hydrolysis, 132–136
  microbial, 123, 124, 125
  oxidation, 136–138
  phosphorylation, 130–131
  reduction, 136
  sulfoxidation, 138–139
Antibiotics from *actinomycetes*, 86–89
Arzberger, F., 14, 15
Ascorbic acid, 21
Aureomycin, 33

B

Bacitracin, 33
Brew in Great Britain
  analysis, 7
  history, 2–6
  training, 7, 8

C

Celesticetin,
  biological properties, 186, 187
Cell wall composition as taxonomic tool, 57–59
Cephalosporins, 34, 35, 37, 38
Chemotaxonomy
  algae, 48, 49
  bacteria, 50
  fungi, 49
  higher plants, 47, 48
Citric acid, 23–25
Commercial Solvents Corporation, 14, 16
Continuous fermentations, 42–44

D

Deoxyribonucleic acid base composition as taxonomic tool, 55, 56
Distillers company, 15, 18, 19

E

Enzymes, 38–40

F

Fermentation equipment
  analytical instrumentation
    fermentor gas effluent, 243
    gas analysis, 240, 241, 242
    rheology, 240
  aseptic metering pumps, 234
  computer techniques, 232, 235
  mechanical seals, 233
  sterilization of culture medium, 238
  temperature control, 236
  temperature indication, 237
Fermentation patents, 35, 38
Fermentation wastes, 19

G

Griseofulvin, 33

H

Hydrocarbon fermentation

SUBJECT INDEX

alkane transformation products, 252
alkenes, 271, 272
  amino acids, 272, 273
    glutamic acid, 274-280
    homoserine, 285, 286
    isoleucine and serine, 284, 288
    lysine and diaminopimelic acid, 281-283
    ornithine and citrulline, 285, 289
    phenylalanine, 283, 287
    threonine and valine, 284, 286-288
    tyrosine and tryptophan, 284, 289
  cell yields, 314, 315, 316, 317
  enzymes, 314
  epoxides, 272
  esters, 268, 269
  fatty acids, 253-264
  glycolipids, 264, 265, 266, 267
  ketones, 270, 271
  nucleic acid—related products, 298-302
  organic acids
    citric acid, 293, 294-296
    dipicolinic acid, 298
    fumaric and malic acids, 298
    isocitric acid, 295, 297, 298
    α-ketoglutaric acid, 290-293
    sugars and polysaccharides, 311, 312, 313
  vitamins and pigments
    biotin, 303, 304
    carotenoids, 305, 309, 310
    gibberellin, 305, 311
    phenazinecarboxylic acid, 305, 311
    porphyrins, 305, 308, 309
    pyocyanine, 305, 310
    riboflavin, 304, 306, 307
    vitamin B_{12}, 304, 305, 307, 308
    vitamin B_6, 305, 308
  history, 250, 251

products from cyclic hydrocarbons, 318, 319
  biosynthetic products, 319, 320
  polynuclear hydrocarbons, 338
    cycloalkanes, cyclic monoalkanes, terpenes, 357-375
      higher polycyclic aromatic hydrocarbons, 356
      methylnaphthalenes, 355, 356
      naphthalaene, 339, 344-354

  transformation products, 320
    alkylbenzenes, 336, 337, 340-343
    benzene, 320-324
    methyl substituted benzenes, 321, 325-335
    polynuclear hydrocarbons, 338

I

Industrial alcohol production, 8-11
Industrial fermentations, Great Britain, 1

L

Lactic acid, 23
Lincomycin
  biological properties, 185, 187
  chemical characteristics, 186
  1'-demethylthio-1-ethylthio-, 188
  1'-demethyl-1-demethylthio-1'-ethyl-1-ethylthio-, 188
  1'-demethyl-, 188
  1'-demethyl-1-demethylthio-1-ethylthio-, 188
  esters, 189
    2-acylates and 2-carbonates, 190-192
    3-acylates, 193
    7-acylates and 7-carbonates, 193, 194
    4-hexanoate, 193
    2-phosphate, 190, 191
  microbial modification, 224
    1'-demethylation, 227
    phosphorylation, 227
    sulfoxide formation, 225
  synthetic, 195
    antimalarial activity of 4'-alkyl-4'-depropyl-1-demethylclindamycins, 218-219, 226
    modification of proline, 207-217
    1-substituted, 204-207
    7-substituted, 195-203
    total synthesis, 220-224
Lipid composition as taxonomic tool, 60-62

M

Molasses, 11, 12, 15

N

Nocardia

biosynthesis of hydrocarbons, 94–96
classification, 93, 94
cooxidation of hydrocarbons, 99–105
degradation of hydrocarbons, 96, 97
nonhydrocarbon oxidation, 105–111
transformation of steroids, 111–119

P

Penicillin, 25–32, 35
Penicillin V, 36
Phenazines, 51–53
Phenoxazinones, 51
Prodiginines, 53

R

Riboflavin, 20

S

Streptomycin, 32, 33

V

Viomycin, 33
Vitamin B_{12}, 21, 22

W

Walker, T. K., 13, 34
Weizmann, C., 12, 13, 14

CONTENTS OF PREVIOUS VOLUMES

Volume 1

Protected Fermentation
Miloš Herold and Jan Nečásek

The Mechanism of Penicillin Biosynthesis
Arnold L. Demain

Preservation of Foods and Drugs by Ionizing Radiations
W. Dexter Bellamy

The State of Antibiotics in Plant Disease Control
David Pramer

Microbial Synthesis of Cobamides
D. Perlman

Factors Affecting the Antimicrobial Activity of Phenols
E. O. Bennett

Germfree Animal Techniques and Their Applications
Arthur W. Phillips and James E. Smith

Insect Microbiology
S. R. Dutky

The Production of Amino Acids by Fermentation Processes
Shukuo Kinoshita

Continuous Industrial Fermentations
Philip Gerhardt and M. C. Bartlett

The Large-Scale Growth of Higher Fungi
Radcliffe F. Robinson and R. S. Davidson

AUTHOR INDEX — SUBJECT INDEX

Volume 2

Newer Aspects of Waste Treatment
Nandor Porges

Aerosol Samplers
Harold W. Batchelor

A Commentary on Microbiological Assaying
F. Kavanagh

Application of Membrane Filters
Richard Ehrlich

Microbial Control Methods in the Brewery
Gerhard J. Hass

Newer Development in Vinegar Manufactures
Rudolph J. Allgeier and Frank M. Hildebrandt

The Microbiological Transformation of Steroids
T. H. Stoudt

Biological Transformation of Solar Energy
William J. Oswald and Clarence G. Golueke

SYMPOSIUM ON ENGINEERING ADVANCES IN FERMENTATION PRACTICE

Rheological Properties of Fermentation Broths
Fred H. Deindoerfer and John M. West

Fluid Mixing in Fermentation Processes
J. Y. Oldshue

Scale-up of Submerged Fermentations
W. H. Bartholemew

Air Sterilization
Arthur E. Humphrey

Sterilization of Media for Biochemical Processes
Lloyd L. Kempe

Fermentation Kinetics and Model Processes
Fred H. Deindoerfer

Continuous Fermentation
  W. D. Maxon

Control Applications in Fermentation
  George J. Fuld

AUTHOR INDEX — SUBJECT INDEX

Volume 3

Preservation of Bacteria by Lyophilization
  Robert J. Heckly

Sphaerotilus, Its Nature and Economic Significance
  Norman C. Dondero

Large-Scale Use of Animal Cell Cultures
  Donald J. Merchant and C. Richard Eidam

Protection Against Infection in the Microbiological Laboratory: Devices and Procedures
  Mark A. Chatigny

Oxidation of Aromatic Compounds by Bacteria
  Martin H. Rogoff

Screening for and Biological Characterizations of Antitumor Agents Using Microorganisms
  Frank M. Schabel, Jr., and Robert F. Pittillo

The Classification of Actinomycetes in Relation to Their Antibiotic Activity
  Elio Baldacci

The Metabolism of Cardiac Lactones by Microorganisms
  Elwood Titus

Intermediary Metabolism and Antibiotic Synthesis
  J. D. Bu'Lock

Methods for the Determination of Organic Acids
  A. C. Hulme

AUTHOR INDEX — SUBJECT INDEX

Volume 4

Induced Mutagenesis in the Selection of Microorganisms
  S. I. Alikhanian

The Importance of Bacterial Viruses in Industrial Processes, Especially in the Dairy Industry
  F. J. Babel

Applied Microbiology in Animal Nutrition
  Harlow H. Hall

Biological Aspects of Continuous Cultivation of Microorganisms
  T. Holme

Maintenance and Loss in Tissue Culture of Specific Cell Characteristics
  Charles C. Morris

Submerged Growth of Plant Cells
  L. G. Nickell

AUTHOR INDEX — SUBJECT INDEX

Volume 5

Correlations between Microbiological Morphology and the Chemistry of Biocides
  Adrien Albert

Generation of Electricity by Microbial Action
  J. B. Davis

Microorganisms and the Molecular Biology of Cancer
  G. F. Gause

Rapid Microbiological Determinations with Radioisotopes
  Gilbert V. Levin

The Present Status of the 2,3-Butylene Glycol Fermentation
  Sterling K. Long and Roger Patrick

Aeration in the Laboratory
  W. R. Lockhart and R. W. Squires

Stability and Degeneration of Microbial Cultures on Repeated Transfer
Fritz Reusser

Microbiology of Paint Films
Richard T. Ross

The Actinomycetes and Their Antibiotics
Selman A. Waksman

Fusel Oil
A. Dinsmoor Webb and John L. Ingraham

AUTHOR INDEX — SUBJECT INDEX

Volume 6

Global Impacts of Applied Microbiology: An Appraisal
Carl-Göran Hedén and Mortimer P. Starr

Microbial Processes for Preparation of Radioactive Compounds
D. Perlman, Aris P. Bayan, and Nancy A. Giuffre

Secondary Factors in Fermentation Processes
P. Margalith

Nonmedical Uses of Antibiotics
Herbert S. Goldberg

Microbial Aspects of Water Pollution Control
K. Wuhrmann

Microbial Formation and Degradation of Minerals
Melvin P. Silverman and Henry L. Ehrlich

Enzymes and Their Applications
Irwin W. Sizer

A Discussion of the Training of Applied Microbiologists
B. W. Koft and Wayne W. Umbreit

AUTHOR INDEX — SUBJECT INDEX

Volume 7

Microbial Carotenogenesis
Alex Ciegler

Biodegradation: Problems of Molecular Recalcitrance and Microbial Fallibility
M. Alexander

Cold Sterilization Techniques
John B. Opfell and Curtis E. Miller

Microbial Production of Metal-Organic Compounds and Complexes
D. Perlman

Development of Coding Schemes for Microbial Taxonomy
S. T. Cowan

Effects of Microbes on Germfree Animals
Thomas D. Luckey

Uses and Products of Yeasts and Yeast-like Fungi
Walter J. Nickerson and Robert G. Brown

Microbial Amylases
Walter W. Windish and Nagesh S. Mhatre

The Microbiology of Freeze-Dried Foods
Gerald J. Silverman and Samuel A. Goldblith

Low-Temperature Microbiology
Judith Farrell and A. H. Rose

AUTHOR INDEX — SUBJECT INDEX

Volume 8

Industrial Fermentations and Their Relations to Regulatory Mechanisms
Arnold L. Demain

Genetics in Applied Microbiology
S. G. Bradley

Micotoxins
Alex Ciegler and Eivind B. Lillehoj

Microbial Ecology and Applied Microbiology
Thomas D. Brock

The Ecological Approach to the Study of Activated Sludge
Wesley O. Pipes

Control of Bacteria in Nondomestic Water Supplies
Cecil W. Chambers and Norman A. Clarke

The Presence of Human Enteric Viruses in Sewage and Their Removal by Conventional Sewage Treatment Methods
Stephen Alan Kollins

Oral Microbiology
Heiner Hoffman

Media and Methods for Isolation and Enumeration of the Enterococci
Paul A. Hartman, George W. Reinbold, and Devi S. Saraswat

Crystal-Forming Bacteria as Insect Pathogens
Martin H. Rogoff

Mycotoxins in Feeds and Foods
Emanuel Borker, Nino F. Insalata, Colette P. Levi, and John S. Witzeman

AUTHOR INDEX — SUBJECT INDEX

Volume 9

The Inclusion of Antimicrobial Agents in Pharmaceutical Products
A. D. Russell, June Jenkins, and I. H. Harrison

Antiserum Production in Experimental Animals
Richard M. Hyde

Microbial Models of Tumor Metabolism
G. F. Gause

Cellulose and Cellulolysis
Brigitta Norkrans

Microbiological Aspects of the Formation and Degradation of Cellulosic Fibers
L. Jurášek, J. Ross Colvin, and D. R. Whitaker

The Biotransformation of Lignin to Humus — Facts and Postulates
R. T. Oglesby, R. F. Christman, and C. H. Driver

Bulking of Activated Sludge
Wesley O. Pipes

Malo-lactic Fermentation
Ralph E. Kunkee

AUTHOR INDEX — SUBJECT INDEX

Volume 10

Detection of Life in Soil on Earth and Other Planets. Introductory Remarks
Robert L. Starkey

For What Shall We Search?
Allan H. Brown

Relevance of Soil Microbiology to Search for Life on Other Planets
G. Stotzky

Experiments and Instrumentation for Extraterrestrial Life Detection
Gilbert V. Levin

Halophilic Bacteria
D. J. Kushner

Applied Significance of Polyvalent Bacteriophages
S. G. Bradley

Proteins and Enzymes as Taxonomic Tools
Edward D. Garber and John W. Rippon

Transformation of Organic Compounds by Fungal Spores
Claude Vézina, S. N. Sehgal, and Kartar Singh

Microbial Interactions in Continuous Culture
Henry R. Bungay, III and Mary Lou Bungay

Chemical Sterilizers (Chemosterilizers)
Paul M. Borick

Antibiotics in the Control of Plant Pathogens
M. J. Thirumalachar

AUTHOR INDEX—SUBJECT INDEX

CUMULATIVE AUTHOR INDEX—CUMULATIVE TITLE INDEX

Volume 11

Successes and Failures in the Search for Antibiotics
Selman A. Waksman

Structure–Activity Relationships of Semisynthetic Penicillins
K. E. Price

Resistance to Antimicrobial Agents
J. S. Kiser, G. O. Gale, and G. A. Kemp

Micromonospora Taxonomy
George Luedemann

Dental Caries and Periodontal Disease Considered as Infectious Diseases
William Gold

The Recovery and Purification of Biochemicals
Victor H. Edwards

Ergot Alkaloid Fermentations
William J. Kelleher

The Microbiology of the Hen's Egg
R. G. Board

Training for the Biochemical Industries
I. L. Hepner

AUTHOR INDEX—SUBJECT INDEX

Volume 12

History of the Development of a School of Biochemistry in the Faculty of Technology, University of Manchester
Thomas Kennedy Walker

Fermentation Processes Employed in Vitamin C Synthesis
Miloš Kulhánek

Flavor and Microorganisms
P. Margalith and Y. Schwartz

Mechanisms of Thermal Injury in Nonsporulating Bacteria
M. C. Allwood and A. D. Russell

Collection of Microbial Cells
Daniel I. C. Wang and Anthony J. Sinskey

Fermentor Design
R. Steel and T. L. Miller

The Occurrence, Chemistry, and Toxicology of the Microbial Peptide-Lactones
A. Taylor

Microbial Metabolites as Potentially Useful Pharmacologically Active Agents
D. Perlman and G. P. Peruzzotti

AUTHOR INDEX—SUBJECT INDEX

Volume 13

Chemotaxonomic Relationships Among the Basidiomycetes
Robert G. Benedict

Proton Magnetic Resonance Spectroscopy — An Aid in Identification and Chemotaxonomy of Yeasts
P. A. J. Gorin and J. F. T. Spencer

Large-Scale Cultivation of Mammalian Cells
R. C. Telling and P. J. Radlett

Large-Scale Bacteriophase Production
K. Sargeant

Microorganisms as Potential Sources of Food
Jnanendra K. Bhattacharjee

Structure–Activity Relationships Among Semisynthetic Cephalosporins
M. L. Sassiver and Arthur Lewis

Structure–Activity Relationships in the Tetracycline Series
Robert K. Blackwood and Arthur R. English

Microbial Production of Phenazines
J. M. Ingram and A. C. Blackwood

The Gibberellin Fermentation
E. G. Jefferys

Metabolism of Acylanilide Herbicides
Richard Bartha and David Pramer

Therapeutic Dentifrices
J. K. Peterson

Some Contributions of the U.S. Department of Agriculture to the Fermentation Industry
George E. Ward

Microbiological Patents in International Litigation
John V. Whittenburg

Industrial Applications of Continuous Culture: Pharmaceutical Products and Other Products and Processes
R. C. Righelato and R. Elsworth

Mathematical Models for Fermentation Processes
A. G. Fredrickson, R. D. Megee, III, and H. M. Tsuchiya

AUTHOR INDEX — SUBJECT INDEX